Cross-Sector Leadership for the Green Economy

Cross-Sector Leadership for the Green Economy

Integrating Research and Practice on Sustainable Enterprise

Edited by
Alfred Marcus, Paul Shrivastava, Sanjay Sharma,
and Stefano Pogutz

CROSS-SECTOR LEADERSHIP FOR THE GREEN ECONOMY
Copyright © Alfred Marcus, Paul Shrivastava, Sanjay Sharma, and Stefano Pogutz, 2011.

All rights reserved.

First published in 2011 by
PALGRAVE MACMILLAN®
in the United States—a division of St. Martin's Press LLC,
175 Fifth Avenue, New York, NY 10010.

Where this book is distributed in the UK, Europe and the rest of the world, this is by Palgrave Macmillan, a division of Macmillan Publishers Limited, registered in England, company number 785998, of Houndmills, Basingstoke, Hampshire RG21 6XS.

Palgrave Macmillan is the global academic imprint of the above companies and has companies and representatives throughout the world.

Palgrave® and Macmillan® are registered trademarks in the United States, the United Kingdom, Europe and other countries.

ISBN: 978–0–230–11940–6

Library of Congress Cataloging-in-Publication Data

Cross-sector leadership for the green economy : integrating research and practice on sustainable enterprise / by Alfred Marcus...[et al.].
    p. cm.
ISBN 978–0–230–11940–6 (hardcover)
 1. Green technology. 2. Technological innovations—Environmental aspects. 3. Business enterprises—Environmental aspects. 4. Sustainable development. 5. Social responsibility of business. 6. Greenhouse gas mitigation. I. Marcus, Alfred Allen, 1950–

HC79.E5C765 2011
658.4'092—dc23   2011023353

A catalogue record of the book is available from the British Library.

Design by Newgen Imaging Systems (P) Ltd., Chennai, India.

First edition: December 2011

10 9 8 7 6 5 4 3 2 1

Printed and bound in Great Britain by
CPI Antony Rowe, Chippenham and Eastbourne

CONTENTS

List of Figures		vii
List of Tables		ix
Introduction Alfred Marcus, Paul Shrivastava, Sanjay Sharma, and Stefano Pogutz		xi

Part I Institutions Matter

One	The Role of Institutions in the Implementation of Wind Energy *Michelle Bernard, Michael Craig, and Itai Sened*	3
Two	How Regulatory Uncertainty Impedes the Reduction of Greenhouse Gas Emissions *Jens Hamprecht, David C. Sprengel, and Volker H. Hoffmann*	19
Three	Renewable Energy Investment Decisions under Policy Risk: An Adaptive Conjoint Analysis (ACA) Approach *Sonja Lüthi and Rolf Wüstenhagen*	37
Four	Why Some Managers Expect to Benefit from Public Policies and Others Do Not *Alfred Marcus, Susan Cohen, and Kathleen Sutcliffe*	53

Part II Innovation Matters

Five	Rethinking Sustainability, Innovation, and Financial Performance *Timo Busch, Bryan T. Stinchfield, and Matthew S. Wood*	81
Six	What Kinds of Photovoltaic Projects Do Lenders Prefer to Finance? *Florian Lüdeke-Freund and Moritz Loock*	107
Seven	Path Dependence and Creation in Venture Capital Investment *Alfred A. Marcus, Shmuel Ellis, Joel Malen, Israel Drori, and Itai Sened*	125

Eight	High-Tech Cluster Revolution from an Organizational Ecology Perspective *Deborah E. de Lange*	141
Nine	Shecopreneuring: Stitching Global Ecosystems in the Ethical Fashion Industry *Kim Poldner, Oana Branzei, and Chris Steyaert*	157

Part III Global Supply Chains Matter

Ten	Institutionalizing Proactive Sustainability Standards in Supply Chains: Which Institutional Entrepreneurship Capabilities Matter? *Jörg H. Grimm, Joerg S. Hofstetter, Martina Müggler, and Nils J. Peters*	177
Eleven	Supply Chain Structure as a Critical Driver of Sustainable Supplier Practices *Amrou Awaysheh and Robert D. Klassen*	195
Twelve	Going Green by Exporting *Emilio Galdeano-Gómez, Eva Carmona-Moreno, and José Céspedes-Lorente*	221
Thirteen	Internationalization, Innovativeness, and Proactive Environmental Strategy among Small and Medium Enterprises *Javier Aguilera-Caracuel, M. Ángeles Escudero-Torres, Eulogio Cordón-Pozo, and Nuria Esther Hurtado-Torres*	241

Part IV New Forms of Cross-Sector Cooperation Matter

Fourteen	Why Do Private Firms Invest in Public Goods? *Gurneeta Vasudeva and Hildy Teegen*	263
Fifteen	Voices from the Field: The Green Economy Partnership Process *Jack Hogin and Georgia Rubenstein*	277
Sixteen	Moving the Green Economy Forward: Conclusions from Research and Practice *Alfred Marcus*	295

About the Contributors	305
Index	315

FIGURES

1.1	Percent of total electricity generated by renewable resources in the United States, Germany, and Spain	11
3.1	Willingness to accept a certain duration of the administrative process, correlated with €ct/kWh	47
3.2	Willingness to accept a loose cap or a tight cap	48
3.3	Willingness to accept a certain number of policy changes	48
4.1	Why firms decide to be politically active	55
5.1	The moderating effect of innovation on ESP and CFP	87
5.2	Mediated model of ESP, innovation, and short-term CFP	94
5.3	Mediated model of ESP, innovation, and long-term CFP	95
5.4	Time-dependent inverse-U-shaped relationship between a given level of investments in ESP-related resources and CFP	98
6.1	Graph showing relative importance of PV business model attributes to project lenders	117
7.1	The uneven paths in the development of solar power, biofuels, wind power, and energy efficiency portfolios for 15 countries, 2003 to 2009	130
7.2	The impact of initial conditions and subsequent events	131
7.3	A full model of the factors that affect path dependence	131
7.4	Growth of solar power and concurrent economic and policy data, 2003–2009	134
7.5	Growth of wind power and concurrent economic and policy data, 2003–2009	134
7.6	Growth of biofuels and concurrent economic and policy data, 2003–2009	135
7.7	Growth of energy efficiency and concurrent economic and policy data, 2003 to 2009	135
8.1	Two scenarios for green technology clusters. In which case are new firms more likely to survive and experience more rapid growth?	146
8.2	Organizational ecology concepts and niche conditions	148
8.3	First-generation (old) and second-generation (new) industry interactions	149

9.1	Practices of ecosystem design	170
11.1	Conceptual model of supply chain structure	202
12.1	Productivity component estimations: Technological change *(TEC)*, efficiency change *(EFC)* and environmental productivity *(EP)*	233
13.1	Proposed model of the effect of international diversification and innovation environmental strategy	248
13.2	Structural model estimation (standardized solution)	252

TABLES

0.1	Overview of chapters and their topics, theory, and methodology used.	xxiv
1.1	Added and cumulative wind power capacity and national rank of Minnesota (MN) and Texas (TX), 2007–2009	6
1.2	Electricity generated from wind and from all sources, from 2006 to 2008, in Minnesota (MN) and Texas (TX)	6
1.3	Types of renewable energy policies enacted by Germany, Spain, and the United States	12
2.1	Data sample, by industry and region.	24
2.2	Means, standard deviations (s.d.), and correlations of the companies studied	27
2.3	Results of the regression analysis testing Hypotheses 1 and 2	28
2.4	Nine sources of pressure to reduce GHG emissions	32
2.5	Questions to assess the level of perceived regulatory uncertainty	32
2.6	Which items did your company pursue in response to pressure to reduce CO_{2e} emissions?	33
3.1	Potential attributes for ACA survey	42
3.2	Attributes and attribute levels used in the ACA experiment.	43
3.3	Average part-worth utility estimates and standard deviations by attribute levels (hierarchical Bayes models with all normally distributed part-worth utilities)	46
3.4	Description of project developers' likelihood of investing in different policy frameworks (holdout tasks) and SMRT simulation	49
3.5	Investment likelihood simulations for changes in the PV policy framework of Spain in 2007	49
4.1	Summary of investments in energy efficiency and renewable energy in the US 2009 Economic Stimulus Act	54
4.2	Cognitive template for managers' views of public policy benefits	61
4.3	Descriptive statistics and zero-order correlation ($N = 66$)	65
4.4	Results of the regression analyses	67

5.1	Descriptive statistics and correlations	92
5.2	Regression results for short-term CFP (Tobin's q for 2004)	93
5.3	Regression results for long-term CFP (Tobin's q scores, average of years 2005 through 2007)	95
6.1	Photovoltaic business model attributes and levels.	113
6.2	ACBC (Adaptive Choice-Based Conjoint) analysis—Heirarchical Bayes (HB) summary of results	116
6.3	Overview of simulation results	118
9.1	Micro- and macro-practices of ecosystem design: definitions and illustrations.	164
10.1	Theme analysis of intraorganizational and interorganizational capabilities needed for compliance to PSCSS	180
11.1	Descriptive statistics and correlation table	207
11.2	Regressions results	208
12.1	Definitions of variables, and their sources	231
12.2	Descriptive statistics and correlations	232
12.3	Simultaneous equation GMM (Generalized Method of Moment) parameter estimates	234
13.1	Descriptive statistics and correlations for the constructs	251
13.2	Standardized factor loadings, Average Variance Extracted, and reliability estimates	251
13.3	Structural model results (standardized structural coefficients)	253
14.1	Specific benefits to investing firms and governments, and to the global public	271
14.2	Hierarchy and interplay of benefit types from investing in a collaborative technology development venture	272

INTRODUCTION

Alfred Marcus, Paul Shrivastava, Sanjay Sharma, and Stefano Pogutz

A green economy seeks to lower the impact of economic activities on the environment while striving to maintain or achieve high levels of material well-being. A green economy has many components including: efficient energy use; renewable energy (such as solar, wind, and biofuels); clean transportation (such as hybrids, electric vehicles, and public transit); recycling; efficient use of water; organic agriculture; and reforestation. It has the potential to create jobs and ensure economic growth, while at the same time preventing pollution, environmental degradation, and resource depletion. It can be an important source of innovation and economic revitalization, which bring about substantial changes in the way we live and work.

A green economy must also be socially just, as human dignity is as precious as the natural world. Thus, it rests on the belief that all people should have access to an acceptable living standard and opportunities for personal and social development. The diffusion of sustainable business enterprises that adhere to its principles is important. It is important for many reasons, including human dignity, resource scarcity, and the risks of climate change if sustainable business enterprises don't prevail.

To create sustainable economic systems, we need to show more respect for ecosystem integrity. Ecosystems that are critical for our survival have finite resources. These basic facts are well borne out by recent reports. The Stern Review (Stern, 2006) provides detailed costs regarding the impact of climate change on the global economy. It concludes that 1.5 percent of total GDP invested in climate mitigation measures now can avert an over 20 percent decline in economic production over the next few decades. The 2010 Economics of Ecosystem and Biodiversity (TEEB) Report (Sukhdev, Wittmer, Schröter-Schlaack, Nesshöver, Bishop, Ten Brink, Gundimeda, et al., 2010) provides methodologies for valuing the natural capital that is represented by ecosystem services. It points out that until now humans have been using many ecosystems services without paying for them or ensuring their renewability. Overusing these services has caused ecosystem collapse in many instances. Most recently, the UNEP report, "Towards a Green Economy" (United Nations Environment Program, 2011) shows the way to creating a green economy that would innovate across energy, industrial, financial, and built environment sectors, to usher in long-term sustainability.

Cross-Sector Leadership

Research and practice suggest that cross-sector leadership can facilitate the growth and development of a green economy. The message of this book is that in order to break impasses and hasten the introduction of a green economy, cross-sector leadership is needed.

By themselves, institutions are insufficient, innovation and resources are limited, and global supply chains do not provide an adequate enough push to bring into being a green economy. Top-down initiatives that come from government will not get the job done, nor will bottom-up market mechanisms motivated by financial gain.

To organize a global supply chain around green products and practices and human dignity is a formidable task. Critical work has to be done by local, regional, national, and international organizations that are neither part of government nor agents of individual firms. These organizations need to tie together the loose ends. They must bring together governments, corporations, and nonprofits and must create new and reliable institutions that people in the business community can depend on to make long-term investments and spread green practices across the global supply chain.

To accomplish these tasks, research is needed about the rules that will engage and inspire these heterogeneous actors to cooperate. These diverse actors must overcome their long-standing differences and antagonisms, not in order to create perfect harmony, but to forge a common interest in a more sustainable world. No doubt progress is being made. Many organizations and many institutions are already attempting to do this. The roles they should be playing, however, must be better understood. Highlighting successes that should be emulated is a worthy endeavor. Nonetheless, a deeper understanding of the barriers to additional progress also is needed.

At its core, cross-sector leadership depends on the rules—both tacit and explicit—that allow people from diverse sectors to work together. Rules that permit people from diverse sectors to cooperate with one another are difficult to bring into being. Research into achieving a green economy must start with a better understanding of these rules.

More integrated transdisciplinary research is also needed. Sustainability as a global challenge cuts across multiple disciplines and areas of study and action. A sustainable economy requires a holistic understanding of global economics and of enterprise sustainability. This means going beyond functional area thinking and developing integrated cross-sector theories and approaches. Current research is highly fragmented and lacks this integration. For example, research in business schools has evolved into silos of management, accounting, finance, information systems, operations, and marketing. In each of them, there is some interest in sustainability. Beyond business research, there are many areas of the sciences, engineering, law, and the arts that are pertinent to understanding enterprise sustainability. Through this book, we seek to create cross-cutting discourse. We understand that this is only a first step. Scholars and practitioners still must break the silos and develop relationships with each other to evoke a broader more holistic and systemic understanding of the green economy.

University administrators can catalyze the process of developing a cross-cutting discourse by encouraging and incentivizing this type of research and by encouraging ongoing interactions between academics and practitioners. University research centers and schools that have taken steps to cut across disciplines and involve multiple stakeholders such as practitioners, policymakers, and civil society are playing a major role, like the University of Minnesota's Center for Integrative Leadership and the SDA Bocconi School of Management in Milan, which sponsored the conferences where most of the chapters in this volume were first presented.

We also wish to acknowledge a generous grant we received from the Institute for Renewable Energy and the Environment that assisted us in carrying out some of the research that is presented in this book and the contributions made by the Center for New Institutional Social Sciences at Washington University in St. Louis.

Issues in Moving toward a Green Economy

Moving toward a green economy will yield many benefits, including regional economic development, financial returns for firms, implementation of the best practices for sustainability, greater legitimacy for actors in different sectors of society, and additional possible spillovers and learning. The issues that should engage researchers and practitioners who are moving toward a green economy include the following:

1. Institutions: The Role of Regulation
    a. public policy and regulatory certainty
    b. public policy and regulatory predictability
    c. public policy and regulatory enforcement and compliance

2. Innovation and Resources
    a. the role of investors and capital markets
    b. the role of business groups
    c. the role of entrepreneurial and business strategies

3. The Global Supply Chain
   a. the impact of activist pressures, stakeholder partnerships, and social networks
   b. voluntary actions such as information disclosure
4. Cross-Sector Leadership
   a. mobilization strategies, trust, and coordination
   b. coalitions, consensus, and conflict management
   c. bargaining and negotiating
   d. power, control, and influence
   e. collective action

With this much complexity, researchers and practitioners will have to increasingly rely on systems theory to recognize unintended consequences. They will have to be aware of path dependencies and prevent premature lock-in to technologies that are outliving their usefulness. Three matters deserve further attention: strong and multiple path dependence; global impacts; and the need to create new knowledge.

Strong, Multiple Path Dependence. An obstacle to the development and diffusion of the green economy is related to strong and multidimensional path dependence. Our infrastructure system is a burden that limits the possibility for a rapid change to more sustainable technologies. The structure of the electric energy grid, for example, based on large, centralized power generation plants, slows down the diffusion of a decentralized model based on small-scale renewable energy production. At the same time, the very long lifetime of conventional power plants places obstacles in the diffusion of clean technologies. In the case of cars, the automotive and oil industries have huge sunk costs in production plants, logistics, and distribution systems, which impede the growth of a market for electric or fuel cell vehicles.

Path dependence also conditions our pattern of consumption. Desire and need are satiated by goods and services, and materiality is mistakenly equated with the quality of life. Individualism and possessiveness dominate our culture, and material goods have become important in defining our social position, in the construction of our self-identity, and in signaling our affiliation to a group of peers. Other forms of path dependence engage the political and financial community, where short-term thinking dominates over long-term approaches, impeding investments that have returns in the long run, like many green technologies. Research must address these challenges and investigate the many forms of path dependence and the type of cross-sector collaborations involving practitioners and academics that are needed to favor the fast diffusion of sustainable models of production and consumption.

None of this will be easy, as the green economy is likely to challenge such assumptions as the notion of eternal growth. The current rates of consumption, especially in the industrialized West, are already well over the biocapacity of the respective countries. Collectively, the world is currently consuming more than the total productive biocapacity of the Earth ecosystem. According to the World Resource Fund's 2010 Living Planet Report, we need one and a half planets to satisfy our wants. Yet, there are millions of people whose consumption levels are below basic human requirements. A green economy will need to find forms of alternative, dematerialized, environmentally anchored, recyclable, socially equitable growth that balances out developed country overconsumption with underconsumption in the rest of the world.

Global Impacts. The Brundtland Commission Report (1987), which popularized the term "sustainable development," has argued that improvement of ecosystem health without the alleviation of poverty, the redistribution of economic opportunity, community self-sufficiency, and human freedom, can only lead to incremental results. Solutions to global environmental problems such as climate change involve the development and transfer of technologies to developing countries and the facilitation of grassroots economic capacity building. A U.S. energy company cannot tackle climate change without tackling global poverty. This is because marginalized and poor societies often survive by burning wood and cow dung, and using slash-and-burn cultivation, all of which exacerbate climate change around the world.

Attempts at reducing greenhouse gas emissions in the West by themselves will not make a dent in the global climate change that affects the entire world. The spread of the Internet has allowed groups and individuals to communicate and find common cause to pressure firms and to address their social and environmental impacts in an integrated manner. Coalitions of NGOs and individuals—smart mobs—have made it increasingly difficult for firms to deal separately with each social, ecological, or economic issue in one country or region to the exclusion of other related issues in another country or region. Integrated ecosystems and social sustainability require cross-sector approaches that will generate fundamental changes in business practice, which will be based on disruptive rather than incremental innovation and on patient investments rather than short-term returns.

Thus, an important issue that must be addressed is the relation between industrialized and emerging countries in the race toward a green economy. Unlike industrialized economies, emerging ones in many cases are lacking in infrastructure, have alternative financial systems—the shareholder-based company model is almost absent—and firm governance also differs. These degrees of freedom are an immense opportunity for designing sustainable communities that favor a quicker diffusion of cleaner technologies than can be done in developed countries. Moreover, these countries can learn from their local capabilities and shape new and innovative business models that fit better with their ecosystem requirements and societal needs.

To manage the global commons, some degree of global governance as well as local, regional, and national governance is essential. Even when they operate locally, organizations are embedded in a global context. Thus, an integrated cross-sector social and environmental strategy is critical for their future survival and competitive advantage. Future research must address this challenge.

Creating Knowledge. Knowledge about the green economy, i.e., technologies, impacts, regulations, and social preferences, is constantly evolving on multiple fronts, and organizations need to connect with the processes of knowledge creation. Cross-sector and cross-disciplinary innovations that address environmental and social problems require organizations to develop capabilities that will enable them to engage multiple stakeholders, generate a diversity of information, transform this information into relevant knowledge via higher-order learning and cross-functional teams, and continuously innovate to find sustainable solutions to building a green economy. A prolific research team on environmental strategy has found that proactive solutions to environmental problems are more likely to be found by organizations that are able to build and deploy organizational capability for stakeholder integration, higher-order learning, and continuous innovation. These capabilities enable firms not only to work collaboratively with their suppliers and channels of distribution, but also to engage competitors, industry associations, NGOs, governments, and local communities to collaboratively seek innovative solutions.

We would be remiss if we did not come back to the role of universities and business schools in this process: How much do management disciplines help in building the green economy? There definitely is a need for promoting cross-sector fertilization among the disciplines and for breaking the boundaries of our theoretical fields. The development of a green and sustainable economy depends on innovative outputs emerging from many scientific domains (the sciences, social sciences, technology, law, and so on). However, a decision-making and problem-solving "mantra" prevails in many business schools. This approach tends to favor a limited-risk consultancy-style perspective more than long-term vision, breakthrough thinking, and holistic knowledge, therefore limiting the entrepreneurial innovativeness needed for a green and sustainable society. Thus, business schools need to encourage breaking down functional and disciplinary silos, yielding cross-fertilizations among disciplines, empowering pragmatic creativity, and promoting a long-term vision.

What This Book Is About

This book is a compilation of chapters about the importance of cross-sector leadership for the green economy.[1] It is divided into four major parts, which correspond to its main contentions that for the green economy to flourish institutions, innovation and resources, global supply chains, and new forms of cross-sector leadership matter.

INTRODUCTION   xvii

Part I: Institutions Matter

In Chapter 1, "The Role of Institutions in the Implementation of Wind Energy," Michelle Bernard, Michael Craig, and Itai Sened argue that much of the debate about the implementation of strategies that would enhance growth in the use of new energy-producing technologies has been couched in terms of technological and economic feasibility. The common argument is that many of these technologies are not sufficiently developed for practical applications, and to the extent that they are, they impose excessive costs on consumers that cast a doubt on whether they are economically feasible. The authors maintain that though wind power is both technologically feasible and economically competitive, the market has not jumped on the opportunity to bring about widespread implementation. A significant reason is found in the institutions that regulate the market. The issue is not the extent to which the government intervenes—conventional energy sources are heavily regulated. The question is how government regulates. Chapter 1 provides an analysis of the institutional structures that regulate the introduction of wind energy and relates the industry's progress to differences in these structures. It concludes that for wind implementation to grow, governments must ensure proactive transmission expansion, reduce conflict among stakeholders, guarantee markets, and secure increased market certainty.

In Chapter 2, "How Regulatory Uncertainty Impedes the Reduction of Greenhouse Gas Emissions," Jens Hamprecht, David C. Sprengel, and Volker H. Hoffmann examine how institutional pressures and regulatory uncertainty influence the response strategies of companies to greenhouse gas (GHG) emissions. Their results are based on a global survey that demonstrates how uncertainty has impeded the reduction of greenhouse gases. Previous research has provided conflicting results on the question of how uncertainty influences response strategies. This is because most scholars understand an organization's response strategy as being of one specific type. This chapter holds that this simplification has contributed to conflicting findings. The authors understand a strategy to be a combination of response measures with different strategic directions. Organizations are more likely to reduce their greenhouse gas emissions as institutional pressure increases; however, as uncertainty regarding future regulations increases, they also increase the scope of their responses and organizations engage in multiple, potentially conflicting strategies simultaneously. This increased response scope prevents the emergence of role models and common practices in an uncertain environment. Various market stakeholders (including the financial community, customers, and suppliers) build pressure on companies to reduce their GHG emissions, but such pressure is not sufficient to bring about change in dominant industry practices if regulatory uncertainty exists. Reducing uncertainty about the future of the Kyoto Protocol therefore must be a pivotal task of policymakers.

In Chapter 3, "Renewable Energy Investment Decisions under Policy Risk: An Adaptive Conjoint Analysis (ACA) Approach," Sonja Lüthi and Rolf Wüstenhagen examine the influence of policy risks in solar energy project investment decisions. Solar photovoltaic (PV) technology has particular potential for contributing to a future low-carbon energy supply, but given the current cost of PV, market diffusion crucially depends on policy incentives such as feed-in tariffs (FITs), which have been implemented in a number of countries. FITs have been praised for their effectiveness, but have also received mixed reviews when it comes to assessing their efficiency. A key empirical puzzle is why similar FITs lead to differing outcomes in terms of newly installed PV capacity in different countries. Previous research suggests that answering this question requires a better understanding of policy risk, rather than just the level of return. This chapter examines the influence of policy risk on the decision of a PV project developer to invest in a given country. Based on an original data set of 1575 choice decisions by 63 PV project developers, it demonstrates that risk matters in policy design and that a price tag can be attached to it. By providing an adequate compensation for risk, governments can build on the results to design policies that will be effective in attracting private PV investment.

In Chapter 4, "Why Some Managers Expect to Benefit from Public Policies and Others Do Not," Alfred Marcus, Susan Cohen, and Kathleen Sutcliffe investigate the heuristics that managers rely on to anticipate the benefits of public policies. Understanding these heuristics can help explain why some firms seek government assistance while others do not. The authors analyze the cognitive frameworks and heuristics that managers apply to discern whether public policies are likely to benefit their firms. Based on qualitative interviews and a survey of managers in energy efficiency and renewable energy businesses, the authors propose that managers who expect that their firms will benefit from public policies believe they compete in an unstable environment in which they have little ability to influence market forces, yet they believe they can predict and influence government. In contrast, managers who expect that their firms will not benefit from public policies believe they compete in stable environments in which they can influence market forces but cannot predict and influence government.

Part II: Innovation and Resources Matter

In Chapter 5, "Rethinking Sustainability, Innovation, and Financial Performance," Timo Busch, Bryan T. Stinchfield, and Matthew S. Wood examine the question of the extent to which there is a financial payoff for addressing ecological and social issues. They include a time component for corporate financial performance (CFP) and a firm's innovativeness and ask "When does addressing ecological and social issues pay?" Combining such a contingency perspective with a resource-based view of the firm clarifies the relationship between corporate environmental and social performance (ESP) and CFP. The full benefits of ESP take time to pay off financially. Further, the authors of Chapter 5 find support for the moderating effect of innovation on the relationship between the ESP and short-term CFP. In the long term, innovation mediates the ESP–CFP relationship, they find. This result suggests that innovation should be considered as a long-term investment required to unlock the full potential of ESP initiatives.

In Chapter 6, "What Kinds of Photovoltaic Projects Do Lenders Prefer to Finance?" Florian Lüdeke-Freund and Moritz Loock argue, as previous authors have done, that the diffusion of innovative renewable energies is more than a technical challenge. Although governments, researchers, and industries around the globe are developing new solutions for the provision of environmentally friendly energies, the availability of capital has become a major concern. Innovative photovoltaic (PV) projects will only be realized if project developers and sponsors are able to convince financiers, especially banks, of the profitability and robustness of their project designs. Chapter 6 examines the preferences that financing experts have, in order to gain insights into the factors they find favorable to PV project financing; this information will help to further the diffusion of photovoltaic power. The authors of Chapter 6 found a bias in financing PV projects: lenders prefer PV projects with premium-brand technology rather than low-cost technology. Based on the results from a conjoint experiment with 43 German PV project financing experts, the authors carried out market simulations that suggest that lenders use premium brands as a crucial decision heuristic during project assessments: lenders chose inferior proposals with comparably lower Debt Service Cover Ratios (DSCR), as long as these projects included premium-brand technologies. This effect, which the authors call "debt for brands," can be an important consideration in the financing of renewable energy technology.

In Chapter 7, "Path Dependence and Creation in Venture Capital Investment," Alfred Marcus, Shmuel Ellis, Joel Malen, Israel Drori, and Itai Sened investigate the extent to which venture capital investments in innovative solar power, wind power, biofuels, and energy efficiency start-up firms stuck to or altered their paths in response to market conditions and other external perturbations that took place during the time period 2003 to 2009. Rather than finding a single path common to all of these innovative start-ups, the authors found that each had separate paths. Their separate paths can be explained by their relative price competitiveness with conventional fossil fuel and nuclear power generating technologies. Start-ups in less mature technologies, like solar power, generated investor excitement, but were subject to higher levels of volatility and had more sensitivity to shifts in the external environment than start-ups in more mature energy efficiency technologies.

In Chapter 8, "High-Tech Cluster Revolution from an Organizational Ecology Perspective," Deborah E. de Lange argues that innovation in high-tech industrial clusters consists of revolutionary second-generation change; that is, the entry and growth of a new population of green firms will impinge on the niches of first generation high-tech firms, such as information and communications technology and/or biotech. From an organizational ecology perspective, are the new green firms better off locating in new places or in existing clusters? This chapter investigates second-generation firms' location choices. It argues that existing clusters are like incubators; they can reduce the liability of newness that second-generation firms face. However, the original firms in a cluster may treat the new green firms as rivals, thus creating a challenging environment for the new green firms, which then face the double liability of both newness and greater competition. As well as competing with each other, deLange points out, second-generation firms can help revitalize first generation and there can be synergy between the two.

In Chapter 9, "Shecopreneuring: Stitching Global Ecosystems in the Ethical Fashion Industry," Kim Poldner, Oana Branzei, and Chris Steyaert attempt to broaden our understanding of how women entrepreneurs in the innovative ethical fashion industry, called *shecopreneurs*, begin to change the world by means of a feminine duty of care. Drawing on three multi-sensorial ethnographies, the authors suggest that more sustainable ecosystems can be first reimagined and then progressively realized, as shecopreneurs deliberately create and harness a dialectic tension between the imaginary and the real in their everyday work and by combining three distinct sets of practices: romanticizing, being precise, and generalizing.

Part III: Global Supply Chains Matter

In Chapter 10, "Institutionalizing Proactive Sustainability Standards in Supply Chains: Which Institutional Entrepreneurship Capabilities Matter?" Jörg H. Grimm, Joerg S. Hofstetter, Martina Müggler, and Nils J. Peters focus on the institutionalization of proactive supply chain sustainability standards. Based on four case studies, the chapter identifies five key capabilities needed to bring about the institutionalization of a company's proactive supply chain sustainability standards, namely: (1) interfirm dialogue, (2) risk management, (3) external stakeholder collaboration, (4) cross-functional integration, and (5) continuous improvement. These capabilities are associated with greater supply chain partner compliance.

In Chapter 11, "Supply Chain Structure as a Critical Driver of Sustainable Supplier Practices," Amrou Awaysheh and Robert D. Klassen explore the integration of social issues in supply chain management from an operations management perspective. They rely on a large-scale survey of plant managers in three industries in Canada and develop multi-item scales for dimensions of socially responsible supplier practices that they empirically validate. These dimensions are supplier human rights, labor practices, codes of conduct, and social audits. The authors find that increased transparency, as reflected in greater product visibility by the end-consumer, is related to the increased protection of the human rights of the suppliers' workforce, which in turn helps to protect a firm's brands. Organizational distance, as measured by the total length of the supply chain (number of supply chain tiers) is related to the increased use of multiple socially responsible supplier practices. When plants are positioned further upstream in the supply chain, managers increase their use of supplier codes of conduct.

In Chapter 12, "Going Green by Exporting," Emilio Galdeano-Gómez, Eva Carmona-Moreno, and José Céspedes-Lorente focus on the relations between environmental performance and exports. Can environmental performance explain differences in export intensity among firms? Does a firm's environmental performance improve as a firm's export activity increases? The composite equation model for exporting firms in the agro-food industry in southeast Spain, upon which the authors rely, shows that export intensity and productivity components, which includes environmental productivity, are in fact jointly determined. The findings point out that the different components of total firm productivity affect export intensity and that exporting increases productivity, including environmental productivity.

In Chapter 13, "Internationalization, Innovativeness, and Proactive Environmental Strategy among Small and Medium Enterprises," Javier Aguilera-Caracuel, M. Ángeles Escudero-Torres, Eulogio Cordón-Pozo, and Nuria Esther Hurtado-Torres analyze the influence that the internationalization of small and medium-sized enterprises has had on innovativeness and on the adoption of proactive environmental strategies. Using direct interviews with the CEOs of 155 Spanish export firms in the food industry, the authors found that international knowledge, derived from the firm's activities in foreign markets, had a significant impact on the level of innovativeness and development of green innovation. More specifically, they showed that a high degree of international diversification and an international learning orientation leads firms to easily acquire new and valuable knowledge about foreign markets that reinforces their innovativeness. Moreover, internationalizing firms with a high level of innovativeness are more willing to undertake an advanced and proactive environmental strategy, which they apply in the different locations where they are active.

Part IV: New Forms of Cross-Sector Cooperation Matter

In Chapter 14, "Why Do Private Firms Invest in Public Goods?" Gurneeta Vasudeva and Hildy Teegen provide a framework for understanding private participation and investment in the creation of global public goods like a clean environment. Based upon observations of and interviews with executives representing an important multinational, collaborative, pooled equity technology development venture, the authors suggest that firms share incentives to establish such a venture in collaboration with other firms and governments when the venture yields private benefits that are particular to individual firms, common benefits that are shared by the collaborating firms, and public goods that are shared by global publics. The authors argue that firms do not choose among these various benefit types, but instead seek to maximize the net advantage resulting from the interplay among these benefits.

In Chapter 15, "Voices from the Field: The Green Economy Partnership Process," Jack Hogin and Georgia Rubenstein describe how in 2009, the Minnesota Environmental Initiative (MEI) was asked by the cities of Minneapolis and Saint Paul, the Blue Green Alliance, and the state of Minnesota to convene a stakeholder process, the Green Economy Partnership Process (GEPP), to design a partnership model to support and further develop the green economy of the Twin Cities and greater Minnesota. Chapter 15 describes this effort, and others, to form partnerships for green economic development.

In Chapter 16, "Moving the Green Economy Forward: Conclusions from Research and Practice," Alfred Marcus provides nine conclusions from academic research about what is needed for the green economy to flourish. Then he turns to a practitioner's perspective and provides six key challenges that practitioners typically believe must be confronted and overcome. The nine conclusions from academic research are:

1. The number and type of organizations needed to induce takeoff must be large, their activities many, and the relationships among them must be dense and complicated.
2. The influences of government, social movements, and natural capital are large.
3. Organizations must engage in joint problem-solving.
4. There must be positive feedback loops.
5. Both new entrepreneurs and incumbent (existing) firms in an industry have important roles to play.
6. The burden on new entrepreneurs to develop markets is high.
7. By themselves, governments do not have the power to induce a long-term takeoff.
8. Venture capital can be very influential.
9. Managerial perceptions are critical.

The six challenges that practitioners believe must be confronted are in the areas of:

1. cooperation and conflict
2. focus vs. diversity
3. government vs. business vs. academia
4. egoism vs. altruism
5. local vs. national vs. international undertakings
6. long-term vs short-term thinking

In Sum

The chapters in this book provide interesting clues to better understand the multidimensional challenges related to creating a green economy (see Table 0.1). The authors illustrate the many dimensions of this transformation and explore changes that are needed in several sectors and industries. The broad focus on renewable energies and clean technologies is a dominant feature that characterizes many of the chapters: Chapter 1, by Bernard et al.; Chapter 3, by Lüthi & Wüstenhagen; chapters 4 and 7 by Marcus et al.; Chapter 8, by de Lange; Chapter 6, by Lüdeke-Freund and Loock; and Chapter 14, by Vasudeva and Teegen.

Renewable energies and cleantech represent a setting still not sufficiently explored in managerial and innovation theory. The growth of these sectors has accelerated over the last few years to the point that in 2008, for the first time, investments in renewable capacity were greater than investments in conventional energy both in Europe and in the United States. This feature makes the investigation of these industries interesting not only from the perspective of scholars in the field of organizations and the natural environment, but also to practitioners. Two chapters from Spanish scholars, namely Chapter 12 by Galdeano-Gómez et al. and Chapter 13 by Aguilera-Caracuel et al., address the agro-food industry and investigate how firms from these industries develop green strategies. Chapter 9, by Poldner et al., focuses on the design and fashion industry, exploring the development of a social movement that attempts to lead the change to more sustainable consumption by promoting a new idea of what is fashionable. In this volume, there are also cross-industry chapters, dealing with sectors such as retail, packaging, and publishing (Chapter 10, by Grimm et al.); and with food, chemicals, and transportation equipment (Chapter 11, Awaysheh & Klassen). There are also chapters not set in specific industries that explore cross-cutting topics. Recurring actors found in the book's chapters are finance and public sectors, where partnerships are needed with the business community in order to develop a sustainable economy. The lesson is that a number of value chains are involved in the process of change, and that the boundaries of a green economy are extremely broad.

With regard to the theories and methodologies used in the chapters, the book offers a mosaic of different approaches. Institutional, resource-based, and stakeholder theory are used by many scholars to analyze the diffusion of environmental strategies and practices, the relation with regulators and policymakers and NGOs, and the establishment of green partnerships. Other chapters are grounded in innovation and entrepreneurial theory or apply consumer theory and investigate the formation of preferences and the choice of clean technologies. Different research methodologies are used, including field case studies and secondary data and surveys, coupled with different statistical tools. The chapters prove

Table 0.1 Overview of chapters and their topics, theory, and methodology used

Chap. #/Authors	Title	Topic	Theory	Methodology
1. Michelle Bernard, Michael Craig, Itai Sened	The Role of Institutions in the Implementation of Wind Energy	Government, renewable and wind energy, market development	Institutional theory	Exploratory study, US, Germany and Spain
2. Jens Hamprecht, David C. Sprengel, Volker H. Hoffmann	How Regulatory Uncertainty Impedes the Reduction of Greenhouse Gases	Regulatory uncertainty, institutional pressure and strategic responses by firms	Institutional theory and uncertainty	Empirical global survey, data from DJSI
3. Sonja Lüthi and Rolf Wüstenhagen	Regulatory Risk and Cleantech Investment Decisions	PV projects, market diffusion, regulatory/policy frameworks	Consumer theory, preference investigation	Empirical, choice experiment
4. Alfred Marcus, Susan Cohen, and Kathleen Sutcliffe	Why Some Managers Expect to Benefit from Public Policies and Others Do Not	The role of and need for government in the emergence of new energy efficiency and renewable energy businesses	Managerial cognition and heuristics	Qualitative interviews and quantitative survey of energy efficiency and renewable energy innovators and entrepreneurs
5. Timo Busch, Bryan T. Stinchfield, Matthew S. Wood	Rethinking Sustainability, Innovation, and Financial Performance	The role of innovation as a mediator/moderator between ESP and CFP. The importance of contingency	"Does it pay to be green?" stream and innovation theory	Empirical, data from KLD, global survey
6. Florian Ludeke-Freund and Moritz Loock	What Kinds of Photovoltaic Projects Do Lenders Prefer to Finance?	PV projects, project finance, innovative business models	Project finance theory	Empirical, survey and interviews with experts, conjoint experiment, Germany
7. Alfred Marcus, Shmuel Ellis, Joel Malen, Israel Drori, and Itai Sened	Path Dependence and Creation in Venture Capital Investment	Venture capital and cleantech, renewable (solar, wind and biofuels) and energy efficiency, path dependency	Path dependence	Exploratory analysis, 15 countries
8. Deborah E. de Lange	High-Tech Cluster Revolution from an Organizational Ecology Perspective	High-tech clusters and innovation, green firms, newcomers vs. incumbents	Organizational ecology, innovation theory	Theoretical study

Continued

Table 0.1 Continued

Chap. #/Authors	Title	Topic	Theory	Methodology
9. Kim Poldner, Oana Branzei, Chris Steyaert	*Shecopreneuring: Stitching Global Ecosystems in the Ethical Fashion Industry*	Female entrepreneurship, ecosystems, fashion and design	Social entrepreneurship, social imaginary	Empirical, ethnography, three case studies, UK and Canada
10. Jörg H. Grimm, Joerg S. Hofstetter, Martina Müggler, Nils J. Peters	*Institutionalizing Proactive Sustainability Standards in Supply Chains: Which Institutional Capabilities Matter?*	Stakeholders, sustainable supply chain strategies, entrepreneurship	Institutional theory, institutional entrepreneurship	Empirical, four case studies based on interviews and secondary data, Swiss, Germany, and Ireland
11. Amrou Awaysheh, Robert D. Klassen	*Supply Chain Structure as a Critical Driver of Sustainable Supplier Practices*	Supply chain management and operations, responsible and social practices	Operations and supply chain management: transparency, dependence and distance	Empirical, survey, focus on Canadian companies
12. Emilio Galdeano-Gómez, Eva Carmona-Moreno, José Céspedes-Lorente	*Going Green by Exporting*	Internationalization, environmental performance and export activity	International trade theory, productivity	Empirical, panel data and survey, focus on Spanish companies
13. Javier Aguilera-Caracuel, M. Ángeles Escudero-Torres, Eulogio Cordón-Pozo, Nuria Esther Hurtado-Torres	*Internationalization, Innovativeness, and Proactive Environmental Strategy among Small and Medium Enterprises*	Internationalization, innovation and environmental strategies, SMEs	Internationalization and innovativeness, learning	Empirical, survey with direct interviews, focus on Spanish companies
14. Gurneeta Vasudeva, Hildy Teegen	*Why Do Private Firms Invest in Public Goods?*	Private sector, public goods, carbon capture technologies	Public goods theory	Exploratory study and direct interviews
15. Jack Hogin, Georgia Rubinstein	*Voices from the Field: The Green Economy Partnership Process*	Green economy, stakeholders, partnership, cities	Practitioner perspective	Exploratory study, focus on US
16. Alfred Marcus	*Moving the Green Economy Forward: Conclusions from Research and Practice*	Green economy, development paradoxes	Practitioner perspective	Theoretical study

that the field of organizations and the natural environment is lively, rich, and productive.

The number of geographical perspectives offered to the reader is a distinctive feature of this book. Authors come from several countries and represent many organizations and universities in Europe, the United States, Israel, and Canada. This is another relevant issue when considering the breadth of interest in the green economy.

Note

1. The chapters in the book originally were presented at two conferences held in 2010. The first, *Cross-Sector Leadership for the Green Economy*, was sponsored by the University of Minnesota's Center for Integrative Leadership. This conference was held on April 29–30, 2010 (see http://www.leadership.umn.edu/news/annual_conferences.html). The second conference, *Corporate Sustainability, Innovation, and Ecosystems in a Globalized World: Addressing the Green Challenge*, was sponsored by the Group on Organizations and the Natural Environment (GRONEN) and was held at SDA Bocconi School of Management, Milan, Italy, on June 23–26, 2010.

   The book benefitted from a grant received from the Institute for Renewable Energy and the Environment at the University of Minnesota. This grant supported the research found in chapters one and seven. The Center for New Institutional Social Sciences at the Washington University in St. Louis likewise supported the research found in these chapters.

References

Bruntdland, G. (1987). *Our Common Future, World Commission for Economic Development.* (NewYork: Oxford University Press).

Stern, N. (2006). *The Stern Review: The economics of climate change.* London: HM Treasury, 2006.

Sukhdev, P., Wittmer, H., Schröter-Schlaack, C., Nesshöver, C., Bishop, J., Ten Brink, P., Gundimeda, H., et al. (2010). *TEEB synthesis report: Mainstreaming the economics of nature: A synthesis of the approach.* Geneva: United Nations Environment Programme.

United Nations Environment Programme (2011). *Towards a green economy: Pathways to sustainable development and poverty eradication.* Geneva: United Nations Environment Programme.

World Resource Fund's 2010 *Living Planet Report*.

PART I

Institutions Matter

Firms play a major role in building a green economy by investing in clean and renewable technologies instead of in conventional and fossil fuel technologies. However, not only do some firms tend to invest in green technologies to a greater extent than other firms, but those in one jurisdiction also tend to invest in such technologies to a greater extent than those in other jurisdictions. Why has Denmark developed a significant lead in wind turbine technology? Why does Germany have such a large percentage of solar energy generation? Why does Spain have one of the highest percentages of wind energy production amongst developed countries? And why have Texas in the United States and Alberta in Canada (both major producers of oil and natural gas) invested much more in wind energy generation than Minnesota and Ontario (both of which have greater wind resources than Texas and Alberta)?

Research has shown that there are many explanatory drivers and influences for investments in green technologies. Natural resources such as sunshine and wind matter; firm strategy matters; firm capabilities matter; managerial perceptions and decisions matter; stakeholder influences matter; societal norms and preferences matter; and of course, institutional influences matter. Institutions shape regulations, investment returns, risk and uncertainty, and give firms incentives to invest.

This volume brings together chapters on several of these drivers of the green economy. Part I of the book focuses on how and why institutions matter. The chapters in Part I help us understand that it is not regulations alone but rather *how* regulations and public policy reduce technical and economic uncertainty and create incentives that drives investments in clean technologies such as renewable energy. Regulatory jurisdictions that foster the growth of renewable technologies are designed to influence investments in support infrastructure (for example, grid interfaces and transmission systems), to guarantee tariffs, and to manage conflicting stakeholder expectations. The chapters also show that institutional uncertainty and risk are not objective, but rather are perceived differently by firms and their managers. Managerial perceptions of an institutional environment as more or less uncertain, risky, or beneficial (munificent) drive managers' analysis of clean technology decisions and affect their subsequent behavior in investing in such projects. The insights from these chapters will help managers and policymakers to develop better regulations, conditions, and support systems that will foster clean technologies and green economies.

CHAPTER ONE

The Role of Institutions in the Implementation of Wind Energy

MICHELLE BERNARD, MICHAEL CRAIG, AND
ITAI SENED

Given the threat of climate change, the 2010 Deepwater Horizon oil spill, volatile natural gas prices, and unrest in the Middle East, implementing clean alternative energies is urgently needed. What drives the adoption of these energies, however, has yet to be fully determined. Neoclassical economists suggest that implementation would be accelerated by reducing regulations and establishing free markets, in which resource scarcity would drive deployment. Others argue that technology propels alternative energy use. North (1990) maintains that markets lead to institutional changes as firms and politically motivated interests induce changes in institutional structures (see also Sened, 1997). What then are the roles of resource availability, technology, and institutions in driving alternative energy utilization? We examine this question by focusing on wind, thereby avoiding controversies surrounding other alternative energies, such as the extent to which they are "clean" and the costs associated with their large-scale production, for there is a consensus that wind is clean and economically viable. We analyze the role of institutions in wind power implementation, comparing its adoption in Minnesota and Texas and then broadening our perspective with a global comparison among Spain, Germany, and the United States. In accord with North (1981, 1990), we maintain that market forces are not the only, or the most important, determinants of wind's performance. Rather, the institutions that govern and regulate wind implementation have a very large impact.

Institutional Theory

Neoclassical economic theory proposes that the main exogenous variable that explains shifts in economic performance is technology (Mokyr,

1990). As long as technologies are constant and preferences stable, everything else is accounted for by supply and demand. In the global economy of the twenty-first century, however, technology is shared worldwide. Wind technology is relatively simple and is commonly understood across different geographical and geopolitical arenas. Though wind conditions vary across geographical regions, this variation cannot completely account for differences in implementation. In fact, wind technology has been more widely implemented in regions with less favorable resources, such as southwest Europe, than in regions with more favorable resources, such as the central United States. Why have some markets heavily invested in wind technology while others have not? North provides an explanation. "The structure of the economy" (North, 1981, 1990) is just as potent an explanatory variable as technology and market forces.

In what follows, we show the connections between institutional arrangements that govern and regulate the production of wind energy and the extent to which it has been implemented. Our method of analysis is known as *analytic narrative* (Bates, Greif, Levi, Rosenthal, & Weingast, 1998). Formal models of mathematical economics are often abstractions of reality with little relevance to actual occurrences in real marketplaces. Analytic narrative connects economic theory with the real world. We assume that technology is constant across our study areas and control for resource availability (wind) using widely available global maps of wind conditions across regions of the globe (3Tier, 2011, U.S. Department of Energy, 2011a, 2011b). From preliminary analysis of data from OECD (Organisation for Economic Co-operation and Development) countries, we further note that competitive markets and production costs are similar for most inputs in the countries we studied. From the results of our study, we found that we could attribute much of the variance in wind power capacity to variance in institutional arrangements.

Wind's Cost-Competiveness

An oft-repeated argument for the minimal deployment of renewable energies is their noncompetitive pricing compared with fossil fuels, because of the underdeveloped technology of renewable energies. Coal- and natural gas-fired power plants, the reasoning goes, produce electricity at a lower cost per kilowatt-hour than renewables such as wind and so coal and natural gas are therefore preferred by profit-maximizing energy utilities. However, recent studies show the situation to be more nuanced, for wind is currently cost-competitive in certain scenarios and it is likely to be even more so in the near future. In 2010, the International Energy Agency (IEA) compared the cost to produce baseload electricity, or the minimum amount of electricity demanded over a 24-hour window, from nuclear, coal, gas, and wind for projects to be brought online by 2015. The IEA found onshore wind power in some cases to be cost-competitive with other sources, depending on local wind conditions. The study assumes a reasonable but highly uncertain price per ton of carbon dioxide, due to an anticipated cap-and-trade policy, which increases the cost of electricity from the carbon-emitting technologies coal and gas. Yet even without such a price, the study concludes that wind would be cost-competitive, albeit to a lesser extent. Two discount rates, or the annual interest rate at which money can be borrowed, are considered. The lower discount rate, 5 percent, favors capital-intensive low-carbon energy sources like nuclear and wind; the higher rate, 10 percent, favors coal and natural gas. It is crucial to emphasize that current interest rate forecasts predict discount rates much lower than 5 percent (International Energy Agency, 2010).

A study conducted in 2009 by the NRC or National Research Council (2009), a division of the U.S. National Academy of Sciences, titled *America's Energy Future,* agrees with the IEA's findings for the United States. The NRC study determined that for new electricity generation facilities, onshore and offshore wind are price-competitive with other energy sources, even in the absence of a price on carbon. Both the IEA's and the NRC's findings admit that the cost of wind-generated electricity varies widely, based on available wind resources. Furthermore, both studies consider prices at the power plant rather than at the system level, which excludes transmission and grid expenditures. These transmission and grid expenditures are often higher for wind farms, because of their isolated locations, but they still only add a few additional cents per kilowatt hour to the cost of electricity and so would most likely not put wind at a significant disadvantage.

The U.S. Energy Information Administration, or EIA, arrived at the same conclusions in 2009 as the previously mentioned two studies, even when factoring transmission and grid expenditures into costs. Areas with high-intensity winds outcompete fossil fuel electricity sources, even with transmission costs incorporated, underscoring the importance of available resources to wind energy adoption (U.S. Energy Information Administration, 2009). These three studies demonstrate that wind is more than cost-competitive with traditional energy sources (such as coal and natural gas) in certain regions. One would therefore expect regions with high-intensity winds to readily adopt wind power as a source of electricity, for the sake of profits as well as the environment. We next investigate whether this occurs in reality and, if not, why.

Resource Availability and Institutions

Although renewable energy technology—and its corresponding effect on electricity prices—does not vary across OECD nations, renewable resources (wind) and institutions do vary. We therefore examined the effects of resource (wind) availability and institutions on wind power implementation in the states of Minnesota and Texas, and then we broadened our investigation to a comparison of the United States, Spain, and Germany. We asked which state or country we would predict to have the greatest wind power capacity based on the available resources, as measured by wind intensity. Then we investigated whether our predictions aligned with reality, and if not, we looked into whether the regions' institutional approaches explained the differences.

Minnesota and Texas—Installed Wind Power Capacity. To gauge the extent of wind power implementation in Minnesota and Texas, we use two sources of electricity data. Table 1.1 summarizes the added and cumulative wind power capacity for Minnesota (MN) and Texas (TX) over 2007, 2008, and 2009 as well as their national rank in cumulative capacity, per the 2007 to 2009 Year End Market Reports of the AWEA or American Wind Energy Association (2008, 2009, 2010). As shown, Texas not only had significantly greater cumulative capacity than Minnesota as of 2009, but also added significant amounts of capacity annually, whereas capacity additions in Minnesota stalled and sharply decreased. Another metric for measuring the size of a state's wind power sector is its actual amount of electricity generated. The U.S. Department of Energy (DOE) provides this information for the electric power industry, as summarized in Table 1.2. Though Minnesota generated a higher percentage of its electricity from wind than Texas, by the raw quantity of wind-generated elec-

Table 1.1 Added and cumulative wind power capacity and national rank of Minnesota (MN) and Texas (TX), 2007–2009

Year	Added Capacity (MW)		Cumulative Capacity (MW)		National Ranking (Cumulative Capacity)	
	MN	TX	MN	TX	MN	TX
2007	405	1618	1299	4356	3	1
2008	455	2671	1754	7118	4	1
2009	56	2292	1809	9410	5	1

Source: American Wind Energy Association, 2008, 2009, and 2010.

Table 1.2 Electricity generated from wind and from all sources, from 2006 to 2008, in Minnesota (MN) and Texas (TX)

Year	Wind (MWh)		Total (MWh)		Wind (%)	
	MN	TX	MN	TX	MN	TX
2006	2,054,947	6,670,515	53,237,789	400,582,878	3.9	1.7
2007	2,638,812	9,006,383	54,477,646	405,492,296	4.8	2.2
2008	4,354,620	16,225,022	54,763,360	404,787,781	8.0	4.0

Source: Data from the U. S. Energy Information Administration, 2010a.

tricity, Texas's wind sector once again dominated Minnesota's, producing nearly four times the electricity in 2008.

Having established that Texas boasts a more vibrant wind power sector than Minnesota does in terms of size and recent growth, we next focus on explaining this difference. We know to a high degree of certainty that technology does not explain the difference, for technology, competitive markets, and production costs are likely constant across the two states, per our prior assumption. Is it true that Texas has greater wind resources than Minnesota? The annual average wind speeds at 80 meters elevation are an indicator for overall wind resources. Wind speed directly determines the potential for wind power in a state. The U.S. Department of Energy (2011a, 2011b) estimates that winds must have annual average speeds of greater than 6.5 m/s (meters per second) to be suitable for wind power development, as noted in the report "Wind Powering America." Another crucial determinant, however, is the location of the wind resources, since wind turbines in isolated gusty regions require large investments in new transmission lines to bring the electricity to its point of consumption, which has been recognized as a prominent barrier to wind power expansion across the United States (Wiser & Barbose, 2008). Therefore, both wind speed and distance from urban centers must be considered in judging the strength of wind resources for a given area.

Although Texas has some of the strongest winds in the United States, its winds of more moderate speeds are much closer to urban centers than its best winds. The state's breeziest region is the Panhandle, the majority of which has winds that average approximately 8.5 m/s in speed. Its largest cities, on the other hand, namely Houston, San Antonio, and Dallas, are about 450, 350, and 240 miles, respectively, southeast of the Panhandle. Although not as windy as the Panhandle, most of the northwestern and central Texas interior is, in fact, also suitable for wind power development, as it has winds measured mostly between 6.5 and 8.0 m/s, although in a patchy distribution. Both Dallas and Austin sit roughly 25 miles from such windy landscapes; Houston, the largest city, lies approximately 150 miles distant.

Minnesota's wind resources mirror those of Texas. Its strongest winds, with average speeds similar to those in the Texas Panhandle, occur in its southwestern corner, about 100 miles from Minneapolis and Saint Paul and 115 miles from Rochester, the third largest city in the state after the Twin Cities. However, these cities are near lands with significant wind resources. The Twin Cities are separated from winds averaging 7.0 to 7.5 m/s by a mere 40 miles, and Rochester is only about 10 miles from winds of speeds between 7.5 and 8.0 m/s.

An examination of the wind speed maps of Minnesota and Texas prove the two states to have similar wind resources with respect to intensity and relative location. For the disparity in the volume of wind power capacity within each state to be explained by available resources alone, Texas would be expected to have either faster or better situated winds than Minnesota. Given that this is not the case, wind resources cannot explain what has propelled Texas to generate more wind electricity than Minnesota. To better understand how each state's institutions affect its wind power sector, we examined the states' political and regulatory institutions. We found that differences between the two states' incentives, rules, policies, and regulatory agencies have better enabled Texas to expand its transmission grid and increase its wind power capacity.

In 2007, Minnesota enacted legislation implementing a statewide Renewables Portfolio Standard or RPS (Minn. Stat. §216B.1691) (Minnesota Office of the Revisor of Statutes, 2007). Minor amendments to this statute have been passed in subsequent years, but the statute remains essentially the same. The legislation mandated Xcel Energy, the largest utility in Minnesota, generate 30 percent of its electricity from renewables by 2020, and all other utilities to generate 25 percent of their electricity from renewables by 2025. Furthermore, 25 percent of Xcel's electricity derived from renewables had to be from wind or solar, with a maximum of 1 percent from solar. Thus, a high mandate was set for wind energy, as well as for other renewable technologies, including solar, thermal, landfill gas, and biomass. Unfortunately, Minnesota's RPS failed to set a high mandate for the construction of additional transmission lines, only requiring utilities to "make a good-faith effort" (see Minnesota Office of the Revisor of Statutes, 2007, §216B.1691, Subd. 2).

Passage of the RPS in Minnesota spurred a nearly five-fold increase in the capacity of proposed wind projects to the Midwest Independent Transmission System Operator (MISO), the regional transmission operator (RTO) responsible for managing the transmission grid in Minnesota along with twelve other states and Manitoba (ISO/RTO Council, 2010). If realized, these proposals would have exceeded the ultimate mandated amount of wind power generation by 340 percent (Marcus, 2010). Unfortunately, none of these proposals has begun operations, because of an extremely lengthy backlog of projects awaiting MISO's approval (Minnesota Department of Commerce, 2010). As of August 27, 2010, the most recent wind turbine project to leave the MISO waiting list for approval with completed reports first joined the list in April 25, 2006 (Midwest Independent Transmission System Operator, Inc. [MISO], 2010). This extraordinary slowness in the approval process is reflected in the wind capacity statistics for Minnesota (see Table 1.1), particularly in capacity additions in 2009. Indeed, on August 25, 2008, the Federal Energy Regulatory Commission (FERC) approved a plan submitted by MISO to reform its waiting list by, among other changes, moving from a first-come, first-served basis to prioritizing projects based on their likelihood of approval, and adding a fast-track option (Federal Energy Regulatory Commission, 2008). Unfortunately, the rewards of this reform have yet to be reaped.

Besides the multifarious policies dealing directly with renewable energies and the sluggish approval process for projects, wind power capacity growth in Minnesota also depends on the transmission grid. The highest wind potential in the state exists along its western and southwestern borders, far from the Twin Cities and Rochester. New wind projects in these areas would therefore require extensive construction of transmission lines to transport the generated electricity. To this end, 11 transmission-owning utilities, including Xcel Energy, have formed a joint initiative named CapX2020, which as of mid-2011 has three project proposals pending judgment by the Minnesota Public Utility Commission (PUC) and two that have been approved. However, only one of these projects is targeted to be in service in 2011; the four others are forecast to be completed between 2013 and 2015 (CapX2020, 2010). Once again, it is evident that Minnesota's institutions have obfuscated the path to quick, widespread wind energy deployment.

A collective action problem among grid stakeholders has further hindered the expansion of the transmission grid, a phenomenon documented by Marcus (2010). In 2007, a diverse coalition consisting of environmental groups, citizens, utilities, and regulators (e.g., MISO and the Office of Energy Security) formed to advocate for the passage of the RPS. Although they collaborated successfully in this case, the coalition began to weaken during the implementation phase of the RPS, as conflicting views on unforeseen issues created tension among members. The construction of new transmission lines, in particular, spread discord among stakeholders, with environmental groups and citizens opposing them for fear of their effects on wildlife and properties, respectively, while the utilities and regulators strongly advocated for them, given their fundamental role in electricity generation. To add to the confusion, various stakeholders randomly entered and exited the debate, making collective action even more difficult. For instance, the U.S. Department of Energy (2008) released a report, "20% Wind Energy by 2030," that included a plan for transmission grid expansion in Minnesota that did not align with the one proposed by other stakeholders, further obscuring the facts and stakes. Consequently, Minnesota has not been as successful in building the controversial infrastructure required for large-scale wind power generation as it has in passing universally appealing legislation like its Renewables Portfolio Standard, and the result has inhibited wind power growth.

Like Minnesota, Texas has enacted a Renewables Portfolio Standard, which was first passed in 1999 and then amended in 2005 to update the mandated minimum renewable energy capacity to 10,000 MW by 2025 (§39.904; Texas Constitution and Statutes, 2005b), or roughly 5 percent of the state's total electricity production. Remarkably, Texas achieved the mandated levels of electricity generation in early 2010 (ERCOT, 2010), although, considering the rapid expansion wind power in Texas has undergone in recent years (see Table 1.1), this should come as no surprise. In addition to the electricity generation requirement, the 2005 amendment to the RPS, Senate Bill 20 (Texas State Legislature, 2005), grants the Public Utility Commission (PUC) of Texas the authority to order an electric utility or distribution and transmission utility to expand or construct new transmission facilities (§39.203(e); Texas Constitution and Statutes, 2005a), a marked institutional departure from the setup in Minnesota. This authority gives the PUC of Texas considerable power to ensure that grid expansion occurs in a timely and efficient manner. Indeed, another part of Senate Bill 20 ordered the PUC of Texas to designate, after consultation with related entities such as the Energy Reliability Council of Texas (ERCOT), Competitive Renewable Energy Zones (CREZs), defined as the best areas for wind energy development because of their exceptional wind resources and available lands. Once the Texas PUC selected the CREZs, it would then commission utilities to expand the transmission grid to these areas, thereby ensuring adequate transmission capacity, regardless of the status of wind projects in the region. In 2008, the Texas PUC selected five areas as CREZs, two in the Panhandle and three in West Texas, and it has already assigned billions of dollars to transmission projects (see Public Utility Commission of Texas, 2010).

In addition to rapidly installing new transmission lines, Texas continues to bring new wind projects online. This is at least partly due to the nature of Texas's regional transmission operator, ERCOT, which only operates within Texas. Specifically, ERCOT covers approximately 75 percent of Texas, including the major urban centers and the southernmost portion of the windy Panhandle. Thanks to ERCOT's intrastate nature, it is only subject to state authority and legislation and so avoids a host of federal regulations (Fleisher, 2008), allowing it to operate freely. Furthermore, since ERCOT's jurisdiction coincides with that of the PUC of Texas, the two entities can collaborate more efficiently than other states with multilateral RTOs. The result: steady and rapid expansion of wind power capacity in Texas, as evidenced by the state's continually growing wind power capacity (see Table 1.1).

Although the political and regulatory institutions of Minnesota and Texas are similar in many respects, key differences exist between them, particularly with respect to the states' RPSs and regulatory processes. Texas has rapidly and, at times, proactively built new transmission lines, and is not hindered by a long backlog of project approvals as Minnesota is. Ultimately, it has been these institutional differences, rather than technology or resource availability, that have produced the divergence in wind power capacity between Texas and Minnesota.

United States, Spain and Germany—Installed Wind Power Capacity. In this section of the chapter, we examine whether the same explanation behind wind power adoption might hold in an analysis of the wind power in the United States, Spain, and Germany. Figure 1.1 reports the percent of total electricity generated by renewable sources in Germany, Spain, and the United States for the year 2009. As can be seen, Germany and Spain have far exceeded the United States in their renewable energy implementation. An investigation into the causes of these disparities in wind-derived electricity reveals that they are driven not by available resources, but rather by national institutions.

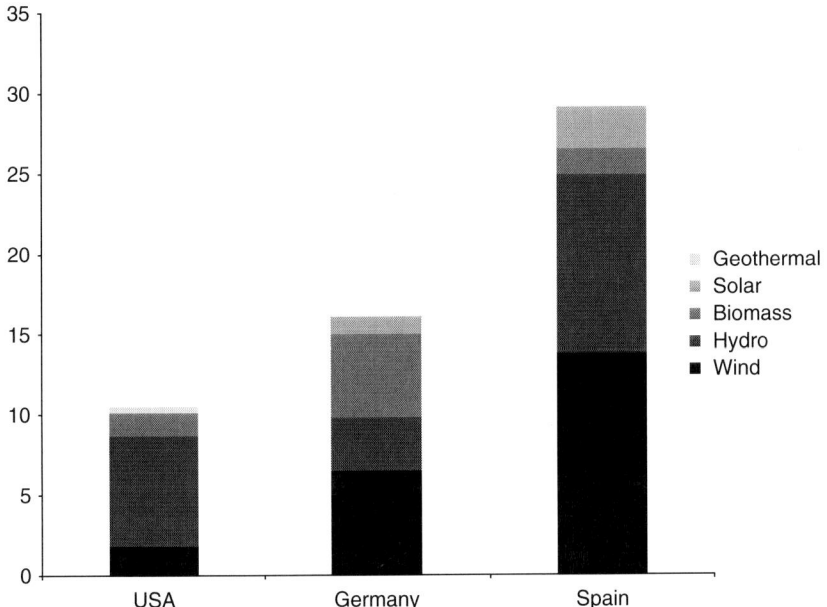

Figure 1.1 Percent of total electricity generated by renewable resources in the United States, Germany, and Spain.

Sources: U.S. Energy Information Administration, 2010a; Germany Working Group on Renewable Energy Statistics for the Federal Ministry for the Environment, Nature Conservation and Nuclear Safety, 2010; Réd Eléctrica de España, 2010.

To better understand the root of the gap in wind-derived electricity between the United States and its European counterparts, Spain and Germany, we first examined the available wind resources in each nation (Archer & Jacobson, 2005). The maps clearly show the United States to have considerably greater wind intensity, particularly in the Midwest, where winds attain speeds of between 6.9 m/s and 8.6 m/s, and along its coasts, which have areas with greater than 9.4 m/s wind speeds. Germany, on the other hand, has fairly moderate wind resources, which are concentrated mostly in its northern half and are mostly below 9.4 m/s in speed. Spain appears to have very limited amounts of wind above 5.9 m/s. With respect to the location of these high-intensity winds, the United States does not have a clear advantage, for the wind-rich Midwest is far removed from the densely populated coasts, although the coastal regions do have high-intensity onshore and offshore winds of their own. Conversely, many of Germany's most populated cities, including Berlin, Hamburg, and Bremen, sit in the northern part of the nation, close to the nation's best winds. Determining how well sited Spain's high-intensity winds are relative to its most populous areas, unlike the situation in the other two nations, is not useful on a broad basis, because Spain's high-intensity winds are not concentrated in a given area. From these analyses, one might safely conclude that, considering the combination of availability and location of wind resources alone, the United States would rank as the most favorable for wind power development, Germany as the second, and Spain as the least. However, this ranking is exactly the inverse of the actual percentage of electricity generated in each country from wind power.

The institutions, political and otherwise, in Spain, Germany, and the United States that affect renewable energy are incredibly diverse, with substantial similarities and differences existing among them. Table 1.3 shows some of the types of policies that each nation has enacted regarding renewable energy. Perhaps the single greatest policy difference among the United States and Spain and Germany is that the latter two nations have adopted a feed-in tariff, passed and revised most recently in Spain in 1997 and 2004 and in Germany in 1991 and 2009. Spain's is the Renewable Energy Plan; Germany's is the Renewable Energy Sources Act. Feed-in tariffs essentially guarantee fixed payments per unit of energy for electricity generated by renewable energies, which gets rid of the uncertainty of what price utilities would pay for completed renewable projects (Sawin, 2004). The efficacy of feed-in tariffs relative to other policies aimed at promoting renewable energies has been well established (Butler & Neuhoff, 2004; Mitchell, Bauknecht, & Connor, 2006), although some question its long-term impact (e.g., Sijm, 2002). Although the feed-in tariffs of Spain and Germany differ in some respects (see Ragwitz & Huber, 2005), they both, in addition to ensuring minimum payments for renewably generated electricity, guarantee grid access to renewable projects, a promise enforceable by state orders to transmission operators (Ragwitz & Huber, 2005). The United States has no such national mechanism in place, and so has no such power to force the expansion of its transmission grid. This same scenario emerged in the case study of Minnesota and Texas, where the state that had the authority to force utilities to expand the grid for renewable projects—Texas—also had higher renewable power capacity.

Another point of comparison between the international and between-state case studies with respect to the transmission grid is the complexity

Table 1.3 Types of renewable energy policies enacted by Germany, Spain, and the United States

	Feed-in tariff	RPS	Capital subsidies, grants, or rebates	Investment or tax credits	Public investment, loans, or financing	Energy production payments/ tax credits	Sales, energy, excise tax, or VAT reduction
Germany	Yes	Yes	Yes	Yes	Yes	No	Yes
Spain	Yes	Yes	Yes	Yes	Yes	No	Yes
United States	No	Yes*	Yes	Yes	Yes	Yes	No

Source: REN21 (Renewable Energy Policy Network for the 21st Century). (2010).
While no national RPS (Renewable Performance Standard) exists in the United States, the majority of states have adopted an RPS or similar legislation.

of their grid operations. As of 2010, the United States has a total of seven RTOs, and they don't even include a large portion of the nation. Germany and Spain, on the other hand, each have a less complicated system and it covers the entire nation. In Germany, there are four grid operators, whereas Spain only has one, the Red Eléctrica de España. Rank-ordering the nations by number of transmission operators yields a list that is the inverse of the list rank-ordered by wind power capacity. In other words, the more transmission operators at the national level, the less wind power results. We also found this pattern in the previous case study, for Texas had its own exclusive transmission operator, ERCOT, whereas Minnesota was controlled by a transmission operator that sprawled across 13 states and provinces.

Spain presents a particularly interesting example, for it has consolidated all of its transmission operation under a single private entity. This has enabled the nation to develop the Control Center for Renewable Energies, which controls all generation of electricity from wind in real-time (within 15 minutes) and is integrated with the broader Power Control Center, which balances power generation from all sources (Domínguez de la Torre, Juberías, Prieto, Rivas, & Ruiz, 2008). This setup allows for maximal incorporation of wind-derived electricity in two ways. First, because these centers operate nationwide, wind-derived electricity can be distributed across a vast geographical region, moderating any localized spikes in wind power and so functioning more as baseload power. Additionally, in times of abnormally high national wind power generation, the nation's baseload generators, e.g., coal or natural gas plants, can be modified accordingly, to make room on the grid for clean energy. Conversely, expanding wind power in the United States is coming into increasing conflict with grid capacity and largely inflexible baseload generators that can't handle surges of electricity generation, often resulting in the temporary curtailment of energy produced by wind farms. Germany, which has a more unified grid than the United States, may at least partly enjoy a similar advantage to Spain's (Frontier Economics, 2009).

The extensive legacy of the German and Spain feed-in tariff policy hints at an even deeper disparity among the United States and Germany and Spain: the extent of institutional uncertainty. Businesses, including renewable energy ones, have been shown to react negatively to perceived institutional uncertainty by shifting investment to more stable jurisdictions (see e.g., Lüthi & Wüstenhagen, 2010), a notion reflected in the renewable energy capacities of our three study nations. Germany and Spain have had meaningful policies, including an RPS and feed-in tariff, aimed at increasing renewable energy capacity since 1997 (Germany) and 1991 (Spain). Such long-term commitments have indicated strong support for renewable energy and decreased institutional uncertainty, which is reflected in high renewable power capacity. The United States, on the other hand, does not have, and has not ever had, a feed-in tariff. Furthermore, although it does have a panoply of financial incentives aimed at renewable energies, their existence has historically been highly uncertain. For instance, one of the major financial incentives of the United States, the Production Tax Credit (PTC), expired three times between 1994 and 2009, in 1999, 2001, and 2003. Each year following the tax credit's expiration, renewable energy power capacity growth has all but ceased, dropping to about 20 percent of the previous year (Combs, 2008). Such vacillation in policymaking is but one of many things that have plagued the U.S. renewable power industry, hence the nation's relatively low wind-power capacity as a percent of total power capacity.

Referring to another PTC lapse expected in 2010, George Sterzinger, executive director of the Renewable Energy Policy Project at the U.S. Department of Energy, explained in an interview on National Public Radio, "Right now, if you build a wind project, the government will, essentially, cut you a check for 30 percent of the cost. But that incentive is running out of rope and scheduled to expire at the end of 2010. That deadline prompted a lot of activity last year. Everybody moved their projects forward into 2009 to take advantage of it, but now some developers are waiting to see if the credit will be extended" (Brady, 2010). A lapse of the PTC in 2010 was averted when President Obama extended it for 2 years, to 2012, in the American Recovery and Reinvestment Act (U.S. Internal Revenue Service, 2009). Nonetheless, sensitive to the point of view expressed by Sterzinger, legislators have in fact pursued long-term legislation to create market certainty for years, but have had little success to date.

Conclusion

In this chapter, we utilized analytic narratives to study the discrepancies in the growth of wind power adoption across two U.S. states and between the US and two other OECD countries. We started with the neoclassical economics argument that in a global economy where renewable energy technologies spread rapidly, faltering evolution of wind power in some countries, particularly in the United States, poses a puzzle. Given the fact that wind power can offer competitive costs of electricity production as a result of technological development, as well as the environmental and political advantages of wind power, it should enjoy wider implementation in U.S. electricity markets. To investigate this anomaly, we restricted our analysis to OECD countries, to control for preferences and trends of consumption, as well as for average income per capita. After ruling out those variables, we found that institutional structure (North, 1981, 1990) is a potential exogenous variable that might help us sort out our empirical puzzle.

We examined whether institutions could significantly affect wind power implementation by using analytic narrative to examine wind power in two states of the United States and in three countries, namely,

Minnesota, Texas, the United States, Germany, and Spain. Our main analytical argument comes from a commonly made proposition in the neoclassical economic literature that high transaction costs and high uncertainty delay economic growth, whereas lower transaction costs and lower uncertainty enhance it. For the institutional structures of our cases, we specified the exact ways in which each institutional structure affected economic performance. We supplemented our analysis with available data and validated our findings by relying on other studies that point in the same direction.

Ultimately, we found that institutions matter. Our analytic narratives clearly indicate that institutional control of the power grid and of its expansion, coupled with some key policies, such as feed-in tariffs, explain most of the variance in the degree of expansion of the industry across states and across countries. Although for methodological reasons we have limited ourselves to wind power in OECD countries, we have every reason to believe that our findings apply beyond wind power and beyond OECD nations.

In addition to its academic value, our research underscores the importance to wind power implementation of a number of policies. Adequate expansion of electricity transmission should be assured via policies that, for instance, mandate the connection of renewable energy facilities in a proactive and timely manner. More homogeneous control over grid operations, be it of transmission expansion or power generation, can lower some barriers to significant renewable energy implementation—for example, the inability of grids to accommodate the inherently variable wind or solar electricity generation. Long-term institutional commitments by governments to renewable energies should be adopted in order to decrease uncertainty and assure continuous rather than sporadic construction. Such actions can greatly change the nature of renewable energy power generation, allowing cleaner and cheaper generators of electricity like wind to play a more significant role in this sector.

In calculating the costs and benefits involved with renewable energies, we only accounted for the supply side. There is every reason to believe that the spread of alternative energy production will do much more good than what our analysis suggests by, for instance, improving our environment, creating new jobs, reducing the price of electricity to consumers, and helping to revitalize the economy.

References

American Wind Energy Association (AWEA). (2008). *AWEA 2007 Market Report.* Retrieved August 1, 2010, from http://www.awea.org/learnabout/publications/upload/4Q07.pdf

American Wind Energy Association (AWEA). (2009). *American Wind Energy Association Annual Wind Industry Report Year Ending 2008.* Retrieved August 1, 2010, from http://www.awea.org/learnabout/publications/upload/AWEA-Annual-Wind-Report-2009.pdf

American Wind Energy Association (AWEA). (2010). *AWEA Year End 2009 Market Report*. Retrieved August 1, 2010, from http://www.awea.org/learnabout/publications/upload/4Q09.pdf

Archer, C. L., & Jacobson, M. Z. (2005). Evaluation of global wind power. *Journal of Geophysical Research, 110*, D12110.

Bates, R., Greif, A., Levi, M., Rosenthal, J. L., & Weingast, B. (1998). *Analytic narratives*. Princeton, NJ: Princeton University Press.

Brady, J. (2010, September 6). Wind power wanes with fading federal incentives. NPR (National Public Radio— from All Things Considered). Retrieved from http://www.npr.org/templates/story/story.php?storyId=129632055

Butler, L., & Neuhoff, K. (2004). *Comparison of feed in tariff, quota and auction mechanisms to support wind power development* (CMI Working Paper 70). Retrieved August 9, 2010, from http://www.dspace.cam.ac.uk/bitstream/1810/131635/1/ep70.pdf

CapX2020. (2010) Transmission line projects. Retrieved August 22, 2010, from http://www.capx2020.com/Projects/index.html

Combs, S (2008). Wind energy. Chapter 11 of *The energy report 2008*. (May 6, 2008). Retrieved August 1, 2010, from http://www.window.state.tx.us/specialrpt/energy/pdf/96-1266EnergyReport.pdf

Domínguez, T., de la Torre, M., Juberías, G., Prieto, E., Rivas, R., & Ruiz, E. (2008). *Renewable energy supervision and real time production control in Spain*. Paper 302, presented at International Conference on Renewable Energies and Power Quality, Santander, Spain, March 12, 13, 14, 2008. Retrieved August 7, 2010, from http://www.icrepq.com/icrepq-08/302-DOMINGUEZ-summary.pdf

ERCOT (Electric Reliability Council of Texas). (2010).Texas posts record increase in voluntary renewable energy credits. (May 14, 2010). Retrieved October 29, 2010, from http://www.ercot.com/news/press_releases/2010/nr-05-14-10

Federal Energy Regulatory Commission. (2008).124 FERC ¶61,183, Docket No. ER08-1169-000, Order conditionally accepting tariff revisions addressing queue reform. http://www.ferc.gov/EventCalendar/Files/20080825213947-ER08-1169-000.pdf

Fleisher, J. M. (2008). ERCOT's jurisdictional status: A legal history and contemporary appraisal. *Texas Journal of Oil, Gas and Energy Law, 3*(1), 5–21.

Frontier Economics. (2009, March). *Options for the future structure of the German electricity transmission grid*. http://www.bmwi.de/English/Redaktion/Pdf/options-future-structure-german-electricity-transmission-grid,property=pdf,bereich=bmwi,sprache=en,rwb=true.pdf (accessed Aug. 18, 2011)

Germany Working Group on Renewable Energy Statistics for the Federal Ministry for the Environment, Nature Conservation and Nuclear Safety. (2010, March). *Development of renewable energy sources in Germany 2009*. http://www.germany.info/contentblob/2674862/Daten/748533/BMU_Development_of_Renewable_Energy_Sources_in_Germany_2009_DD.pdf

International Energy Agency (2010). The projected costs of generating electricity: 2010 edition. Executive summary. (2010). Retrieved March 25, 2010, from http://www.iea.org/Textbase/npsum/ElecCost2010SUM.pdf

ISO/RTO Council. (2010). ISO RTO Operating Regions. Retrieved September 6, 2010, from http://www.isorto.org/site/c.jhKQIZPBImE/b.2604471/k.B14E/Map.htm

Jaccard, M. (2004). Renewable portfolio standard. *Encyclopedia of energy*, vol. 5, 413–421. Amsterdam, UK: Elsevier Inc.

Lüthi, S., & Wüstenhagen, R. (2010). The price of policy risk – Empirical insights from choice experiments with European photovoltaic project developers (Project no. 031569). *DISTRES*. Retrieved March 2, 2010, from http://www.alexandria.unisg.ch/publications/61086

Marcus, A. (2010). Institutional leadership in renewable energy development: filling the institutional void. Presented at the Center for New Institutional Social Sciences, St. Louis, Missouri, March 22, 2010.

Midwest Independent Transmission System Operator, Inc. (MISO). Midwest ISO generator interconnection queue— August 27, 2010. (2010). *Midwest ISO*. Retrieved August 27, 2010, from https://www.midwestiso.org/Planning/GeneratorInterconnection/Pages/InterconnectionQueue.aspx

Minnesota Department of Commerce, Office of Energy Security. (2010). Minnesota's electric transmission system–Now and into the future. Retrieved January 15, 2010, from http://www.state.mn.us/mn/externalDocs/Commerce/Minnesota_Electric_Transmission_System_Report_2009_011510034109_ElecTransmissionSystemReport2009.pdf

Minnesota Office of the Revisor of Statutes (2007). §216B.1691 renewable energy objectives. *Minnesota Statutes*. Retrieved August 27, 2010, from https://www.revisor.mn.gov/statutes/?id=216B.1691&year=2007

Mitchell, C., Bauknecht, D., & Connor, P. M. (2006). Effectiveness through risk reduction: A comparison of the renewable obligation in England and Wales and the feed-in system in Germany. *Energy Policy, 34*, 297–305.

Mokyr, J. (1990). *The lever of riches: Technological creativity and economic progress*. New York and London: Oxford University Press.

National Research Council. (2009). *America's energy future*. Washington, D.C.: National Academies Press.

North, D. C. (1981). *Structure and change in economic history*. New York: Norton.

North, D. C. (1990). *Institutions, institutional change and economic performance*. New York: Cambridge University Press.

Public Utility Commission of Texas. (2010). PUCT– CREZ home page. *CREZ [Competitive Renewable Energy Zone] Transmission Program Information Center*. Retrieved August 19, 2010, from http://www.texascrezprojects.com/default.aspx

Ragwitz, M., & Huber, C. (2005). Feed-in systems in Germany and Spain and a comparison. *Fraunhofer Institute for Systems and Innovation Research*. Retrieved August 11, 2010, from http://www.bmu.de/files/english/pdf/application/pdf/langfassung_einspeisesysteme_en.pdf

Red Eléctrica de España. (2010). The 2009 Spanish electricity system summary. Retrieved March 27, 2010, from http://www.ree.es/ingles/sistema_electrico/pdf/infosis/sintesis_REE_2009_eng.pdf

REN21 (Renewable Energy Policy Network for the 21st Century). (2010). Renewables 2010 global status report. Retrieved August 22, 2010, from http://www.ren21.net/Portals/97/documents/GSR/REN21_GSR_2010_full_revised%20Sept2010.pdf

Sawin, J. L. (2004, January). National policy instruments. International Conference for Renewable Energies, Bonn. June 1–4, 2004. Bonn, Germany. Co-chairs Heidemarie Wieczorek-Zeul and Jürgen Trittin.

Sened, Itai (1997). *The political institution of private property*. Cambridge: Cambridge University Press.

Sijm, J. P. M. (2002). The performance of feed-in tariffs to promote renewable electricity in European countries (ECN no. 7.7748). From Energy Research Centre of the Netherlands, November 2002. Retrieved September 1, 2010, from http://www.ecn.nl/docs/library/report/2002/c02083.pdf

Texas State Legislature. (2005). Senate Bill 20 (SB20) Texas Senate, 79(1) session. Retrieved September 2, 2010, from http://www.capitol.state.tx.us/BillLookup/Text.aspx?LegSess=791&Bill=SB20

Texas Constitution and Statutes. (2005a). §39.203 Transmission and distribution service. *Texas Utilities Code*. Retrieved August 28, 2011, from http://www.statutes.legis.state.tx.us/Docs/UT/htm/UT.39.htm

Texas Constitution and Statutes. (2005b). §39.904 Goal for renewable energy. *Texas Utilities Code*. Retrieved August 28, 2011, from http://www.statutes.legis.state.tx.us/Docs/UT/htm/UT.39.htm#39.904

3Tier. (2011). Global mean wind speed at 80m. Retrieved August 17, 2011, from http://www.3tier.com/static/ttcms/us/images/support/maps/3tier_5km_global_wind_speed.jpg

U.S. Department of Energy. (2008). 20% wind energy by 2030: Increasing wind energy's contribution to U.S. electricity supply. Retrieved August 23, 2010, from http://www.nrel.gov/docs/fy08osti/41869.pdf

U. S. Department of Energy. (2011a). Wind powering America: Minnesota wind map and resource potential (2011).. Retrieved March 19, 2011, from http://www.windpoweringamerica.gov/wind_resource_maps.asp?stateab=mn

U. S. Department of Energy. (2011b). Wind powering America: Texas wind map and resource potential. (2011). Retrieved March 19, 2011, from http://www.windpoweringamerica.gov/wind_resource_maps.asp?stateab=tx

U.S. Energy Information Administration. (2009). 2016 levelized cost of new generation resources. From the *Annual energy outlook 2010*. Retrieved February 12, 2010, from http://www.eia.doe.gov/oiaf/aeo/electricity_generation.html

U.S. Energy Information Administration. (2010a). Electric power monthly July 2010. [DOE/EIA-0226 (2010/07)]. Retrieved August 2, 2010, from http://www.eia.doe.gov/ftproot/electricity/epm/02261007.pdf

U.S. Internal Revenue Service. (2009, April). Energy provisions of the American Recovery and Reinvestment Act of 2009. Retrieved August 18, 2011, from http://www.irs.gov/newsroom/article/0,,id=206871,00.html

Wiser, R., & Barbose, G. (2008). *Renewables portfolio standards in the United States: A Status Report with Data Through 2007*. Published by Lawrence Berkeley National Laboratory. Retrieved January 27, 2010, from http://eetd.lbl.gov/ea/EMS/reports/lbnl-154e

CHAPTER TWO

How Regulatory Uncertainty Impedes the Reduction of Greenhouse Gas Emissions

JENS HAMPRECHT, DAVID C. SPRENGEL, AND
VOLKER H. HOFFMANN

> *It is completely unclear how things will develop beyond 2012. Nobody knows what a [climate] regulation will look like then. We are moving in the dark and we are waiting for someone to turn on the lights.*
> —VP of a European electricity provider on climate regulations

Empirical evidence exists on the response strategies of organizations to institutional pressures (Goodstein, 1994); however, how organizations respond to institutional pressures when regulatory uncertainty is high remains disputed. Although some authors suggest that in times of high uncertainty organizations shift toward conformity to institutional pressures (Oliver, 1991), others found response strategies diverge and become less focused (George, Chattopadhyay, Sitkin, & Barden, 2006; Goodrick & Salancik, 1996; Miller & Toulouse, 1998). Early empirical contributions to this debate have been limited as they only used one measure, such as the strategic direction, in order to assess the response strategy of an organization. Such a strategy, however, can be multifaceted and complex. In this chapter we extend previous research by disaggregating an organization's response strategy. We distinguish between the scope, i.e., the number of response measures, and their strategic direction, i.e., the objective of these measures. Our results confirm that organizations adapt the direction of their response strategy as pressure for greenhouse gas reduction increases. They become more likely to avoid the pressure, for example by divesting themselves of greenhouse-gas-intensive product lines. However, as regulatory uncertainty increases, organizations do not pursue a narrow strategic direction. Instead, the scope of their response strategy becomes broader, and they may simultaneously engage in conflicting activities, such as adjusting to the institutional pressure and lobbying against it in the political process. We argue that these contradictory activities hinder the emergence of industry role models and best practices.

A common understanding is that new environmental regulation gradually transforms itself into a norm and then into a taken-for-granted industry standard (Hoffman, 1999). In the case of the Kyoto Protocol, however, this transition is obscured by uncertainty. Policymakers should not underestimate how significantly the uncertainty regarding future greehouse gas (GHG) regulations limits the emergence of common practices for GHG reduction. Our findings contradict the frequently held viewpoint that industry will resolve the issue of GHG reduction without regulation. If there is no future regulation on GHG reduction, it is unlikely that the best practices for greenhouse gas reduction will be widely adopted.

We start with an overview of previous research about our line of inquiry. Then, we develop hypotheses about how institutional pressures and regulatory uncertainty should influence the strategic direction and scope of organizational responses. We test the hypotheses with data from a global survey of companies' response strategies to pressures to reduce their greenhouse gas emissions. At the end of the chapter, we summarize our contributions and provide suggestions for future research.

Responses to Institutional Pressures

Organizations can respond to institutional pressures in four ways: adjust to it, influence it, avoid it, or simply ignore it. Adjusting to institutional pressure implies changes to the organization itself. Here, the objective is to bring the organization's activities or structure in line with institutional pressure by complying with rules, norms, and standards (Oliver, 1991). Such a strategy is pursued by Volkswagen, for example, which aims to systematically reduce the fuel demand and resulting greenhouse gas emissions of its automobiles. Measures to *influence* the organization's environment intend to "actively change or exert power over the content of the expectations themselves or the sources that seek to express or enforce them" (Oliver, 1991, p. 157). Such a strategy has been pursued by Exxon, which has challenged assumptions that there is a link between greenhouse emissions and global warming, funded the election campaign of George W. Bush, and sought to exert pressure on the U.S. government not to sign the Kyoto Protocol (Guardian, 2005). An organization may also employ measures to *avoid* institutional pressure (Engau & Hoffmann, 2010). For example, the steel manufacturer ThyssenKrupp describes emission trading schemes as a threat to its production sites in Germany (Finanznachrichten, 2008) and has shifted its investments outside of Europe to Brazil where it has invested over five billion euros, which do not fall under strict emission reduction regulation as investments in Europe do. (Financial Times Deutschland, 2010). Finally, there is the option of *ignoring* institutional pressure, which entails no concrete action with reference to the pressure.

Because a direct link between inaction and institutional pressures cannot always be assumed, we exclude this from further analysis.

The above strategies should not be regarded as monolithic responses. As Lounsbury observes (2007, p. 290), "[too little] is known about the *variety* of ways in which organizations respond to institutional pressures." We do not assume that organizations follow just one *direction* at a time when pursuing their response strategies (Goodstein, 1994; Oliver, 1991). Instead, they can simultaneously engage in a number of different response measures with the intention of adjusting, influencing, or avoiding institutional pressures, each to a different extent. We further recognize that a response strategy may vary in its *scope:* the number of response measures and strategic directions the organization engages in at a time will vary (Miller & Toulouse, 1998). These distinctions allow us to develop more fine-grained hypotheses on how organizations respond to institutional pressures and regulatory uncertainty concerning greenhouse gas emissions.

Institutional pressure can be perceived differently by managers in different organizations (Goodstein, 1994). For example, the executives of a large and prominent oil and gas company are likely to perceive higher levels of institutional pressure from nongovernmental organizations (NGOs) and subsequently also from other stakeholders than their smaller competitors perceive (Buckley, 1968; Miller & Toulouse, 1998). Based on Oliver's (1991) categorization, we distinguish three strategic directions: *adjusting to, avoiding,* and *influencing* institutional pressure. These three directions can be characterized in terms of the active agency required by the organization. Adjusting to institutional pressure typically demands only a low level of activity, as the organization merely complies with the pressure. Similarly, measures to avoid institutional pressure are not likely to be that demanding. However, measures to avoid require more effort than adjusting to pressures. In the case of avoiding pressure, the organization does not actually comply with the pressure, but rather circumvents it. Influencing institutional pressure is the response strategy that requires agency by an organization as the organization seeks to gain control over the institutional pressure and change it in its own interest (Oliver, 1991).

Goodstein (1994) analyzed the relationship between an increase in institutional pressure and conformist responses. In his study, he did not determine whether organizations influence institutional pressure (Goodstein, 1994, p. 365). However, his results suggest that more conformist actions are likely if institutional pressure is high. This observation implies that measures to influence institutional pressure should be lower when institutional pressure is high. The conceptual research of Oliver (1991) indicates that the long-term survival chances of organizations are higher when they respond in a conformist manner to very high pressure. This is because the risk of losing their license to operate becomes too high if they continue opposing the institutional pressure.

Hypothesis 1a: The higher the level of perceived institutional pressure, the more an organization's response strategy consists of measures to adjust to this pressure.

Hypothesis 1b: The higher the level of perceived institutional pressure, the more an organization's response strategy consists of measures to avoid this pressure.

Hypothesis 1c: The higher the level of perceived institutional pressure, the less an organization's response strategy consists of measures to influence this pressure.

Although there are consistent findings on the effects of institutional pressure on response strategies, the effects of environmental uncertainty remain disputed. Environmental uncertainty can be understood as the inability of organizations and individuals to predict the future state of the environment (Hoffmann, Trautmann, & Hamprecht, 2009; Hoffmann, Trautmann, & Schneider, 2008). Several studies indicate how uncertainty can influence the activities of an organization (Amit & Schoemaker, 1993; Aragón-Correa & Sharma, 2003; Dess & Beard, 1984). However, few empirical contributions have assessed how *regulatory* uncertainty influences the organization's response to pressure in the institutional environment. From the perspective of transaction cost economics (TCE), companies should be expected to limit their activities and to focus on the one strategic direction they regard as most promising (Williamson, 1981). TCE suggest that companies should minimize switching costs between different strategic directions as well as the costs for enforcing each strategic direction in the organization. On the contrary, game theory suggests that organizations can successfully manage uncertainty by hedging their bets and pursuing multiple strategies at the same time (Harsanyi, 1973).

George et al. (2006) follow the latter line of thinking. They suggest in a conceptual paper that environmental uncertainty leads decision-makers to bet on several future outcomes. In another conceptual contribution, Miller (1993) also suggests that companies broaden the scope of their responses in an uncertain situation. However, as there is a lack of empirical contributions, it remains unclear if the logic of transaction economics or game theory is more relevant for predicting the scope of an organization's strategy. For our research we assume that the rationale of transaction cost economics (i.e., resource optimization through focus on one strategic direction) is valid as long as organizations have the necessary information to evaluate strategic options. However, in the presence of uncertainty, organizations minimize risk by pursuing several options. Hence, as long as there is no environmental uncertainty, the organizations tend to invest all resources in the most promising strategy. As uncertainty increases, the key issue becomes engaging in the right strategy at all. In that case the motivation to minimize transaction costs in strategy implementation becomes secondary. In such an uncertain situation, wasting resources on a strategy that may ultimately turn out to be inappropriate is less of a concern. Instead, organizations minimize risk by increasing the scope of their strategic response to institutional pressure when the uncertainty increases.

Hypothesis 2: The higher the level of perceived uncertainty, the broader the scope of an organizational response.

To summarize, whereas the level of perceived institutional pressure influences the strategic direction of an organizational response strategy, we hypothesize that regulatory uncertainty increases the scope of an organizational response strategy.

Climate Change

The issue of climate change has only recently entered the business environment (Hoffman, 2005; Hoffman & Woody, 2008; Kolk & Levy, 2001; Kolk & Pinkse, 2005). This context lends itself very well to analyzing how regulatory uncertainty influences the response strategies of organizations. First, companies in industries that are greenhouse gas emission-intensive face increasing regulative pressure to reduce their GHG emissions. The increased implementation of GHG emission regulations around the world, such as the European Union Emission Trading Scheme (EU ETS), demonstrates the mounting regulative pressures put on GHG emission-intensive companies (Hoffmann, 2007; Kolk & Hoffmann, 2007; Kolk & Pinkse, 2007). In addition, other institutions have begun to screen organizations' GHG emissions more intensively (Kolk & Pinkse, 2007). For example, investors demand the disclosure of companies' GHG emissions. Likewise, customers pay attention to the amount of GHG emitted throughout the life cycles of the products they purchase.

Second, companies in GHG emission-intensive industries face substantial uncertainty regarding the pressure for GHG reduction. The Kyoto Protocol is a major driver for pressure to reduce GHG emissions. However, a successor to the Kyoto Protocol for the period following the year 2012 was still uncertain at the time of our research (Hoffmann et al., 2008). In addition, organizations are uncertain about industry coverage, reduction targets, and implementation details of regional GHG regulations (Engau & Hoffmann, 2010, 2011; Hoffmann, 2007). Consequently, the future of regulatory pressure at a regional or country level also remains uncertain. Moreover, as companies find it difficult to predict the future requests of investors, customers, and NGOs regarding climate change and GHG emission reductions, the future of normative pressure is also uncertain (Kolk & Pinkse, 2007).

Hypotheses Testing

The hypotheses developed above are tested by means of response data from a global survey of companies in GHG emission intensive industries.

To test the hypotheses, we apply linear regression analysis using ordinary least squares. To test Hypotheses 1 and 2, we perform separate linear regression analyses for the share of *adjust, avoid,* and *influence* measures. We subsequently illustrate the data sample, the variables used, and the sample representativeness.

The questionnaire developed was part of an annual survey conducted by a Swiss-based asset management company, and was sent to companies from eight different GHG intensive industries, all of which were part of the Dow Jones 2500 global index. We focused on the 821 companies in GHG emission intensive industrial sectors only, as we expected GHG emissions and the respective institutional pressures to be critical to these companies. These industries included basic resources, chemicals, construction and materials, oil and gas, aviation, and utilities as they demonstrated a (direct) GHG intensity of at least 500 tons per thousand USD (US dollars) of sales. The industries creating automobiles and parts as well as industrial goods and services were added because of the indirect emissions caused by their products. The questionnaire was sent to the companies' CEOs or, in the case of multidivisional companies, to the heads of the business units operating in GHG emission intensive industries. Respondents could complete the survey online or return it by mail or fax during the response time period from May to July 2007. Out of the 199 companies completing the questionnaire, 81 provided incomplete responses. These companies did not provide specific data that was important to verify our hypotheses, e.g. the level of institutional pressure that the companies are exposed to. We excluded the incomplete responses which resulted in a final sample of 118 companies. Table 2.1 provides an overview of the respondents by industrial sectors and geographical regions.

The *organizational response strategy* variable was measured by the combination of measures that pursue one of the three strategic directions to adjust, avoid, or influence the institutional pressures. These measures

Table 2.1 Data sample, by industry and region.

| | Region | | | |
Industry	Americas*	Europe	Asia Pacific	Total
Automobiles and parts	2	5	1	8
Basic resources	5	7	3	15
Chemicals	5	10	1	16
Construction and materials	0	8	1	9
Industrial goods and services	5	6	13	24
Oil and gas	5	7	3	15
Aviation	0	3	0	3
Utilities	9	15	4	28
Total	31	61	26	118

were derived from a review of theory and empirical studies in the area of climate change research (e.g., Hoffman, 2005; Kolk & Levy, 2001), and were subsequently refined in cooperation with industry analysts from the aforementioned Swiss asset management firm that we partnered with in our research. Eight items were used to represent measures from the three strategic directions in the questionnaire. Respondents were asked whether the company pursued the described response measure or not. Using the method described by Miller and Toulouse (1998), we measure the scope of an organizational response as the sum of measures a company pursues as part of its response strategy. Hence, we computed the scope of an organizational response by the total number of response measures pursued by a company. This variable ranged from 0 to 8. The degree to which a company pursued each of the three strategic directions was determined by the ratio of the number of measures of a strategic direction to total number of measures the company took, i.e., the relative share of response measures of a strategic direction (Miller & Toulouse, 1998). For example, if a company engages in a total of 5 response measures of which 3 pursue the strategic direction of influencing institutional pressures, influence takes on the value 0.6, because 3/5 = 0.6.

Similar to Goodstein's (1994) study, *institutional pressure* is directed at one specific fact, namely the reduction of companies' GHG emissions. Although Scott (2001) postulated keeping regulative, normative, and cognitive pressures in the institutional environment distinct, other views promoted the idea that these three are strongly interconnected and that they persist simultaneously in an organizational field, each to a different degree at a given point in time (Hirsch, 1997; Hoffman, 1999). We follow this line of argument and do not explicitly distinguish the different types of institutional pressure for the purpose of this analysis. Moreover, we follow the line of argument that managers' perceptions of the environment determine the decisions made regarding an organization's strategy (Dill, 1958; Duncan, 1972; Lawrence & Lorsch, 1967; Sharfman & Dean, 1991) and thus we use a perceptual rather than an objective measure to reflect the level of institutional pressure. In line with other empirical research (e.g., Buysse & Verbeke, 2003; Henriques & Sadorsky, 1996), respondents were asked to assess the pressure to reduce GHG emissions put on their company from nine different sources on a five-point Likert scale ranging from "very low pressure" to "very high pressure." (See Table 2.4) We utilized Freeman's (1984) internal and external stakeholder groups with the exception of consumer advocates and special interest groups who did not seem to participate in the GHG emission reduction debate at the time of the analysis. Additionally, we added pressure arising from the direct costs incurred by emitting GHG (i.e., emission allowances or taxes). The overall level of perceived institutional pressure of a company was measured by the sum of items scoring either "high pressure" or "very high pressure" (Cronbach's Alpha of 0.85 for all items). Thus, the variable can take on integers between 0 and 9.

To determine a company's *level of perceived regulatory uncertainty*, respondents were asked to indicate how certain the company was about the future of a possible emission reduction regulation regarding five aspects (see Table 2.5). A 5-point Likert scale ranging from "very uncertain" (value of 5) to "'very certain" (value of 1) was used to differentiate responses. The overall level of perceived uncertainty is represented by the sum of items scoring either "uncertain" or "very uncertain" (Cronbach's Alpha of 0.87). The variable can therefore take on integers between 0 and 5.

In our research we also included control variables for a company's *industry, region, size, profitability*, and *direct GHG emission intensity*. As the sample includes companies from eight different industries, ranging from airlines to utilities, as well as companies from three different regions (see Table 2.1), we controlled for the industry and region by including dummy variables (Dess, Ireland, & Hitt, 1990). We controlled for differences in company size and profitability by using total company sales (in millions of USD) and return on assets (in percent) as of 2006, taken from the Compustat database. Finally, we also controlled for differences in direct GHG intensity. This is the total amount of equivalent carbon dioxide emissions (CO_{2e}) in tons emitted during one year, divided by total sales. It reflects the degree to which company sales rely on direct GHG emissions. These data were also obtained by the questionnaire.

To enhance validity, the explanatory variables were developed in collaboration with industry experts from the asset management company and further tested with company representatives. Moreover, to reduce the risk of common method bias being present, we positioned the items representing the dependent and independent variables in different sections of the questionnaire. All control variables (except for direct GHG emission intensity) were taken from archival sources. To test whether common method bias was still present, we performed Harman's single-factor test. An unrotated principal component factor analysis of all perceptual questionnaire items resulted in three distinct factors with eigenvalues greater than 1.0, rather than one single factor. The three factors together account for 62 percent of the total variance, and the first factor does not account for the majority of the total variance (40 percent). This suggests that common method bias is unlikely in our study (Podsakoff, MacKenzie, Lee & Podsakoff, 2003; Podsakoff & Organ, 1986).

Results

Table 2.2 reports descriptive statistics and Pearson's correlation coefficients among all variables.

In Table 2.3 we show the results of the two models to test the hypotheses.

Model 1 shows a positive coefficient for the level of perceived institutional pressure for the share of measures to *avoid* institutional

Table 2.2 Means, standard deviations (s.d.), and correlations of the companies studied

Variable	Mean	s.d.	1	2	3	4	5	6	7	8	9
1. Scope of organizational response	4.72	1.21									
2. Adjust measures	0.30	0.09	−.29**								
3. Avoid measures	0.15	0.14	.54**	−.23*							
4. Influence measures	0.54	0.13	−.41**	−.30**	−.75**						
5. Level of perceived institutional pressure	3.50	2.55	.32**	.14	.30**	−.39**					
6. Level of perceived uncertainty	1.81	1.74	.32**	.05	.22*	−.23*	.35**				
7. Firm size (in bn USD)	27.28	34.13	.13	−.04	.00	.01	.12	.19*			
8. Firm profitability (in percent)	5.83	4.42	−.24*	.00	−.15*	.29**	−.22*	−.12	−.12		
9. Direct GHG intensity (t per bn USD sales)[1]	1.12	1.96	.03	−.11	−.06	.14	−.04	−.09	−.19*	−.10	
10. Automobiles and parts dummy	0.06	0.25	.08	.02	.11	−.16	.19*	.24*	.55**	−.24*	−.15
11. Basic resources dummy	0.12	0.33	−.12	.03	−.09	.15	−.11	−.04	−.09	.22*	−.08
12. Chemicals dummy	0.13	0.34	−.07	−.09	.04	.00	.00	−.05	−.15	.02	−.09
13. Construction and materials dummy	0.07	0.26	−.09	.06	−.03	−.02	−.15	.01	−.10	−.11	.12
14. Industrial goods and services	0.20	0.40	−.06	.25*	.02	−.15	.00	.01	−.02	.00	−.22*
15. Oil and gas dummy	0.12	0.33	−.14	−.11	−.20*	.27**	−.01	−.01	.12	.40**	−.03
16. Aviation dummy	0.02	0.15	.03	−.07	.05	−.01	.09	−.01	−.04	−.11*	.00
17. Utilities dummy	0.23	0.42	.32**	−.12	.10	−.05	.06	−.06	−.11	−.26**	.40**
18. Americas dummy	0.26	0.44	−.10	−.04	−.13	.13	−.23*	−.26**	−.09	.07	.17
19. Europe dummy	0.51	0.50	.09	−.19*	−.04	.11	−.05	.06	.18*	−.04	−.06
20. Asia Pacific dummy	0.22	0.41	.00	.28**	.18*	−.27**	.31**	.21*	−.11	−.02	−.10

*p < 0.05.
**p < 0.01.
[1] tons per billion USD sales.

Table 2.3 Results of the regression analysis testing Hypotheses 1 and 2

| | Model 1 | | | | | | Model 2 | |
| | Adjust | | Avoid | | Influence | | Scope of organizational response | |
Variable	Coeff.	Sig. (t)	Coeff.	Sig. (t)	Coeff.	Sig. (t)	Coeff.	Sig. (t)
Constant	0.26	0.00**	0.14	0.01**	0.56	0.00**	4.86	0.00**
Level of perceived institutional pressure	0.15	0.15	0.21	0.05*	−0.30	0.00**	0.22	0.03*
Level of perceived uncertainty	−0.03	0.74	0.12	0.24	−0.08	0.41	0.27	0.00
Firm size	−0.04	0.74	−0.05	0.67	0.12	0.26	0.10	0.37
Firm profitability	0.05	0.68	0.04	0.70	0.11	0.28	−0.06	0.59
Direct GHG intensity	−0.07	0.50	−0.09	0.36	0.18	0.05*	−0.06	0.54
Automobiles and parts dummy	0.08	0.52	0.00	0.97	−0.06	0.59	−0.25	0.03*
Basic resources dummy	0.08	0.52	−0.17	0.14	0.16	0.14	−0.30	0.02*
Chemicals dummy	−0.02	0.87	−0.05	0.66	0.06	0.57	−0.26	0.01**
Construction and materials dummy	0.16	0.13	−0.05	0.60	−0.04	0.63	−0.22	0.03*
Industrial goods and services dummy	0.20	0.13	−0.15	0.25	0.01	0.91	−0.27	0.03*
Oil and gas dummy	−0.06	0.63	−0.28	0.03*	0.24	0.03*	−0.32	0.01**
Aviation dummy	−0.03	0.72	0.00	0.99	0.04	0.65	−0.09	0.32
Americas dummy	0.09	0.42	0.01	0.95	−0.06	0.52	−0.01	0.93
Asia Pacific dummy	0.19	0.10	0.13	0.27	−0.15	0.15	−0.11	0.30
F-Value	1.37		1.63†		3.55**		3.07**	

†$p < 0.10$.
*$p < 0.05$.
**$p < 0.01$.

pressure at the 5 percent significance level. Similarly, the predicted negative effect on the share of measures to *influence* the institutional pressure is confirmed at the 1 percent significance level. This implies that as the pressure for GHG reduction increases, companies engage less in activities such as offsetting GHG emissions or other *influence* measures as listed in Table 2.6. Furthermore, companies engage less in the political process for future GHG regulations in such circumstances. The F-tests for each of the regression analyses for Hypothesis 1b (*avoid*) and Hypothesis 1c (*influence*) are significant at the 10 percent (1b) and 1 percent (1c) significance levels. Due to the statistically significant coefficients, the hypotheses can be robustly confirmed. However, we observe no effect of the level of perceived institutional pressures on the share of measures to *adjust* to the institutional pressures; therefore Hypothesis 1a cannot be confirmed. Overall, these results indicate that the higher the level of perceived institutional pressure, the more a company engages in *avoid* measures and the less in *influence* measures. This implies that as the level of perceived institutional pressures to reduce GHG emissions increases, companies shift away from *influence* measures and toward *avoid* measures, while the share of *adjust* measures does not significantly change. This is an interesting finding, as it indicates that an increase in institutional pressure is not related to an increased adjustment of companies to that pressure. Instead, companies choose the more drastic alternative—to avoid the pressures altogether. Also, this means that as institutional pressure to reduce GHG emissions rises, companies would actually rather reduce their level of GHG emissions (requiring more internal change) than try to influence the institutional pressure. Activities such as offsetting GHG emissions or image-building do not seem appropriate responses to companies when they are confronted with very pronounced and explicit pressure to reduce their emissions.

We controlled for the level of perceived uncertainty, but our results suggest that it has no significant effect on any specific strategic direction. However, Model 2 shows a positive influence of perceived uncertainty on the scope of organizational response (number of simultaneously implemented measures). The F-test for this regression analysis is significant at the 1 percent level. Hypothesis 2 can be robustly confirmed, indicating that the higher the level of perceived uncertainty, the larger the scope of an organization's response strategy. In our test of Hypothesis 2, we controlled for effects that an increase in institutional pressure might have on the scope of organizational response. In line with previous research (e.g., Oliver 1991), our findings indicate a positive association between these two variables. In summary, the results show that increased institutional pressure is associated with an increase in the scope of the organizational response as well as with changes in the strategic direction of response. An increase in perceived uncertainty, however, is only associated with an increased scope of the response.

Conclusion

Our results help advance our understanding how perceived regulatory uncertainty influences responses to institutional pressure. Previous research has provided conflicting results on this topic (George et al., 2006; Goodrick & Salancik, 1996; Miller & Toulouse, 1998). Whereas some researchers suggested that strategies become more comprehensive in an uncertain environment (George et al., 2006), others suggested that they become more focused (Oliver, 1991). Most scholars understand an organization's response strategy as being of one specific type. We argue that such simplification has contributed to conflicting findings. In this study, we extend previous research, as we understand a strategy as a combination of response measures with different strategic directions. In a first step, we draw on this distinction to show that the strategic direction of an organization's response changes when institutional pressure increases. In such a context, organizations increasingly seek to avoid the pressure and make less of an attempt to influence the pressure in their own interest. In a second step, we show that the level of perceived uncertainty has a positive influence on the scope of the organizational response. As long as organizations do not face an uncertain environment, the scope of their response strategy is narrow. This means that companies act in line with the prescriptions of transaction costs economics (TCE). TCE suggests that companies seek to minimize their search and information costs as well as their costs for implementing a strategy (Williamson, 1981). This is achieved by focusing on few response strategies and by narrowing the scope of the organizational response. Therefore, TCE is useful in predicting the scope of organizational response strategies as long as uncertainty is low.

However, game theory suggests that organizations increase the scope of their response strategy when they are confronted with an uncertain environment. In such a situation, they balance their resource investments on different strategic directions according to their expected likelihood of success (Baldani, Bradfield, & Turner, 2005; Harsanyi, 1973). The prescriptions of game theory match with our observations.

Our findings hold important implications for research that examines how organizational fields mature (Hargrave & Van de Ven, 2006; Hoffman, 1999). Past research has assumed that regulative pressure for protecting the environment can gradually transform into normative and cultural-cognitive pressure (Hoffman 1999). Finally, the pressure is thought to become taken for granted and only becomes evident when an organization does not comply with the pressure. For example, such a gradual transformation of pressure could be observed with the chemical DDT. Initially, usage was banned by a regulation and today it is taken for granted that the chemical is not offered by chemical companies. However, regulatory uncertainty obscures this gradual change of practices in an entire industry. We observe two mechanisms that hinder the emergence of a new dominant practice like GHG reduction. First, organizations increase the *scope* of their responses in an uncertain environment. As a result, the variety of practices increases and it is less likely that a dominant organizational action emerges. Second, we observe no link between regulatory uncertainty and any specific strategic direction of the organizational responses. Instead, organizations only increase the scope of their responses as uncertainty increases. This makes it difficult for an organization to assess which of the numerous responses of its competitors is worth copying in order to be judged as legitimate in an uncertain environment.

Our findings extend research on the factors that can hinder the maturing of an organizational field. Previous studies have demonstrated how a set of beliefs in a single organization (Delmas & Toffel, 2008) or an entire industry (Lounsbury, 2007) can pose a barrier to the establishment of a new dominant practice. The regulatory uncertainty that we observe, however, leaves an organization with a puzzling variety of possible response options. There seems to be a lack of joint understanding in the industry with relation to which strategic direction may become most common, and therefore legitimate.

The fact that we have only collected data at one point in time is a limitation of our study. Future research should conduct a longitudinal analysis, in order to provide a broader basis for these findings. Despite this limitation, we believe our study holds important implications for policymakers. Our observations question the possible understanding that the self-regulation of the markets is a viable approach in order to reach a global reduction of greenhouse gases. Our research does show that various market stakeholders (including the financial community, customers, and suppliers) can build up pressure on companies to reduce their GHG emissions. Still, such a pressure is not sufficient to lead to a change in the dominant practices of an industry. Our data suggests that regulatory certainty is an additional and necessary precondition to reach ambitious targets for greenhouse gas reduction. Reducing uncertainty on the future of the Kyoto Protocol should therefore be a pivotal task for policymakers.

Appendix Tables

Questionnaire Items Measuring Level of Perceived Institutional Pressure

Companies were asked to indicate on a 5-point Likert scale how intense the pressure to reduce direct CO_{2e} emissions was from each of the stakeholders shown in Table 2.4.

Table 2.4 Nine sources of pressure to reduce GHG emissions

- Financial community (e.g., analysts)
- Public opinion (e.g., media, society)
- Customers (e.g., demand for low-emission products)
- Government (e.g., regulation)
- Cost of emissions (e.g., allowance price)
- Suppliers (e.g., green initiatives across value chain)
- Competitors (e.g., competitor actions)
- Employees/unions (e.g., initiatives)
- NGOs (e.g., publications)

Questionnaire Items Measuring Level of Perceived Regulatory Uncertainty

Companies were asked to indicate on a 5-point Likert scale how certain the company was about the features of a possible future emissions reduction regulations, shown in Table 2.5.

Table 2.5 Questions to assess the level of perceived regulatory uncertainty

- How certain is your company of what the future of a global agreement on the reduction of CO_2e emissions will be after the expiration of the Kyoto Protocol in 2012?
- How certain is your company of what the future of a possible regulation to reduce CO_2e emissions of your company will be after 2012?
- How certain is your company of what the design and details of a possible future regulation to reduce CO_2e emissions will be after 2012?
- How certain is your company of what the impact of a possible future regulation to reduce CO_2e emissions on your industry as a whole will be?
- How certain is your company of what the impact of a possible future regulation to reduce CO_2e emissions on your company in specific will be?

Questionnaire Items Measuring Organizational Response Strategy

Companies were asked which of the items in Table 2.6 they pursued in response to pressure to reduce their direct CO_{2e} emissions.

Table 2.6 Which items did your company pursue in response to pressure to reduce CO_{2e} emissions?

Strategic Direction (not indicated in questionnaire)	Questionnaire Items
Adjust	• Our company increases efficiency, substitutes input factors or modifies products or production processes to reduce our direct CO_2e missions
	• Our company limits the production and sale of CO_2e emission intensive products
Avoid	• Our company engages in activities in order to become largely independent of direct CO_2e emissions
	• Our company explores new markets/environments with lower societal or governmental pressure to reduce CO_2e emissions in order to avoid emission reduction pressure for carbon intensive products
	• Our company outsources CO_2e emission intensive processes or technologies
Influence	• Our company increases the emission limits by offsetting our own emissions (e.g., by engaging in emission reduction projects) or by acquiring additional emission capacity (e.g., by purchasing emission allowances)
	• Our company informs stakeholders such as customers or analysts of our efforts to reduce our direct CO_2e emissions, e.g. by image building, marketing lower emission products, reports or publications
	• Our company engages in the political process regarding a future emission reduction regulation that could potentially include our company

References

Amit, R., & Schoemaker, P. (1993). Strategic assets and organizational rent. *Strategic Management Journal, 14*, 33–46.

Aragón-Correa, J. A., & Sharma, S. (2003). A contingent resource-based view of proactive corporate environmental strategy. *Academy of Management Review, 28*, 71–88.

Baldani, J., Bradfield, J., & Turner, R. W. (2005). *Mathematical economics* (2nd ed.). Mason, OH: Thomson South-Western.

Buckley, W. (1968). Society as a complex adaptive system. In W. Buckley (Ed.), *Modern system research for the behavioural scientist* (pp. 490–513). Chicago: Aldine.

Buysse, K., & Verbeke, A. (2003). Proactive environmental strategies: A stakeholder management perspective. *Strategic Management Journal, 24*, 453–470.

Delmas, M. A., & Toffel, M. W. (2008). Organizational responses to environmental demands: Opening the black box. *Strategic Management Journal, 29*, 1027–1055.

Dess, G. G., & Beard, D. W. (1984). Dimensions of organizational task environments. *Administrative Science Quarterly, 29*, 52–73.

Dess, G. G., Ireland, R. D., & Hitt, M. A. (1990). Industry effects and strategic management research. *Journal of Management, 16*, 7–27.

Dill, W. R. (1958). Environment as an influence on managerial autonomy. *Administrative Science Quarterly, 2*, 409–443.

Duncan, R. B. (1972). Characteristics of organizational environments and perceived environmental uncertainty. *Administrative Science Quarterly, 17*, 313–327.

Engau, C., & Hoffmann, V. H. (2010). Corporate response strategies to regulatory uncertainty: Evidence from uncertainty about post-Kyoto regulation. *Policy Sciences, 44*, 53–80.

Engau, C., & Hoffmann, V. H. (2011). Strategizing in an unpredictable climate: Exploring corporate strategies to cope with regulatory uncertainty. *Long Range Planning, 44*, 42–63.

Financial Times Deutschland. (2010, December 10). Brasiliens Justiz nimmt ThyssenKrupp ins Visier [Justice in Brazil takes a hard line with ThyssenKrupp], *Financial Times Deutschland*.

Finanznachrichten. [Financial News] (2008, May 27). ThyssenKrupp-Betriebsrat: EU-Pläne zum Emissionshandel gefährden Standort, *Finanznachrichten*. [Work council of ThyssenKrupp: EU plans on emission trading threaten production site]

Freeman, R. E. (1984). *Strategic management: A stakeholder approach*. Marshfield, Massachusetts: Pitman Publishing.

George, E., Chattopadhyay, P., Sitkin, S. B., & Barden, J. (2006). Cognitive underpinnings of institutional persistence and change: A framing perspective. *Academy of Management Review, 31*, 347–365.

Goodrick, E., & Salancik, G. R. (1996). Organizational discretion in responding to institutional practices: Hospitals and cesarean births. *Administrative Science Quarterly, 41*, 1–28.

Goodstein, J. D. (1994). Institutional pressures and strategic responsiveness: Employer involvement in work–family issues. *The Academy of Management Journal, 37*, 350–382.

Guardian. (2005, June 18). Revealed: How oil giant influenced Bush, *The Guardian*.

Hargrave, T. J., & Van de Ven, A. H. (2006). A collective action model of institutional innovation. *Academy of Management Review, 31*, 864–888.

Harsanyi, J. C. (1973). Games with randomly disturbed payoffs: A new rationale for mixed strategy equilibrium points. *International Journal of Game Theory, 2*, 1–23.

Henriques, I., & Sadorsky, P. (1996). The determinants of an environmentally responsive firm: An empirical approach. *Journal of Environmental Economics & Management, 30*, 381–395.

Hirsch, P. M. (1997). Sociology without social structure: Neoinstitutional theory meets brave new world. *American Journal of Sociology, 102*, 1702–1724.

Hoffman, A. J. (1999). Institutional evolution and change: Environmentalism and the U.S. chemical industry. *Academy of Management Journal, 42*, 351–371.

Hoffman, A. J. (2005). Climate change strategy: The business logic behind voluntary greenhouse gas reductions. *California Management Review, 47*, 21–46.

Hoffman, A. J., & Woody, J. G. (2008). *Climate change: What's your business strategy?* Cambridge, MA: Harvard Business School Press Books.

Hoffmann, V. H. (2007). EU ETS and investment decisions: The case of the German electricity industry. *European Management Journal, 25*, 464–474.

Hoffmann, V. H., Trautmann, T., & Hamprecht, J. (2009). Regulatory uncertainty – a reason to postpone investments? Not necessarily. *Journal of Management Studies, 46*, 1227–1253.

Hoffmann, V. H., Trautmann, T., & Schneider, M. (2008). A taxonomy for regulatory uncertainty—application to the European Emission Trading Scheme. *Environmental Science & Policy, 11*, 712–722.

Kolk, A., & Hoffmann, V. H. (2007). Business, climate change and emissions trading: Taking stock and looking ahead. *European Management Journal, 25*, 411–414.

Kolk, A., & Levy, D. L. (2001). Winds of change: Corporate strategy, Climate change and oil multinationals. *European Management Journal, 19*, 501–509.

Kolk, A., & Pinkse, J. (2005). Business responses to climate change: Identifying emergent strategies. *California Management Review, 47*, 6–20.

Kolk, A., & Pinkse, J. (2007). Towards strategic stakeholder management? Integrating perspectives on sustainability challenges such as corporate responses to climate change. *Corporate Governance, 7*, 370–378.

Lawrence, P. R., & Lorsch, J. W. (1967). *Organization and environment: Managing differentiation and integration.* Reading, MA: Harvard Business School.

Lounsbury, M. (2007). A tale of two cities: Competing logics and practice variation in the professionalizing of mutual funds. *Academy of Management Journal, 50,* 289–307.

Miller, D. (1993). The architecture of simplicity. *The Academy of Management Review, 18,* 116–138.

Miller, D., & Toulouse, J.-M. (1998). Quasi-rational organizational responses: Functional and cognitive sources of strategic simplicity. *Canadian Journal of Administrative Sciences, 15,* 230.

Oliver, C. (1991). Strategic responses to institutional processes. *Academy of Management Review, 16,* 145–179.

Podsakoff, P. M., MacKenzie, S. B., Lee, J.-Y., & Podsakoff, N. P. (2003). Common method biases in behavioral research: A critical review of the literature and recommended remedies. *Journal of Applied Psychology, 88,* 879–903.

Podsakoff, P. M., & Organ, D. W. (1986). Self-reports in organizational research: Problems and prospects. *Journal of Management, 12,* 531.

Scott, W. R. (2001). *Institutions and organizations* (2nd ed.). London: Sage Publications.

Sharfman, M. P., & Dean, J. W. (1991). Conceptualizing and measuring the organizational environment—A multidimensional approach. *Journal of Management, 17,* 681–700.

Williamson, O. E. (1981). The economics of organization: The transaction cost approach. *The American Journal of Sociology, 87,* 548–577.

CHAPTER THREE

Renewable Energy Investment Decisions under Policy Risk: An Adaptive Conjoint Analysis (ACA) Approach

SONJA LÜTHI AND ROLF WÜSTENHAGEN

Solar energy is a promising energy source for the future. During the past few years, the installed photovoltaic (PV) capacity, a form of solar energy, has been increasing, especially in Germany and Spain. However, the contribution of solar power to total power production is still negligible. The barriers slowing this transition process are manifold, but to a large extent are related to current high prices of this technology. PV technology is still in an early stage, and the transition from central power production to distributed power production brings along transition costs. The cost disadvantage of PV technology is also influenced by subsidies for conventional, nonrenewable energy sources and a lack of internalization of external costs for those sources. Furthermore, the investment profile for PV is different than competing technologies (it has a higher initial cost, lower operating cost, and lower fuel price risk). Other barriers to diffusion of solar energy are related to path dependencies (e.g., market power of incumbent energy businesses) and cognitive factors (e.g., valuation methods that favor large-scale power plants).

Because of these barriers, the PV market is not yet self-sustaining, but is dependent on policy. To facilitate the emergence of this clean technology industry and to reach a self-sustaining market, effective policies and financing mechanisms are required. Thanks to effective incentives for PV systems by national and local governments, countries like Germany have become front runners in the adoption of PV panels (Jacob, Beise, Blazejcak, Edler, Haum, and Jänicke, 2005). But there is controversy about the effectiveness of Germany's incentives and it remains unclear what effective financing schemes are and how an effective PV policy should be designed. To date, the literature has rarely studied the effectiveness of policy schemes from the point of view of renewable energy companies' investment decisions. A notable exception is the work of Wiser & Pickle (1998) who analyzed, by means of case studies, the influence of renewable energy policies on the financing process and on financing costs. To analyze renewable energy companies' point of view is of importance, because these companies are transfer agents (Jacob et al., 2005). By entering new countries, they transfer products that are successful in their home markets to markets worldwide. A company will, however, only enter a market that provides an adequate policy framework. Motivated by this fact, this chapter addresses the question of policy effectiveness by analyzing the PV project developers' point of view. Specifically, it aims at identifying the most relevant policy-related factors in the location decision. The argument we make is that investment income (which is influenced by the level of the feed-in tariff) is not of higher importance than noneconomic policy risks. We calculated investors' willingness to accept such policy risks. The questions analyzed in the chapter are addressed by a multistage methodological approach, consisting of qualitative expert interviews and a quantitative adaptive conjoint analysis (ACA). Expert interviews provide in-depth understanding, and the ACA data allow statistical precision and generalization.

PV Project Developers

The expert interviews were conducted with PV project developers and other solar or project development experts. These market professionals were asked to recount their location decision process and to explain the different influencing factors. In this way, we identified their business models and reviewed the roles of host country characteristics as determinants in PV location patterns, especially in regard to the PV policy factors. Qualitative interviews with 8 experts confirmed that the policy conditions are currently key factors in a PV project location decision. These policy conditions include the public financial incentive schemes, the application procedure, policy targets for renewable energy, and support for policy stability.

The most common and effective incentive scheme in Europe was reported to be the feed-in tariff (FIT). Here, the level, duration, and yearly reduction of the tariff, as well as the presence of a limitation of the promoted power (the existence of a cap) are taken into account in a location choice. Sometimes other incentive schemes, such as investment subsidies and tax exemptions, provide additional support. Regarding the application procedure, the duration and the complexity of the approval procedure are of primary importance. A project developer is interested in starting a project as soon as possible. If the procedure to get the necessary permissions is long and complicated, and especially if it is uncertain when the permissions are forthcoming, or if they will be forthcoming at all, project developers hesitate to invest.

Furthermore, developers take PV policy stability into account. If significant unexpected changes in the policy have occurred frequently, a project developer hesitates to enter the market, because planning security is not provided. Most of the countries have fixed policy targets for renewable energies, and sometimes for solar power in particular. If the gap between the actual amount of PV used and the targeted amount is large, there is a high probability that the country will make stronger efforts to promote renewable energy in subsequent years. This is, however, a long-term process and thus of low importance for project developers.

Legal factors are to a large extent linked to political conditions. Legal conditions include regulatory requirements, a legally regulated FIT, legal backup of the FIT repayment, and law enforcement. A FIT can only be guaranteed if power utilities are obliged to accept feed-in power (there is a power purchase agreement). The interviewees also mentioned the security of private property rights as a factor. The interviewed experts are all active only in European countries, where legal security is provided and consequently is not decisive in the location choice.

In addition to policy conditions, the amount of solar irradiation is another influencing factor. However, because the current level of FIT in countries where PV project developers are active is relatively high, the solar resource is of minor importance in explaining the differences in return from one country to another.

Economic conditions are currently also of secondary importance, because the PV market is still strongly dependent on public policy and is not yet self-sustaining. The market demand and potential are thus artificially created through FITs. However, as soon as grid parity is reached and the market is self-sustaining, these factors will increase in importance. Grid parity refers to the point where the cost of renewable electricity generation is on par with the cost of electricity generation from conventional energy sources.

The Optimal Features of Projects

Upon the background of the expert interviews, we conducted an adaptive conjoint analysis (ACA) (Hartmann & Sattler, 2002). This is a well-established market research technique to determine the optimal features of projected, but as yet undeveloped, products and services. ACA belongs to the family of conjoint experiment methods. The conjoint experiment was initiated by mathematical psychologists (Anderson, 1970; Kruskal, 1965; Luce & Turkey, 1964), and was introduced in marketing research in the early 1970s (Green & Srinivasan, 1990; Orme, 2007). Since 1990, conjoint experiments have been frequently used by market researchers for elicitation of consumers' preferences (Green & Srinivasan, 1990) and have spread quickly over a wide array of research communities (Shin & Park, 2008). At the beginning, conjoint studies mainly analyzed the importance of product attributes and price. Later, concerns shifted to the simulation of customers' choices, and to the forecast of market responses to changes in the business's products or those of its competitors (Batsell & Lodish, 1981; Ben-Akiva & Gershenfeld, 1998; Louviere & Woodworth, 1983). More recently, conjoint analysis also has been used in environmental and resource economics and in studies on investment behavior (e.g., Franke, Gruber, Harhoff, & Henkel, 2006; Muzyka, Birley, & Leleux, 1996; Riquelme & Rickards, 1992; Shepherd & Zacharakis, 1999; Zacharakis & Meyer, 2000; Zacharakis & Shepherd, 2001). The methodological approach of our present study is novel in that it uses ACA to investigate investor choices among policy frameworks. The conjoint analysis approach suits this study well and alleviates some shortcomings of previous research on location decision-making. Most studies analyzing decision-making used post-hoc methodologies (e.g., Ajami & Ricks, 1981; Cheng & Kwan, 2000; Larimo, 1995; Ulgado, 1996), which may generate biased results (Shepherd & Zacharakis, 1999; Ulgado & Lee, 2004). The respondents using post-hoc methodologies had to evaluate location factors in terms of their importance to the most recent location decision, so they made the location decision at different points in time, with various business resources and constraints, and under different environmental conditions. Also, the location alternatives were different. These variations can significantly affect a factor's importance. Conjoint analysis, however, is a real-time method (Shepherd & Zacharakis, 1999) and respondents have to make their decision based on an identical set of alternatives (Ulgado & Lee, 2004).

Investors surveyed in past studies were often asked to evaluate location attributes individually (e.g., Ajami & Ricks, 1981; Ulgado, 1996). In reality however, businesses evaluate their location alternatives as a group of varying location characteristics. Location decision-makers trade off the different factors in comparing the available alternatives. PV project developers, for example, may want to invest in a certain country even if the return is lower, because the administrative procedure is very short. For that reason, an approach that asks respondents to assess a location site using a combination of attributes is more realistic (Ulgado & Lee, 2004).

Theory. We conducted choice experiments that built on the assumption that project developers make their choices based on their own individual preferences. The foundations underlying the preference investigation are briefly explained in the rest of this paragraph (Ben-Akiva & Lerman, 1985; Hensher, Rose, Ortúzar, & Rizzi, 2009; Louviere, Hensher, & Swait, 2003; Train, 2009). Microeconomic consumer theory provides the foundation for discrete choice experiments. Consumer theory analyzes the economic decisions, especially the consumption decisions, of private households. It states that a consumption decision is based on a cost–benefit comparison of the different product alternatives and that the consumer chooses the product that maximizes his utility. The theory provides the means to transform assumptions about consumers' preferences into a demand function. Lancaster (1966) advanced consumer theory by indicating that products can be considered as bundles of attributes with different levels (or characteristics) and that the utility of a product is the sum of the part-worth utilities of its attributes. This microeconomic consumer theory view of demand is appropriate to situations where the feasible choices are continuous. However, where choices are a selection of one out of a finite set of attributes (as is the case in this chapter), discrete choice theory is appropriate. Discrete choice theory uses the concept of the rational consumer (or, in our case, project developer), but it differs from consumer theory in that it works directly with the utility function, instead of deriving demand functions.

It is not possible to completely describe any product's utility in terms of its attributes; there will always be some unknown or intangible characteristic that may also provide utility. As a result, the other underlying foundation of discrete choice theory is random utility theory (Mansky, 1977), which allows the direct utility function of a person to be broken down into observable (deterministic) and unobservable (stochastic) parts. The utility is thus not an apparent value, but an unobservable random variable. This probabilistic approach accounts for randomness in choice behavior.

Our study is built on the assumption that renewable energy project developers evaluate the different factors influencing their location choice according to the theory described above. They do not choose among different products, but among policy frameworks. The policy framework of a country can be described as a bundle of attributes, analogous to a product with multiple attributes. A renewable energy project developer chooses the location for his or her project by looking for the country with the policy framework that provides the highest utility. As in the case of a choice among products, when choosing among policy frameworks, there is an inevitable trade-off among the different attributes, and any attribute change influences the attractiveness of the respective country for the project developer. A higher level of return, for example, increases the utility and thus the attractiveness of a country, whereas higher policy risks decrease the country's utility.

Sample and Questionnaire. The population of interest for the online survey was European PV project developers who were engaged in or were considering developing PV projects abroad in other European countries. The online survey was conducted in October and November 2008. The PV project developers were invited to participate in the survey by phone and/or e-mail, at a solar industry fair, by an article on the Solarserver website (www.solarserver.de), and by a leaflet in a solar industry journal.

Based on the qualitative pre-study, a questionnaire consisting of two parts was compiled: The ACA experiment about the importance of PV policy attributes, and questions to obtain background information about the experience and activities of the project developers and the companies for which they were working.

In the expert interviews, the main element of the pre-study, the decision-influencing attributes could be identified. Besides being relevant to the location decision, the attributes needed to fulfill some other criteria in order to be included in the ACA survey (Backhaus, Erichson, Plinke, & Weiber, 2006); see Table 3.1. The attributes need to be independent, i.e. the utility of the attribute and the perceived utility of a certain level (characteristic of the attribute) should not interact with other attributes. Further, attributes should be compensatory:, attributes and levels have to be able to substitute for each other in investors' perceptions.

Finally, as the study aimed at giving policy recommendations, attributes could be included in the choice experiments only if they could be influenced by policymakers. This is not the case for the attribute *solar radiation* (the amount of solar radiation) or for economic factors such as market demand and market potential, and was only partly the case for *local production* and for the legal factors such as contract enforcement.

As a result, 5 out of 12 attributes shown in Table 3.1 fulfilled all requirements and were chosen for the ACA experiment: *level of feed-in tariff, duration of feed-in tariff, existence of a cap on feed-in tariff payments* (or the time until the cap is reached), *duration of the administrative process until all permits are obtained,* and *significant unexpected policy changes in the last 5 years* in a location under consideration (support policy stability).

Table 3.2 gives a description of each attribute and the levels used in the survey. These attributes and their respective levels form a collection of 2,800 different combinations. The number of combinations corresponds to the multiplication of the number of attribute levels.

To have a comparable initial position for decision-making, the following framework conditions were predefined in the questionnaire: solar radiation: 1,500 kWh/m^2; installation type and size: Greenfield solar plant of 500 kW.

Table 3.1 Potential attributes for ACA survey

Attributes	Relevant	Can be influenced	Independent	Compensatory
Level of feed-in tariff	X	X	X	X
Duration of feed-in tariff	X	X	X	X
Existence of a cap	X	X	X	X
Duration of the administrative process	X	X	X	X
Support policy stability	X	X	X	X
Gap to political solar target	0	X	X	X
Regulated feed-in	X	X	X	X
Law enforcement	X	0	X	X
Market demand	0	—	—	X
Market potential	0	—	—	X
Local production	0	0	X	X
Solar radiation	0	—	X	X

X = criteria fulfilled; 0 = criteria partly fulfilled; — = criteria not fulfilled.

Table 3.2 Attributes and attribute levels used in the ACA experiment.

Attributes	Description	Attribute levels
Level of Feed-in Tariff (€ct/kWh)	The amount paid per kWh fed into the grid.	31, 35, 38, 41, 45 €ct//kWh
Duration of Feed-in Tariff (years)	Number of years for which the feed-in tariff is guaranteed.	15, 20, 25 years
Existence of a cap	Presence of a market cap limiting the promoted PV capacity, and if a cap exists, the predicted time until it will be reached.	No cap, cap reached in 4 years, cap reached in 1 year
Duration of the administrative process (months)	Predicted time from the project submission until all permits are obtained.	1–2, 3–6, 7–12, 13–18, 19–24 months
Significant unexpected policy changes in the last 5 years	A change is considered as significant if it leads to more than 15% of feed-in tariff reduction.	0, 1, 3 policy changes

The computer-based survey questionnaire proceeds in a fixed order, adapted from Sawtooth Software (2007). At the beginning of the process, the respondents usually rate the levels in regard to their relative preference. We skipped this section in our survey, because in the case of the five attributes selected for the final choice experiment, the preference order for the attribute levels is obvious. Our survey started with the paired-comparison section, where the computer program forms pairs that respondents have to compare. Each question showed descriptions of hypothetical policy framework conditions for two countries composed of different levels including two attributes at the beginning, then three, and then four. Assuming that the conditions were identical in all other ways, respondents had to indicate which country they would choose as the next project location. A 9-point scale was given which covers the range from "strongly prefer left" to "somewhat prefer left" to "indifferent" to "somewhat prefer right" to "strongly prefer right." The number of paired-comparison questions to be asked is equal to

$$3 (N - n - 1) - N$$

where N is the total number of levels and n is the total number of attributes, so that 3 (19 – 5 – 1) – 19 = 20. In the last section of the questionnaire, the software composes a series of calibrating concepts, in which product alternatives are described by levels of all attributes. The respondent is asked to indicate a "likelihood of choosing" between 0 and 100 for each concept presented. To assess the spread, the most unlikely concept is presented first to the respondents and then the most likely one.

The conjoint section was concluded by three so-called holdout tasks. Holdout tasks are constructed as the calibrating concepts, but are not used to estimate part-worth utilities. They are used to assess the quality and performance of the model used for the utility estimations (see below). If the responses to holdout questions can be predicted accurately using estimated part-worth utilities, it lends greater credibility to the model.

Data Analysis and Results. Descriptive statistics were obtained from the background data and responses to the ACA questionnaire. For the background data, information about the respondents (profession, professional experience, and knowledge) and their company (activities, headquarters location, and countries in which it is active) are provided. The data from the ACA questionnaire were used to estimate the part-worth utilities of the different attribute levels[1], the relative importance of each attribute, the investors' willingness to accept certain policy risks, and to perform likelihood-of-purchase simulations. (The part-worth utility is the utility of an attribute level. The utility refers to the total utility of a product, made up of all of the part-worth utilities.) Before estimations with the ACA data could be made, the part-worth utility values needed to be normalized. Initial utility estimates were based on the respondent's desirability ratings for attribute levels, together with ratings of attribute importance. The initial estimates were updated during the experiment. As the initial position of the utility estimation was different for each participant, the utility values first had to be scaled so that utilities could be compared across participants. The utilities were scaled in such a way that the sum of the utility "points" across all levels for a respondent were equal to the number of attributes times 100 (Metegrano, 1994).

There were 135 respondents who logged on to the survey website, and 63 questionnaires were completed. Each project developer completed 25 choice tasks, resulting in a final data set of 1575 choice decisions. The ACA interview was time-efficient; the duration had a median of 20 minutes. Eighty percent of respondents were project developers. About 50 percent worked in vertically integrated firms, i.e. they were involved in the planning and building of PV plants, whereas the other 30 percent were just concerned with planning. The remaining 20 percent were investors and project or business managers, also involved in project location decisions. Giving evidence of the emerging nature of the solar industry, more than 80 percent had less than seven years of experience: 27 percent had one year, 29 percent had two to three years of experience, and 27 percent had four to six years of experience. Forty-four percent of the interviewed persons have been involved in 1 to 10 projects and 38 percent percent in more than 10 PV projects. Three project developers had even worked on more than 100 projects. Thirty-six percent of the realized projects are of a capacity smaller than 100 kW, 22 percent of the projects are between 100 and 500 kW, 38 percent are between 500 kW and 10 MW, and 4 percent are bigger than 10 MW.

Of the respondents' companies, 70 percent are active in Germany, 57 percent in Spain, 49 percent in Italy, 30 percent in Greece, 27 percent in France, 17 percent in Portugal, and 14 percent in Switzerland. Of the interviewed PV project developers, 78 percent indicate a good level of knowledge about the PV policy situation of Germany, 71 percent about Spain, 59 percent about Italy, 43 percent about Greece, 37 percent about France, 19 percent about Portugal, and 13 percent about Switzerland. These numbers show the prominent role of Germany and Spain, and so it is not surprising that more than half of the project developers interviewed (58 percent) work for a company having its headquarters in Germany. The other companies' headquarters are located in Spain (17 percent), Italy (10 percent), and several other countries (Greece, France, Portugal, etc.).

The average part-worth utilities are based on the individual part-worth utilities estimated with the hierarchical Bayes method. Part-worth measures the contribution of attribute levels to the overall utility of a product. The utilities are interval data, meaning they are scaled to an arbitrary additive constant to sum to zero within each attribute. Therefore a negative part-worth value for a certain attribute level does not indicate that this attribute level is unattractive, but it shows that it is less preferred than other levels of the same attribute with a higher part-worth value. Average part-worth utilities and standard deviations for each attribute level are displayed in Table 3.3. The part-worth utility examination confirms that the lowest level of each attribute always had the lowest relevance for all project developers. This makes sense intuitively and supports the validity of the results. Standard deviations are all very low, indicating a narrow distribution. The low distribution is also confirmed by an analysis of the correlations. Correlation coefficients of all respondents were close to one (0.95–0.99). A low average part-worth utility and a low standard deviation indicate that such an attribute level is very unattractive (e.g. "31 €ct/kWh").

From the ACA data, the relative importance of each attribute can be estimated by considering how much difference each attribute could make in the overall utility of the product, i.e., between the highest and the lowest utility value of each attribute (see Table 3.3). That difference is the range in the attribute's utility values. The bigger the range is, the more a variation in the attribute can lead to a variation of the overall utility (Backhaus et al., 2006). The relative importance of each attribute was calculated using Formula 1 (adapted from Clark-Murphy and Soutar 2004).

$$RI_i [\%] = \frac{(MaxU - MinU)i}{\sum (Max - Min)i} \times 100$$

where RI_i is the relative importance of the ith attribute; $MaxU$ the maximum utility of the ith attribute; and $MinU$ is the minimum utility of the ith attribute.

The analysis of the relative importance of the attributes reveals the highest importance for the duration of the administrative process with RI of 26 percent. Almost as important is the level of the FIT (24 percent).

Table 3.3 Average part-worth utility estimates and standard deviations by attribute levels (hierarchical Bayes models with all normally distributed part-worth utilities)

Attribute Level	Part-Worth Utility	Standard Deviation
Level of feed-in tariff		
31 €ct/kWh	−62	6
35 €ct/kWh	−27	6
38€ ct/kWh	0	4
41 €ct/kWh	29	6
45 €ct/kWh	60	9
Duration of the administrative process		
1-2 months	63	7
3-6 months	32	6
7-12 months	1	5
13-18 months	−31	6
19-24 months	−64	9
Duration of feed-in tariff		
15 years of support	−35	10
20 years of support	3	6
25 years of support	33	7
Cap status		
No cap	44	9
Cap reached in 4 years	5	9
Cap reached in 1 year	−49	12
Number of policy changes		
0 policy changes	41	8
1 policy change	6	6
3 policy changes	−47	11

The existence of a cap and PV policy changes are of medium importance, with 19 percent and 18 percent, respectively. The lowest importance (14 percent) is attributed to the duration of the FIT.

PV project developers are thus particularly sensitive to the duration of the administrative procedures, followed by other policy risks (policy changes, existence of a cap). Duration of support is relatively less important. This indicates that a more effective administrative procedure enables a lower FIT, without a loss of attractiveness for PV project developers.

In the next step, part-worth utilities are converted into project developers' implicit willingness-to-accept certain policy risks using Formula 2:

$$WTA_l \left[\frac{ct}{kWh} \right] = -1(U_l - MaxU_l) \frac{\Delta FIT}{MaxU_{FIT}}$$

where WTA_l is the implicit willingness-to-accept of the attribute level l; U_l is the part-worth utility of the attribute level l; ΔFIT is the difference of the level of FIT, i.e., 14 €ct/kWh; and $MaxU_{FIT}$ is the maximum utility of the attribute level of the FIT. Figure 3.1 shows that for every half-year

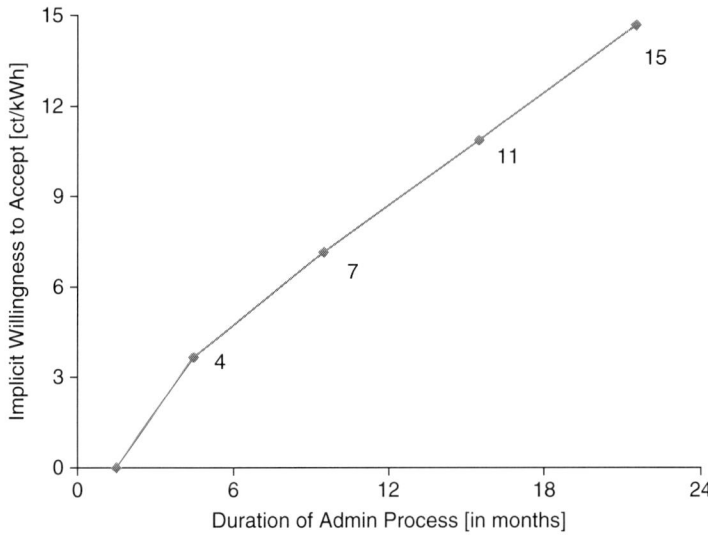

Figure 3.1 Willingness to accept a certain duration of the administrative process, correlated with €ct/kWh

increase in the duration of the administrative process, a government has to pay project developers a FIT premium of about 4 €ct/kWh (all else being equal).

The choice experiments included three attribute levels regarding the existence of a cap: no cap, a cap that is going to be reached in four years (loose cap), and a cap that is going to be reached in one year (tight cap). The analysis shows that removing a loose cap will allow governments to attract the same level of investment at a FIT that is about 5 €ct/kWh lower; removing a tight cap will allow governments to attract the same level of investment at a cap that is about 11 €ct/kWh lower (Figure 3.2). With regard to policy stability, the study estimates that compared to no policy risk conditions, in low-risk conditions (one significant unexpected policy change in the last five years), the FIT needs to be 4 €ct/kWh higher, in high-risk conditions (three significant unexpected policy changes in the last five years), it needs to be 10 €ct/kWh higher to keep its attractiveness (Figure 3.3).

Sawtooth (2007) offers the simulation method Purchase Likelihood (*SMRT Simulation*) to estimate the level of interest for a certain combination of attribute levels. The utilities are scaled so that an inverse logit transformation provides estimates of purchase likelihood. The simulator estimates how each respondent might have answered if presented with a concept with specific levels of attributes similar to those in the calibrating section of the interview. The likelihood projection is given on a 0 to 100 scale. This method can be used to investigate the likelihood of project developers investing in a certain country (i.e., they can investigate a specific combination of attribute levels). As mentioned above, prediction of

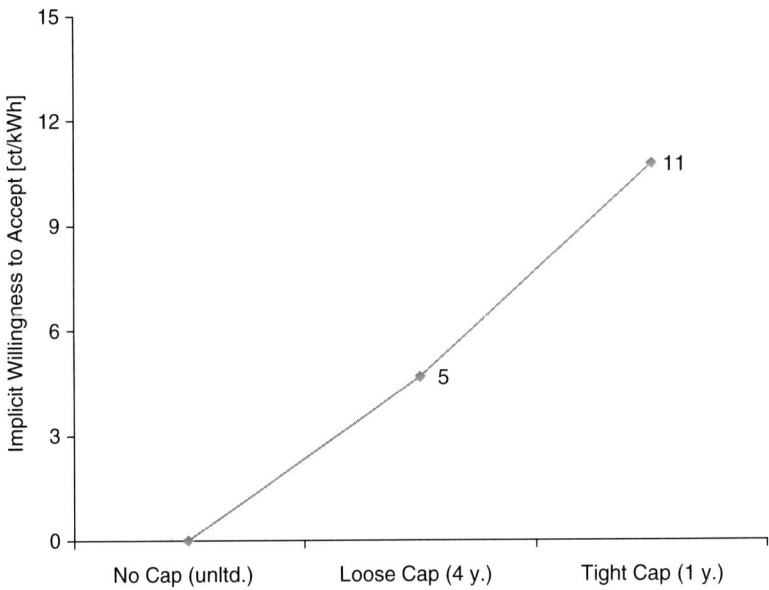

Figure 3.2 Willingness to accept a loose cap or a tight cap.

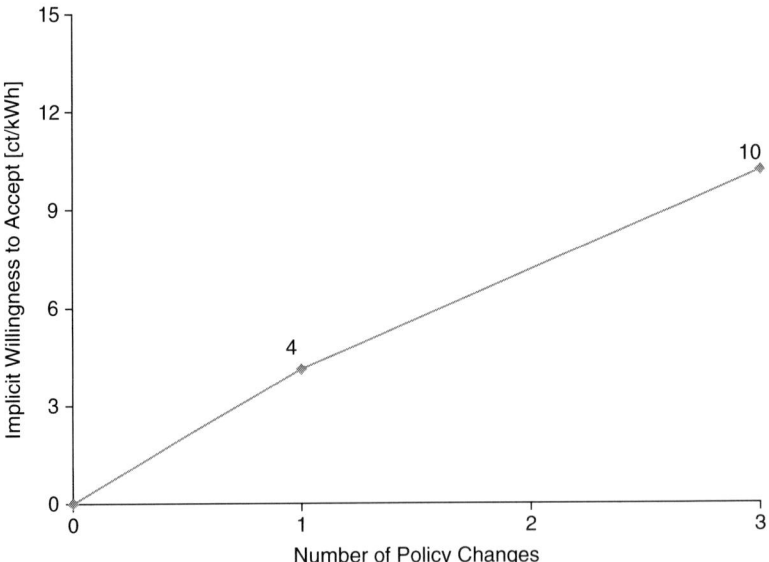

Figure 3.3 Willingness to accept a certain number of policy changes.

choice among holdout tasks can be used to check validity of this simulation method. In the present study, three holdout tasks have been included in the survey (see Table 3.4). Holdout task 1 describes the German PV policy framework in 2007; Holdout task 2 describes the Greek PV policy,

Table 3.4 Description of project developers' likelihood of investing in different policy frameworks (holdout tasks) and SMRT simulation

	Holdout 1 Germany (2007)	Holdout 2 Greece (2007)	Holdout 3 Spain (2007)
Policy Framework			
Duration of admin. process (months)	1–2	19–24	13–18
Level of the FIT (ct/kWh)	35	45	41
Cap situation	No cap	No cap	Cap reached in 1 yr
Number of PV policy changes	0	1	1
Duration of the FIT (years)	20	20	25
Likelihood of Investing (given on a 0 to 100 scale)			
Mean of project developers' likelihood	87	72	38
SMRT Simulation	99	85	39

Table 3.5 Investment likelihood simulations for changes in the PV policy framework of Spain in 2007

	Spain (2007)	Scenario A: Shorter administrative process	Scenario B: Lifting cap	Scenario C: Improved policy stability
Policy Framework				
Admin. process (months)	13 to 18	7 to 12 or 3 to 6	13 to 18	13 to 18
FIT level (€ct/kWh)	41	41	41	41
Cap situation	in 1 year	in 1 year	**in 4 years/ no cap**	in 1 year
Number of PV policy changes in 5 years	1	1	1	**0**
FIT duration (years)	25	25	25	25
Likelihood of Investing (given on a 0 to 100 scale)				
SMRT Simulation	39	66/84	81/95	68

Bold in a cell indicates changes from initial scenario.

and Holdout task 3 describes the Greek PV policy framework in 2007. The investment likelihood of the holdout tasks were predicted very accurately, which indicates the high validity of the results of the study and allows the use of the results to predict investment decisions.

Additionally, this method allows us to conduct simulations to estimate the influence of a hypothetical change in the policy design (e.g., increasing remuneration level, decreasing administrative process duration) on the project developers' likelihood for investing in a certain country. In what follows, the Spanish situation is analyzed (Table 3.5). The attribute levels that were changed from the initial scenario are in bold. One of the policy risks that policymakers can influence to some degree is the duration of the administrative process. *Scenario A: Shorter administrative process* in Table 3.5 reveals that an administrative process that is 6 or 12 months shorter (7 to 12 months long or 3 to 6 months long, instead of 13 to 18 months long) would bring a significantly higher investment likelihood of 66 or 84, respectively, compared to the initial situation (an investing likelihood of 39). In addition to the administrative process, the tight cap is another important issue in Spain. *Scenario B: Lifting cap* in Table 3.5 shows that loosening the cap (reached in four years) or removing the cap (no cap) makes sense to attract project investments, since the likelihood of investing increases to 81 for the 4-year situation and 95 in the no-cap situation.

Finally, the importance of a continuous PV policy is illustrated in *Scenario C: Improved policy stability* in Table 3.5. Having no changes in policy instead of one in the last five years increases the likelihood of investment from 39 to 68.

Conclusion

The transition to a sustainable energy system depends on whether public support policy effectively influences investors' behavior. Applying a sophisticated method from marketing research, this study is one of the first empirical contributions that investigate the influence of renewable energy policies on investors' decisions. By means of different estimations and simulations based on ACA data, the relevance of different policy factors and the costs of different regulatory risks have been quantified. Based on this empirical basis, it is possible to develop specific scenarios that enable policymakers to assess the costs and benefits of reducing various elements of policy risk.

The key finding is that risk matters in PV policy design, and that a price tag can be attached to specific policy risks. More specifically, the attributes "Duration of the administrative process" and "level of feed-in tariff" were perceived as the most important attributes in the location decision. SMRT simulations and willingness-to-accept estimations revealed that a reduction of the administrative process by 6 months enables a 4 €ct/kWh lower feed-in tariff (FIT) without a loss of attractiveness for PV project developers in investing in the given country. Lifting a cap also makes it possible to have a lower FIT: removing a cap that is expected to be reached in one year will allow governments to reduce the FIT by about 11 €ct/kWh and removing a cap that is expected to be reached in four years will allow governments to reduce the FIT by about 5 €ct/kWh. The third policy risk analyzed in this study is policy instability. Compared to having one significant unexpected PV policy change in the prior 5 years, respondents accepted a 4 €ct/kWh lower FIT if the political conditions were stable. These estimations confirm prior research that points to the importance of policy risk and noneconomic barriers—such as duration of the administrative process and political instability— to the deployment of renewable energy.

Governments can build on these empirical results to design policies that will be effective in attracting PV investments while at the same time maintaining efficiency by providing an adequate compensation for policy risk. In particular, policymakers should be aware that long administrative processes and, to a somewhat lesser extent, policy risks related to the existence of a cap and a substantial number of unexpected policy changes, have a cost attached to them that will need to be reflected in a higher level of feed-in tariff in order to attract solar project developers.

As with any piece of research, this study is subject to some limitations, which provide starting points for further research. This study has focused on policy factors; however, there are other factors that influence the location decision of PV project developers, which could be investigated in further studies. Such unobserved factors include language, country size, personal contacts, and the population's attitude towards the new technology.

The study examines the influence of changes in the policy framework on the project developers' investment likelihood. In future studies, the simulation tools that have been developed here can be applied to design scenarios for specific countries and thus can give more detailed policy design recommendations.

Future research could also build on our findings, which are based on stated preferences, and could compare actual valuation of policy risk with revealed preference data. This will become more feasible as an increasing number of countries emerge that provide sufficiently long time series of actual renewable energy investment decisions.

References

Ajami, R. A., & Ricks, D. A. (1981). Motives of non-American firms investing in the United States. *Journal of International Business Studies, Winter*, 25–34.

Anderson, N. H. (1970). Functional measurement and psychophysical judgment. *Psychological Review* (77), 153–170.

Backhaus, K., Erichson, B., Plinke, W., & Weiber, R. (2006). *Multivariate Analysemethoden. Eine anwendungsorientierte Einführung* [Multivariate analyzing methods. A user-oriented translation] (Vol. 11). Berlin: Springer-Verlag.

Batsell, R. R., & Lodish, L. M. (1981). A model and measurement methodology for predicting individual consumer choice. *Journal of Marketing Research, 18*(1), 1–12.

Ben-Akiva, M., & Gershenfeld, S. (1998). Multi-featured products and services: analysing pricing and bundling strategies. *Journal of Forecasting, 17*(3–4), 175–196.

Ben-Akiva, M. E., & Lerman, S. R. (1985). *Discrete choice analysis: Theory and application of travel demand*. Cambridge, Massachusetts: The MIT Press.

Cheng, L. K., & Kwan, Y. K. (2000). What are the determinants of the location of foreign direct investment? The Chinese experience. *Journal of International Economics, 51*(2), 379–400.

Clark-Murphy, M., & Soutar, G. N. (2004). What individual investors value: Some Australian evidence. *Journal of Economic Psychology, 25*(4), 539–555.

Flick, U. (1995). *Qualitative Sozialforschung. Theorien, Methoden, Anwendungen in Psychologie und Sozialwissenschaften [Qualitative research. Theories, methods, applications in psychology and social sciences]*. Reinbeck/Hamburg: Rowohlt.Franke, N., Gruber, M., Harhoff, D., & Henkel, J. (2006). What you are is what you like—similarity biases in venture capitalists' evaluations of start-up teams. *Journal of Business Venturing, 21*(6), 802–826.

Green, P. E., & Srinivasan, V. (1990). Conjoint analysis in marketing: new developments with implications for research and practice. *Journal of Marketing, 54*, 3–19.

Hartmann, A., & Sattler, H. (2002). Wie robust sind Methoden zur Präferenzmessung? *Research Papers on Marketing and Retailing, University of Hamburg, No. 004.*

Hensher, D. A., Rose, J. M., Ortúzar, J. d. D., & Rizzi, L. I. (2009). Estimating the willingness to pay and value of risk reduction for car occupants in the road environment. *Transportation Research Part A: Policy and Practice, 43*(7), 692–707.

Jacob, K., Beise, M., Blazejcak, J., Edler, D., Haum, R., Jänicke, M., et al. (2005). *Lead markets for environmental innovations.* Heidelberg, Germany: Physica Verlag.

Kruskal, J. B. (1965). Analysis of factorial experiments by estimating monotone transformations of the data. *Journal of the Royal Statistical Society, Serie B*(27), 251–263.

Kuhlen, S. (2008). CEO Germany, Research International GmbH. Hamburg.

Lancaster, K. J. (1966). A new approach to consumer theory. *Journal of Political Economy, 74,* 132–157.

Larimo, J. (1995). The Foreign direct investment decision process: Case studies of different types of decision processes in Finnish firms. *Journal of Business Research, 33,* 25–55.

Louviere, J. J., Hensher, D. A., & Swait, J. D. (2003). *Stated choice methods: Analysis and application* (2nd ed.). Cambridge, UK: Cambridge University Press.

Louviere, J. J., & Woodworth, G. (1983). Design and analysis of simulated consumer choice or allocation experiments: An approach based on aggregate data. *Journal of Marketing Research (JMR), 20*(4), 350–367.

Luce, D., & Turkey, J. (1964). Simultaneous conjoint measurement: a new type of fundamental measurement. *Journal of Mathematical Psychology, 1,* 1–27.

Mansky, C. (1977). The structure of random utility models. *Theory and Decision, 8,* 229–254.

Metegrano, M. (1994). *Adaptive conjoint analysis – version 4.0, 1994.* Sawtooth Software, Inc., Evanston.

Muzyka, D., Birley, S., & Leleux, B. (1996). Trade-offs in the investment decisons of European venture capitalists. *Journal of Business Venturing, 11*(4), 273–287.

Orme, B. (2007). *Which conjoint method should I use?*, Research Paper Series. Sawtooth Software Inc., Sequim, WA.

Oschlies, M. (2007). *A behavioral finance perspective on sustainable energy investment decisions.* Unpublished Master Thesis, University of St. Gallen, St. Gallen, Switzerland. Priem, R. L., & Harrison, D. A. (1994). Exploring strategic judgment: methods for testing the assumptions of prescriptive contingency theories. *Strategic Management Journal 15*(4), 311–324.

Riquelme, H., & Rickards, T. (1992). Hybrid conjoint analysis: An estimation probe in new venture decisions. *Journal of Business Venturing, 7*(6), 505–518.

Sawtooth Software. (2007). *The ACA/Web v6.0.* Sequim, WA: Sawtooth Software Inc.

Shepherd, D. A., & Zacharakis, A. (1999). Conjoint analysis: a new methodological approach for researching the decision policies of venture capitalists. *Venture Capital: An International Journal of Entrepreneurial Finance, 1,* 197–217.

Shin, J., Park, Y. (2009). On the creation and evaluation of e-business model variants: The case of auction. *Industrial Marketing Management, 38,* 324–337.

Train, K. (2009). *Discrete choice methods with simulation* (2nd ed.). New York: Cambridge University Press.

Ulgado, F. M. (1996). Location characteristics of manufacturing investments in the U.S.: A comparison of American and foreign-based firms. *Management International Review, 36*(1), 7.

Ulgado, F. M., & Lee, M. (2004). The effects of nationality differences on manufacturing location in the US: a conjoint analysis approach. *International Business Review, 13*(4), 503–522.

Wiser, R., & Pickle, S. (1998). Financing investments in renewable energy: the impacts of policy design. *Renewable & Sustainable Energy Reviews, 2,* 361–386.

Zacharakis, A. L., & Meyer, G. D. (2000). The potential of actuarial decision models: Can they improve the venture capital investment decision? *Journal of Business Venturing, 15*(4), 323–346.

Zacharakis, A. L., & Shepherd, D. A. (2001). The nature of information and overconfidence on venture capitalists' decision making. *Journal of Business Venturing, 16*(4), 311–332.

CHAPTER FOUR

Why Some Managers Expect to Benefit from Public Policies and Others Do Not

ALFRED MARCUS, SUSAN COHEN, AND
KATHLEEN SUTCLIFFE

Governments often provide assistance to firms, especially energy-efficiency and renewable-energy businesses (Marcus, Anderson, Cohen, and Sutliffe, 2010). For example, in the 1970s the U.S. federal government offered subsidies to firms that developed solar panels and tax credits to consumers who purchased these products (Marcus, 1992). There have been a host of such programs, including the 2009 Economic Stimulus Act, which allocated more than $27 billion in direct and indirect aid to energy-efficiency and renewable-energy businesses (See Table 4.1). The government, for example, has provided tax credits to firms producing wind power, enabling them to sell more products and services, or sell them sooner, than otherwise would be economically feasible. Past research suggests that managers in energy efficiency and renewable energy businesses often are the beneficiaries of government policies. By making substitute products and services less attractive, public policies such as mandated performance standards, taxes, and other programs lift public awareness and increase interest in the products and service that energy efficiency and renewable energy businesses offer (Gale & Buchholz, 1987; Russo, 2001; Sine, Haveman, & Tolbert, 2005).

Market–government interactions encourage the development of these businesses (Marcus & Geffen, 1998; Burer & Wüstenhagen, 2008; York & Lenox, 2009). But despite the influence that these policies have, not all firms actively seek government assistance. The question of why some firms actively seek policies that benefit them while others do not has been debated in the literature for some time (e.g., see Stigler, 1971; Salamon & Siegfried, 1977; Mitnick, 1981; Baysinger, 1984; Zardkoohi, 1985; Masters & Keim, 1986; Boddewyn & Brewer, 1994; Hillman.Zardkoohi, and Bierman 1999; Hart, 2004; Holborn & VanDen Bergh, 2008). This

Table 4.1 Summary of investments in energy efficiency and renewable energy in the US 2009 Economic Stimulus Act

---

$6 billion for renewable energy and electric transmission technologies loan guarantees
$5 billion for weatherizing modest-income homes
$3.4 billion for carbon capture and low emission coal research
$3.2 billion toward energy efficiency and conservation grants
$3.1 billion for state programs, help states invest in energy efficiency and renewable energy
$2 billion for manufacturing of advanced car battery (traction) systems and components.
$800 million for biofuel research, development, and demonstration projects
$602 million to support the use of energy efficient technologies in building and in industry
$500 million for training of green-collar workers
$400 million for the geothermal technologies
$400 million for electric vehicle technologies
$300 million for energy efficient appliance rebates
$300 million for state and local governments to purchase energy efficient vehicles
$300 million to acquire electric vehicles for the federal vehicle fleet
$250 million to increase energy efficiency in low-income housing
$204 million in funding for research and testing facilities at national laboratories
$190 million in funding for wind, hydro, and other renewable energy projects
$115 million to develop and deploy solar power technologies
$110 million for the development of high efficiency vehicles
$42 million in support of new deployments of fuel cell technologies

---

Source: http://en.wikipedia.org/wiki/American_Recovery_and_Reinvestment_Act_of_2009

research has analyzed the firm-level motivations for seeking government assistance, which include stimulating demand for a firm's products, lowering operating costs, providing legitimacy for the firm, and raising rivals' operating costs (Kaufman, Englander, & Marcus, 1993; Van de Ven & Garud, 1989; Yoffie, 1987; Getz, 1997). It has postulated that the decision by firms to become active is a function of the attractiveness of political markets (Bonardi, Hillman, & Keim, 2005) and the firms' capabilities for achieving its political goals (Oliver & Holzinger, 2008; Holburn & Zelner, 2010).

However, other than a firm's size (Salamon & Siegfried, 1977; Schuler, 1996; Ungson, James, & Spicer, 1985), it is not clear which firm-level characteristics might be related to the decision to become politically engaged. The results of analyses that have examined the impact of such variables as age, financial condition, and degree of government dependence on the tendency of firms to become politically involved have been mixed (Rudy, 2010; Zardkoohi, 1985; Meznar & Nigh, 1995; Shaffer, 1995; Hillman & Hitt, 1999; Schuler, 1999; Hansen & Mitchell, 2000; Schuler, Rehbein, & Cramer, 2002; Hillman, Keim, & Schuler, 2004; Hillman, 2005; and Lester, Hillman, Zardkoohi, & Cannella, 2008). These mixed results are not surprising, given the complexity of the decision to become politically involved. Managers do not have the time to consider all the possible outcomes of the public policies they might seek, nor are they likely to view the consideration of these outcomes a valuable use of their time. They do not know the precise form that public policies will take, how the public policies will be implemented, and what the unintended consequences of the public policies will be. They cannot easily predict the net gains that the public policies will generate to their firms. Instead, we argue that boundedly rational managers (Simon, 1970), facing numerous demands on their attention including substantial market challenges (Henderson & Stern, 2004), will rely on heuristics, or simple decision rules developed through experience, to inform their expectations about the benefits that the public policies might bring. The domain relevant knowledge that underlies such heuristics is accumulated in path-dependent ways, in which managers scan the environment for salient external opportunities and threats, accumulate knowledge about them, learn to interpret their meaning, and assess their implications (Cyert & March, 1963; Jackson & Dutton, 1988; Ocasio, 1997).

It is surprising that little attention has been devoted to understanding the cognitive frameworks and heuristics that managers apply to discern whether public policies are likely to benefit their firms (Hart, 2004). We think this is a potentially fruitful domain from which to draw new insights, which will move us toward a more complete theory of why firms seek government assistance (see Figure 4.1). The expectation of benefits from public policy, a precondition for corporate political action (Baron, 1995), begin with cognitions (Baron, 2006; Ocasio, 1997; March & Simon, 1958). Hence, it is worthwhile to investigate the heuristics managers use for assessing the opportunities their firms have to gain from public policies.

Although the motivations for seeking benefits from government have been examined, as well as the characteristics and capabilities of firms seeking these benefits, the heuristics that managers use have not been adequately explored. This chapter therefore starts by reviewing the literature

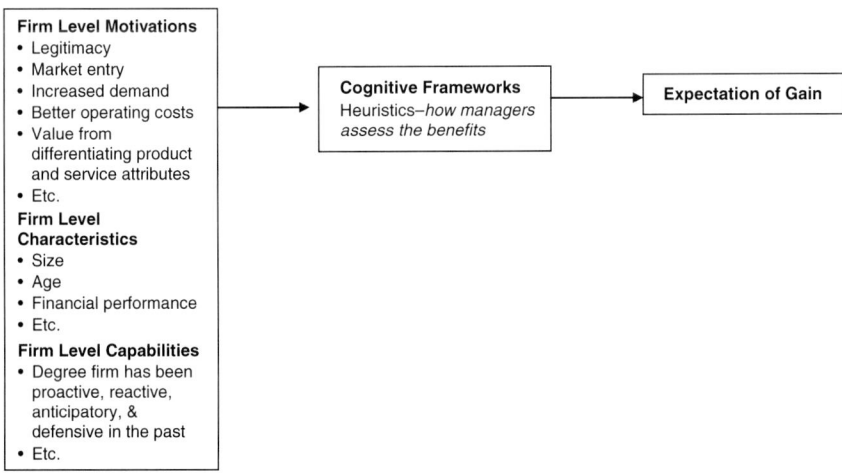

Figure 4.1 Why firms decide to be politically active.

on why managers of some firms actively seek to take advantage of public policies, while others do not. We build on this work to create a framework for describing managerial perceptions of the gains their companies can derive from public policies. We then apply this framework to two sets of overlapping data, qualitative interviews with managers in energy-efficiency and renewable-energy businesses and a survey of managers in this field. Our analysis of these data lead us to the conclusion that managers who view public policies as opportunities for gain believe they are competing in unstable environments. They have little ability to control market forces, but they believe they can predict and control government. Managers who do not view public policies in this way, in contrast, believe that they are competing in stable environments in which they have the ability to control market forces, but that they cannot predict and control government. The implications of our findings are discussed in the conclusion.

Firm–Government Interactions

Government policies influence a firm's performance in many ways. For instance, they alter the distribution of power among substitutes, rivals, suppliers, and customers, and create barriers that slow entry into an industry (Porter, 1980; Baron, 1995). Examples of the kinds of polices that affect firms include price supports, certification requirements, and investment subsidies. Through such mechanisms, governments influence the stability of the industries in which firms compete. They affect demand growth, price competition, and other factors that have important consequences for the firms' investment decisions and competitive strategies. Recognizing the importance of stable conditions to firms contemplating investment decisions, Stigler (1971) maintained that a major reason firms try to influence public policies is to create stability. They solicit the state's coercive powers to limit and control market forces in order to stabilize competititon and reduce uncertainty. However, not all firms that may gain from public policies engage in this behavior (Cook & Barry, 1995; Godwin & Seldon, 2002; Hart, 2004).

Managers are apt to weigh their ability to gain from public policies differently. Thus, as Stigler (1971) admits, there are limitations to his theory that "as a rule" firms try to influence government to achieve their goals (Marcus, 1984). Stigler (1971) takes an industry–level perspective on the reasons some firms are more inclined to seek opportunities for influence than others. His theory has been enriched by the work of Olson (1965) and successors (e.g., Peltzman,1976; Becker, 1983) and by the management scholars previously mentioned who have focused on firm motivations, characteristics, and capabilities that have affect their choices to seek public policies. Neither further refinements of Stigler and Olson's theories (Hart, 2004) nor recent work on corporate political strategies in the management field (Pearce, DeCastro & Guillen, 2008), however, have directly considered the cognitive models and frameworks managers use to evaluate the relative benefits of public policies to their firms. Consequently, we lack a robust theory for why managers perceive the benefits of public policies differently (Bartel & Thomas, 1985; Hart, 2004). What are the heuristics that managers apply to understanding the benefits they expect their firms to achieve from public policies? We propose that the complexity associated with anticipating the net effects of public policies is sufficient to overwhelm the individual managers' cognitive capacities (Hart, 2004; Lau & Redlawsk, 2001; Levinthal & March, 1993). Managers faced with complex decision-making contexts resort to heuristics to process the available information (March & Simon, 1958). They vary in their desire to influence government based on how these heuristics affect their perceptions of the public policies' net benefits.

The Mental Model of Managers

We argue that the mental model of managers who estimate the benefits of public policy has three components. Managers are apt to examine (i) instability, (ii) market adaptation, and (iii) political adaptation before they make choices about government policies.

Instability, the degree and frequency of change in a market, arises from rapid technological change, intense rivalry, fickle consumers, and frequent fluctuations in factor market conditions (Dess & Beard, 1984). Firms competing in unstable markets must make frequent adjustments in their competitive strategies and organizations at the same time that environmental shifts make it difficult for managers to learn how their choices affect firm performance (Levinthal & March, 1993), When the nature of opportunities changes frequently and dramatically, managers must allocate a greater proportion of resources to modifying products and services, changing management practices, and altering how functional activities are carried out. Resources set aside for these purposes can detract from their firm's profitability and raise its short-term risk of failure (Amburgey, Kelly & Barnett, 1993). Such instability makes learning difficult, since the managers cannot easily discern which elements of the strategies they previously pursued (e.g., which product performance criteria they offered customers, which inputs to the production process they employed, and so on) provide future performance advantages (Levitt & March, 1988). Hence, the instability that the managers face increases the costs of operations by imposing high learning-related adjustment costs. Managers facing this kind of instability are likely to expect gains from public policies that they believe can stabilize these turbulent forces. Assuming as Stigler (1971) did that firms seek stability from government, managers of these firms are likely to view public policies as critical to their firms' growth. On the other hand, managers who perceive that their firms' competitive environments are stable are likely to spend less time questioning their current strategies and more time refining their execution (Dess & Beard, 1984; Levinthal & March, 1993). As managers prefer to focus on task domains they believe they can control, those who perceive that the environments of their companies are stable are likely to expect to gain less from public policies than those who perceive that the environments of their companies are unstable (Dutton & Jackson, 1987; Jackson & Dutton, 1988; Ocasio, 1997).

If the managers of firms that compete in unstable markets can frequently predict changes, and/or affect the decisions of market actors upon which the success of their companies depends, they may be able to make market adaptations that will allow them to achieve control over the external environment without resorting to political means. Such foresight and influence depends on the managers of firms developing reliable routines for anticipating and adjusting to environmental shifts (Teece, Pisano & Shuen, 1997; Tushman & O'Reilly, 1997; Eisenhardt & Martin, 2000). An example comes from U.S. machine tool industry. It suffered severe boom and bust cycles for many decades, which tracked the overall health of the economy and the health of the industry's major buyers in the automotive and aerospace industries (Finegold, Brendley, Lempert, Henry, Cannon, Boultinghouse & Nelson, 1994). Although the precise beginning, magnitude, and duration of new cycles could not be pinpointed, the managers of these firms expected that the cycles would regularly recur, and they adopted the practice of backlogging customer orders in an effort to smooth out earnings from year to year.

To the extent that shifts in the competitive environment can be detected in the course of normal operations, managers can learn to spot them, and if particular kinds of fluctuations recur, they can fine-tune the routines of their firms to accommodate these fluctuations (Zollo & Winter, 2002). On the other hand, managers who perceive that they have little ability to predict or influence the decisions of key market actors must resort to trial-and-error learning, such as probing the market with frequent and highly varied product offerings (Sorenson, 2000). While firms that engage in such experimentation might ultimately achieve greater success than those that do not (Ozcan & Eisenhardt, 2010), this path entails more risk (Henderson & Stern, 2004). From the manager's perspective, the hazards and inefficiencies involved in figuring out how to respond to market shifts through trial and error are not likely to be desirable. Managers of firms that can influence the choice of their firm's suppliers, customers, and other market actors require less trial-and-error learning to discern how best to adapt. Their firms incur less adjustment-related inefficiency, and they are more likely to succeed. Hence, we expect that high levels of market adaptation are not likely to be associated with perceptions that public policies are critical to a firm's success; to the extent that managers succeed in predicting and controlling market forces, they are likely to see public policy as a less attractive means to achieve their goals. Rather, under such conditions, a more salient and immediate focus will be to figure out how best to take advantage of market opportunities by reliably and predictably creating the products and services that customers want. In contrast, if managers perceive that markets are hard to predict and control, they are more likely to view government as essential to the future growth of their firms and more likely to perceive public policies as valuable.

Managers who are able to predict and control government's decisions and behaviors are likely to be more certain of the value of public policies. They will adapt politically. When managers can anticipate policy changes and respond to them more effectively than their competitors, they will perceive that they have attractive opportunities to exploit public policies (Mitnick, 1981). If a political party habitually revokes or promulgates certain types of legislation, managers will be able to anticipate the meaning of the change in power for their firms. If they come to know the preferences and strategies of politicians, they can anticipate what the politicians will do and recognize the tactics the politicians use and signals the politicians send, which will help them to anticipate how political disputes will be resolved. If the firm is led by lawyers, as opposed to functional specialists, it may be more politically disposed and confident of its capacity to deal with government, because individuals with backgrounds in law are more accustomed to dealing with laws and rules than their colleagues in marketing and operations are. The political experience of managers provides them with the belief they can predict government and control its behavior (Burris, 2001; Blumentritt, 2003). The better managers are at predicting and controlling the actions and behaviors of government, the less they should need capabilities to deal with changes in the market. The better they can anticipate government's actions and decisions, the more they can develop reliable strategies and invest in the organizational capabilities to respond to public policies and to exploit them to the firm's advantage. The ability to influence and shape the political environment will lead to a perception that government is critical to the firm's growth and valuable, and that public policies are worthwhile to pursue. In contrast, if a firm's managers do not believe they can predict and control government, they are less likely to view government as an attractive domain for increasing their firm's opportunities to grow.

We assume that managers seek growth by whatever means available to them, market and nonmarket (Abernathy & Clark, 1985; Porter, 1980). To the extent that they can predict and control markets and are unable to predict and control governments, they are more likely to rely on market adaptation to increase demand and to be less interested in pursuing opportunities for political gain. On the other hand, to the extent that they can predict and control government and are unable to predict and control market forces, they are more likely to value public policies and to rely on political adaptation to increase demand.

Interviews with Managers of Energy-Efficiency and Renewable-Energy Businesses

We used these concepts to guide interviews with managers in energy-efficiency and renewable-energy businesses and to develop a survey administered to managers in this domain. The sample for the study was drawn from 106 businesses that provide energy-saving products, services, or renewable fuels. These firms were identified by the Alliance to Save Energy (ASE), a nonprofit organization located in Washington, D.C. Each company was contacted by phone to assess the nature of its products and services in order to determine if the firm fit the criteria to be included in a directory that the ASE was constructing. A company was included if its products and services contributed to saving energy and if the company met the following selection criteria. It: (1) manufactures or markets products that are more energy-efficient than existing or traditional products (i.e., offers equal or greater energy service using less energy input); (2) provides services that reduce energy use or are a component of projects to reduce energy use, (3) manufactures or provides services related to renewable energy systems, which conserve traditional energy sources, and (4) is headquartered in or has a facility in the state in which the interviews were carried out (see Appendix A for a more detailed description of the sample).

The advantage of the state-level focus was that we could limit the influences of idiosyncratic and different state policies on managers' perceptions. Energy efficiency and renewable fuels are an attractive context for this study, because firms that distinguish their products in this way must persuade customers to value a product or service attribute (e.g., saving energy or burning cleaner fuels) whose monetary value fluctuates and is affected by myriad government regulations, standards, subsidies, and taxes. Hence, public policy is a salient factor affecting the competitive landscape of these managers. We conducted a series of structured interviews with a chief executive or top manager of each of the firms in our study to validate and fine-tune the framework and to use the findings as the basis for a questionnaire that we subsequently mailed to the entire sample of 106 firms. The questions we asked centered on the following:

- What is the nature of your industry?
- How do you compete?
- How do government policies affect your business?
- Which issues are currently most pressing?
- What is the nature of market evolution and the changes in your industry environment?
- Who are the most salient and influential market and nonmarket stakeholders in your environment whom you believe influence your organization and affect its ability to achieve its objectives?
- Which of these stakeholders are you most and least able to influence?

At least 2 and often 3 of the authors of the study were present at the interviews, and each took copious notes, which we then content-analyzed according to the template found in Table 4.2. We aggregated the responses as described next in terms of eight patterns based on the extent to which the managers expected or did not expect to gain from public policy.

The perceptions of the managers we interviewed divide into the patterns shown in the Table 4.2. We designated the directional influence of statements in each cell, on the concept addressed in that row, with a (+), (–), or (?). Pattern 1 managers expected substantial gain from public policies, while Pattern 8 managers did not; these are the pure types. The in-between states were Patterns 2 through 4 and Patterns 5 through 7. Pattern 2 through 4 managers tilted toward expecting gains from government policies (+/?), while Pattern 5 through 7 managers tilted away from it (–/?). Overall, more of the managers (11) tilted away from expecting government benefits than the opposite. Only one manager fit Pattern 1. His company was involved in energy controls. This manager expressed optimism about his firm's ability to benefit from public policies such as tax credits, contracts, grants for research and development, and recognition programs. In describing his business, the manager commented that, although his company was striving to remain on the leading edge of technology, it found that this quest was quite challenging. Anticipating customer preferences was difficult, as the priority customers placed on minimizing energy costs, relative to other product features, fluctuated substantially according to many factors, including the price of energy, opportunities created by complementary technologies, and new end-user

Table 4.2 Cognitive template for managers' views of public policy benefits

Item	Patterns							
	1	2	3	4	5	6	7	8
Number: How many interviews fit each pattern?	1	5	2	0	2	1	4	4
Instability: Extent to which the managers perceived that the industry environment was unstable	+	+	+	–	–	–	+	–
Market Adaptability: Extent to which the managers perceived that they were able to predict and control market actors	–	–	+	+	–	–	+	+
Political Adaptability: Extent to which the managers perceived that they could predict and control government	+	–	+	+	+	–	–	–
Critical to Growth: Extent to which the managers perceived that public policies were critical to the growth of their firms	+	+/?	+/?	+/?	–/?	–/?	–/?	–

demands. This manager felt that his company's response, offering a broad product line, had not yet been that well received in the market. As a defensive measure, his firm was globalizing rapidly and was offering to provide integrated solutions to the problems its customers encountered, but this approach so far had not made the effort to adapt to rapid market shifts and changes in technology that much easier. In contrast to the problems that this manager perceived that his company was having in adapting to market conditions, he believed his company was having a very strong and direct impact on government through the well-organized trade associations in which it participated. These trade associations not only followed the development of public policies, but helped to forge these policies by setting federal and state standards and shaping energy policy legislation.

Pattern 8 managers (who were in the businesses of consulting, window accessories, heating, and ventilation), in contrast, considered the government unreliable, and they distanced themselves from government. These managers perceived that their companies were in stable business environments, as evidenced by such statements as "energy and pollution awareness grows predictably," to which they successfully had adapted by such means as high-quality products and services, strong distribution, the loyalty of large clients, their global scope, and the customer service they provided. Although Pattern 8 managers were relatively sure of their companies' market abilities, they were anxious about government's impacts, the implementation of its policies (the policies, in the words of one manager, "danced around"), and the difficulties of working with the government's bureaucracy.

The five managers who were ambivalent about public policies (Pattern 2) believed that their companies could gain from policies like rebates, taxes, loans, and standards. Their companies could gain from public policies because they were operating in unstable external environments, caused by such factors as shifting technology, shifting economic and political conditions, and/or substitute products. Their companies also could gain from public policies because the ability of these managers to control market conditions was limited; customers were not yet ready to buy the products and services they offered because of a lack of awareness, a lack of product readiness, high perceived costs, stiff competition from alternative technologies, and other factors. Though these Pattern 2 managers believed that their companies could gain from public policies, they did not have confidence in their companies' capacities to predict or control government. They saw potential value in government policies that might stimulate the growth of their businesses, but they did not believe they could obtain these benefits, because they did not have the political power and were stymied by groups that did have this power.

Pattern 6 and 7 managers (five managers) in Table 4.2 also believed that they did not have much power to predict and control government. But the perceptions of Pattern 6 managers were the opposite of those of Pattern 2 managers with regard to environmental instability and market adaptability and to those of Pattern 7 managers with regard to market adaptability. Pattern 5 managers (2 managers) believed that they had the power to predict and control government, but their need for public policy benefits was limited, since they did not see themselves as operating in an unstable environment. Pattern 3 managers (two managers) also believed they had the power to predict and control government, but they perceived that though they confronted an unstable external environment they had control of market forces and therefore their need for public policy benefits was limited.

The Tilt Away from Government. The literature supplies a number of interrelated reasons that may help to explain the tilt from public policies that we saw among the managers we interviewed (see Vogel, 1978 for an early discussion). Although public policies can be used to influence nongovernment groups such as customers or suppliers indirectly (Baron, 1995), managers are likely to prefer to influence groups like customers and suppliers directly, especially if the benefits of public policies are difficult to appropriate. Public policies rarely apply to individual firms, so individuals firms are shut off from appropriating the benefits entirely to themselves (Olson, 1965). Even when public policies apply to specific firms, if other firms can acquire the benefits offered by the policies, they can obtain the advantages that the policies provide without having invested to obtain these policies themselves (Olson, 1965). In contrast, competitors must invest in their own resources and capabilities, or engage in their own negotiations with customers, suppliers, and other organizations, in order to influence them and benefit from changes in their behavior. As a result, managers should view public policies as a relatively less desirable means of enhancing their growth when they are able to influence customers, suppliers, and competitors directly. Even when managers can influence government directly, they are likely to see public policy as entailing a loss of autonomy (Leone, 1986). Indeed, research shows that public policies often reduce managerial discretion by requiring firms to engage in activities in which they would otherwise not invest or by altering the attractiveness of strategies that they might otherwise have pursued (Ungson, James & Spicer, 1985; Birnbaum, 1984; Carter, 1990). Thus, other things being equal managers are not likely to expect net gains from public policy, because they believe it will reduce the amount of control they have over their operations (Pfeffer & Salancik, 1978).

The Questionnaire

To further validate this framework, we asked three top managers in each company (including the CEO) to complete a questionnaire. We received a total of 66 completed surveys from managers in 43 firms; 62 percent of the firms in our sample returned one or more surveys and at the managerial level our response rate was 21 percent. Many of the firms in this sample were small businesses (48 percent had 10 or fewer employees) in which the proprietor was essentially the only manager in the company. Adjusting our survey counts accordingly brings our response rate to 31 percent. As our unit of analysis is the individual manager, all 66 responses were used to in the analysis that follows.

The dependent variable on which we relied captures the extent to which the managers expected to benefit from public policies. See Appendix A for the specific items in our measures. We asked managers to assess the extent to which four different types of policies would facilitate the growth of their business: taxes, subsidies, public goods, and regulations. Examples of tax policies are fees for energy inefficiency and fuel taxes, such as those designed to account for externality costs. Subsidies include customer rebates, business tax credits, research grants, and low-interest financing. Information dissemination to educate and raise awareness among consumers is a public good. Regulations include energy efficiency standards for certain types of products and certification of service providers. The four policy types were orthogonal and comprehensive. We summed each manager's ratings for each policy type to create an overall index of the extent to which managers expected to benefit from public policies. Thus, this measure reflects the number of different types of policies that managers believed could help their firm, and the extent to which they believed each type of policy would stimulate demand for their products and services.

We created independent variables with questionnaire items adopted or adapted from previous studies (Glick, et. al., 1990). *Perceived industry instability* (see Table 4.3) was measured with a 7-item scale (alpha = .66). We tried to capture the concept of market adaptability with two measures: perceived *market predictability,* which was measured with six items (alpha = .75) (Langenfeld & Silvia, 1993) and perceived *market controllability,* which was also assessed with six items (alpha = .75) adopted from Glick, Huber, Miller, Doty, and Sutcliffe (1990). Similarly, we sought to capture differences in managers' perceptions of their firm's political adaptability using two measures: *perceived government predictability* was measured with two items (alpha = .88), and *perceived government controllability* was assessed with four items (alpha = .81). The items used for all three scales were adopted from Glick, et al. (1990).

As a number of different factors might influence managers' perceptions of their need for public policy, we relied on a variety of control variables. The greater the perceived *munificence* of the market environment, the less pressure managers will feel to look elsewhere for means of achieving their firm's growth objectives. Munificence was measured with a 6-item scale (alpha = .84) adapted from the work of Dess and Beard (1984) and Bourgeois (1985). Along the same lines, managers that perceive their firm to be innovative or entrepreneurial in dealing with their market environment might view public policy as a last resort. The extent to which a firm was a perceived *innovator* was measured with 3 items (alpha = .74). Organizational performance was assessed with a 5-item scale reflecting

Table 4.3 Descriptive statistics and zero-order correlation ($N = 66$)[a]

	Variable	Means	S.D.	1	2	3	4	5	6	7	8	9	10	11	12	13
1.	Expected Policy Benefits	20.80	4.20	—												
2.	Innovator	5.31	1.00	.12	—											
3.	Low Cost	5.76	1.14	.25	.41	—										
4.	Perceived Industry Instability	4.21	.92	.32	.19	.23	—									
5.	Munificence	4.65	1.12	.11	.10	.13	-.04	—								
6.	Market Predictability	4.37	.78	-.01	.23	.05	-.20	.23	—							
7.	Market Controllability	3.83	1.02	-.17	.18	.24	-.03	.13	.33	—						
8.	Gov't. Predictability	3.27	1.40	.28	.13	.19	-.01	.40	.42	.26	—					
9.	Gov't. Controllability	3.06	1.16	.31	.03	-.01	.15	.13	-.07	.18	.12	—				
10.	Energy Efficiency	5.51	1.33	.23	.06	.13	-.08	.02	.05	.14	.03	.05	—			
11.	Salience	4.67	1.42	-.00	.05	-.17	.26	-.25	-.28	.08	-.23	.13	.15	—		
12.	Performance	4.21	1.45	-.14	-.22	-.25	-.13	-.23	-.07	.27	.01	.01	.06	.04	—	
13.	Firm Size[b]	3.78	2.68	-.19	-.09	.15	-.12	-.21	.13	.61	.19	.02	.10	.10	.56	—

Source: Data for table are from Appendix B.
[a] For all $r \geq .20$, $p < .10$; $r \geq .24$, $p < .05$; $r \geq .30$, $p < .01$.
[b] A natural logarithm transformation was applied.

managers' perceptions of their firm's performance relative to other firms in the same industry (alpha = .92), adapted from Glick, et al. (1990). Firm *size* was measured as the logarithm of the number of full-time employees. Managers of better-performing firms ought to perceive less need for and therefore lower benefits from public policies. Larger firms might have less need for, but also a greater ability to appropriate, the benefits of public policy. Because managers' preferences for public policy also might be affected by the firm's dependence on government or the salience of public policy as a strategic instrument (Yoffie, 1987), we controlled for the extent to which managers perceived that the state and federal government influenced their business. This measure was assessed with two items (alpha = .88). Public policy salience would also be influenced by a firm's emphasis on energy savings in its products and services. The extent to which a firm emphasized *energy efficiency* as a feature of their products and/or services was assessed with two items (alpha = .76).

We estimated three regression equations: 1) just the controls; 2) the full model; and 3) an adjusted model. Model 2 was used to determine whether instability, market adaptability, and political adaptability influence managers' expectations of gain, as indicated by our interviews. The descriptive statistics and correlations between the constructs are shown in Table 4.3. Table 4.4 presents the results of the OLS (ordinary least squares) regression analyses. As can be seen, our findings were supportive of the framework. Industry instability was positively related to the benefits that managers expected from public policy ($B = .24$, $p < .05$). The ability to control market forces (*market controllability*), although not to predict it, was negatively associated with the benefits that managers expected ($B = -.36$, $p < .001$); see Table 4.4. On the other hand, the more managers believed they could predict government policies (*perceived government predictability*), the more likely they were to expect benefits ($B = .29$, $p < .05$). Also, the perceived ability to control government (*perceived government controllability*) was positively associated with expected benefits ($B = .31$, $p < .01$). The adjusted full model (Model 3), which included only the significant variables, had an adjusted R^2 of .38).

The survey thus supported our framework for how managers conceptualize opportunities for gain from public policies. Industry instability and the perceived ability to predict and control government policy were positively associated with expected public policy benefits, and the perceived ability to control market forces was negatively associated with the expected benefits. The element in our framework that was not supported was the variable representing the perceived ability to predict the behaviors and decisions of market stakeholders. The fact that market predictability was not significantly related to the expected policy benefits suggests that the capability to adapt to market forces may not be greatly reduced by the capacity to anticipate market fluctuations. Actual control is needed.

The nature of our sample— small firms in a diverse set of industries— and our focus on product markets, rather than industries, made it difficult

Table 4.4 Results of the regression analyses

Variables	Model 1: Control	Model 2: Full Model	Model 3: Adjusted Model
Energy Efficiency	.24*	.24*	.26*
Salience	−.03	.00	
Firm Size	−.15	−.10	
Performance	−.07	.08	
Innovator		−.00	
Low Cost		.24*	.20*
Munificence of Market		−.04	
Market Predictability		.06	
Market Controllability		**−.36***	**−.38***
Perceived Instability		**.24***	**.24***
Perceived Gov't. Predictability		**.29***	**.29****
Perceived Gov't. Controllability		**.31****	**.31****
R^2	.10	.44	.43
Adjusted R^2	.04	.31	.38
F	1.61	**3.49****	**7.56****

a Values shown are the standardized regression coefficients.

$n = 66$; †$p < .10$; *$p < .05$; **$p < .01$; ***$p < .001$. Significant results are shown in bold.

to obtain secondary data for our measures. Future studies could test the robustness of our findings with a sample that enables the use of secondary data to measure the independent variables, particularly firm strategy and resources, but it is not clear that secondary measures of variables such as the predictability and controllability of government and other organizations would necessarily be better if the aim is to examine managers' cognitive schemes. Rather, secondary data are likely to be subject to different sources of measurement error and to capture different dimensions of these constructs.

Implications

Though subject to the limitation of not relying upon secondary data, our findings are important because although many scholars highlight conditions under which firms can profit from public policy, these theories fall short in creating a framework that is fully attuned to managerial perceptions and cogitation about the value of seeking public policies for company gain. Scholars have emphasized that managers' perceptions of political opportunities are influenced by cognitive constraints and incomplete information (Hart, 2004), but to date few studies have directly examined managerial expectations of policy benefits. Rather, researchers have taken for granted that managers expect benefits for their firms based on their reasoning about the relationships between the firm, the industry, and the nature of political issues. Thus, our study provides an important addition to existing research in the collective action and corporate political activity domains. Focusing on the managerial level of analysis provides insights that might not have been gained from prior analyses of industry and firm-level behaviors. Prior studies have examined the motivations for seeking benefits from public policies and the characteristics of the firms seeking these benefits, but not the managerial heuristics— the interpretive frames that managers use to assess the opportunities to gain from public policy,— which are an important antecedent to the decision to be politically active. We provide a more complete picture of why some managers expect to benefit from public policies while others do not.

Overall, this study emphasizes that managerial beliefs about the attractiveness of a firm's opportunities to further its objectives through public policies are shaped by the managers' perceptions' of the firm's inability to achieve these objectives through market means. For any given objective or firm goal, public policies can be viewed as a substitute or complement. That the perceived ability to influence stakeholders other than government affects perceived benefits from public policies suggests that managers regard these as substitutes. This is consistent with, but enriches, the notion that firms attend to more salient stakeholders, and seek to retain managerial discretion and resist external pressure to conform to environmental demands in particular ways (Pfeffer & Salancik, 1978; David, Bloom, & Hillman, 2007). This insight is quite important, and the comments recorded in our interviews are highly supportive of it, but it is a suggestive finding that needs additional validation..

We offer a template of how managers anticipate the value of public policies. Our findings address calls to deepen our understanding of the reasons firms' political behaviors vary so dramatically (Pearce et al., 2008; Hart, 2004; Bonardi et al., 2005). Hart (2004) in particular calls attention to the complexity of governments' influence on businesses and the need to understand decision making structures within organizations as well as the drivers of managers' political preferences. Other fields, such as political science, have demonstrated that individuals rely on heuristics to assess the implications of public policies for their own lives (Lau & Redlawsk, 1997; Miler, 2009; Capelos, 2010). Industry studies reveal that managers' perceptions of the political arena underlie dramatic differences in firms' political activity within industries (Yoffie & Bergenstein, 1985; Suarez, 2000; Martin, 2000). However, we lack a framework for identifying the systematic sources of variation in individual managers' perceptions of the opportunities to gain from public policy. Guidance in this area seems unlikely to come from research that takes the industry as its unit of analysis, given the importance of firm heterogeneity for explaining which public policies ultimately help and hurt the firm, and the infrequency with which we observe coordinated political activity within industries (Smith, 2000). Studies that measure specific firm differences, such as age or size, have failed to offer a robust set of variables explaining inter-firm patterns of political activity (Bonardi et al., 2005). Echoing Pearce et al.. (2008), we suggest it is time to look inside the firm to better understand these differences.

Oliver and Holzinger (2008) move us in this direction by theorizing how dynamic capabilities might influence firms' political strategies. Heinisz (2000) offers a framework for understanding how interactions between a firm and its stakeholders and among those stakeholders lead to different priorities in terms of whether and how to seek government assistance. Both approaches are in the spirit of the framework highlighted in this study.

Our interviews also suggest these new ways to understand managers' heuristics for assessing public policy gain. Specifically, when we compared all of our managers' statements expressing positive expectations toward public policy, and the discussion in which they were embedded, we found certain patterns. First, we saw a structural consistency in managers' descriptions of the kinds of policies toward which they expressed the most optimism. Only when they could envision very simple (i.e. easy to articulate and implement) policies, which they felt had fairly direct and predictable effects, did they conclude that public policies could offer net gains for their firms. Second, many of the managers we interviewed seemed to prioritize the kinds of assessments they made, according to classes of policy implications. For instance, a common approach was to consider how public policies would affect the health of their sector, and only if these concerns could be satisfied, to entertain the potential implications for their specific business. Other managers focused primarily on laying out all the different ways in which public policies could affect their customers, and only when these seemed mostly positive, did they focus on the overall implications for their business.

Further research could examine how general these heuristics and the framework we have presented are. The importance of government in affecting firms' fortunes and shaping their resource allocation decisions has been recognized for a long time in the strategy field and authors have described large, established firms' entrepreneurial efforts to secure private benefits (Yoffie & Bergenstein, 1985; Hart, 2004). A few studies have illustrated that firms develop political capabilities over time through repeated experience with particular kinds of institutional pressures, and that these capabilities affect their strategic choices, such as which markets to expand into and what political initiatives to get involved in, and can be a source of competitive advantage and superior financial performance (Delmas & Montes-Sancho, 2010; Frynas, Mellahi, & Pigman, 2006; Hillman, Zardkoohi, & Bierman, 1999). Our study and much of the literature we review suggest that the way managers assess the likely consequences of public policies for their firm and industry affects their responses to it and hence is a critical antecedent to the kinds of political capabilities that firms develop.

In particular, we described market adaptability and political adaptability in terms of managers' perceptions of their firms' ability to influence and predict market and nonmarket actors. The ability to influence and predict market and nonmarket actors, we suggest, offsets or augments, the policy benefits that managers anticipate. Martin's (2000) study of corporate political capacity suggests some of the paths through which researchers can connect cognitive and structural mechanisms to further develop the idea that firms possess unique capabilities (Oliver & Holzinger, 2008) for political action and for exploiting the opportunities that public policies create. Her careful work documents how the awareness of issues in different parts of the firm, coupled with hierarchical and informal structures, influences what messages get communicated to those individuals and parts of the firm with responsibility for taking action. Managers' cognitive heuristics and the expectations they shape are critical elements in the information filters and communication channels that ultimately govern managerial behavior. Suarez (2000) reveals path-dependent tendencies in how firms learn from their efforts to influence and exploit public policies. Here again, managerial heuristics, which guide and get updated through experimentation and are shaped by a firm's communication and decision-making structures, play a critical role.

APPENDIX A: Description of the Sample

The energy efficiency/renewable fuels sector is composed of firms that manufacture and sell a wide variety of products and services. The firms can be categorized as belonging to one of four groups, according to broad similarities in customers and product/service characteristics, as follows:

1. *Products that contribute to saving energy in residential or commercial buildings (e.g., used in construction or to improve the energy efficiency of existing buildings).* Examples include: energy-efficient windows and lighting components, insulation materials, and energy-efficient appliances. Customers include general, mechanical, and insulating contractors, architects, builders, homeowners, building owners and managers. This sector included 29.6 percent of the respondents.
2. *Products that contribute to saving energy in industrial processes or settings.* Examples include: process controls, thermostats, heat recovery systems, and ventilators. Customers include energy-intensive industries (chemical, food, chapter) and agricultural organizations. This sector included 27.8 percent of the respondents.
3. *Energy-efficiency services that affect energy use in commercial buildings and/or in industrial processes.* Examples of services include: designing energy systems and demand-side management programs, conducting energy audits, training, and developing software for energy systems. Customers include utilities, manufacturers, and nonprofit organizations such as schools and hospitals. This sector included 27.8 percent of the respondents.

4. *Producers of renewable energy or alternate fuel products.* These include photovoltaic products, wind power systems, whole tree biomass systems. This sector included 14.8percent of the respondents.

APPENDIX B: Survey Items

Dependent Variable

Expected Policy Benefits

To what extent do you think each of the following policy types could facilitate the growth of your business in energy efficiency products/services and/or renewable fuels? (1 = not at all; 4 = to a moderate degree; 7 = to a great extent)

1. Subsidies (e.g., customer rebates, business tax credits, research grants)
2. Taxes (e.g., on energy inefficiency/waste, or fuel taxes)
3. Provision of a public good (such as education, dissemination of information, raising awareness)
4. Regulation (e.g., energy efficiency standards, product performance standards, certification)

Independent Variables

Instability

How strongly do you agree with each of the following statements? (1 = very strongly disagree; 4 = neither agree nor disagree; 7 = very strongly agree)

1. Customer demand and preferences are relatively stable in your industry.
2. Your firm must frequently change the way it produces goods or services in order to be competitive.
3. The total value of assets for the firms in your industry varies a lot from year to year.
4. The actions of your major suppliers (including materials, equipment, and labor suppliers) change little from one year to the next.
5. Public/political attitudes toward your industry and its products/services are relatively stable.
6. The volume of sales for firms in your industry fluctuates very little from year to year.
7. Your firm frequently changes its technology to keep up with competitors.

Market Predictability

To what extent are you able to predict the decisions and behavior of each of the following? (1 = never predict; 4 = sometimes predict; 7 = always predict)

1. Competitors
2. Suppliers
3. Distributors
4. Customers
5. Research institutes/consortia
6. Trade/professional associations

Market Controllability

To what extent is your firm able to influence the decisions and behavior of each of the following? (1 = minimal influence; 4 = moderate influence; 7 = tremendous influence)

1. Competitors
2. Suppliers
3. Distributors
4. Customers
5. Research institutes/consortia
6. Trade/professional associations

Government Controllability

To what extent is your firm able to influence the decisions and behavior of each of the following? (1 = minimal influence; 4 = moderate influence; 7 = tremendous influence)

1. Federal government
2. State and local government
3. *How accurate are the following statements?* (1 = not very accurate; 4 = somewhat accurate; 7 = very accurate)
4. By collaborating with other firms within our industry, we can have a great deal of influence over the development of public policy related to energy efficiency.
5. By collaborating with other firms from other industries, we can have a great deal fo influence over the development of public policy related to energy efficiency.

Government Predictability

To what extent are you able to predict the decisions and behavior of each of the following? (1 = never predict; 4 = sometimes predict; 7 = always predict)

1. Federal government
2. State and local government

Controls

Innovator

To what extent is the strategy of your organization targeted to . . . (1 = not at all; 4 = to some extent; 7 = to a great extent)

1. Developing new products and services
2. The development of new markets

To what extent is your organization currently characterized by . . . (1 = not at all; 4 = to some extent; 7 = to a great extent)

1. A strong entrepreneurial orientation

Low Cost

To what extent is the strategy of your organization targeted to . . . (1 = not at all; 4 = to some extent; 7 = to a great extent)

1. Providing low cost products and services

Munificence

How accurate are the following statements? (1 = not very accurate; 4 = somewhat accurate; 7 = very accurate)

1. Demand for the products/services of your principal industry has been growing and will continue to grow.
2. The investment or marketing opportunities for firms in your principal industries are very favorable at the present time.
3. The opportunities for firms in your principal industry to expand the scope of their existing products/markets is extremely limited.
4. In your industry, sales have been growing and are likely to grow.
5. The total value of assets for firms within your industry have been declining and will continue to decline.

Capital expenditures in your firm's principal industry have been growing and will continue to grow.

Energy Efficiency

How strongly do you agree or disagree with the following statements? (1 = very strongly disagree; 4 = neither agree nor disagree; 7 = very strongly agree)

1. We differentiate our products/services from those of our competitors on the basis of energy efficiency.

2. Energy efficiency is an important feature of our major products and/or services.

Salience

To what extent do these groups have the ability to influence your firm either directly (through specific demands or requests) or indirectly (as when internal decisions are constrained by your knowledge of these group's preferences or goals)? (1 = minimal influence; 4 = moderate influence; 7 = tremendous influence)

1. Federal government
2. State and local government

Performance

Compared to the other firms in your principal industry, over the past two years, the following was (1 = very low; 4 = average; 7 = very high)

1. After-tax return on your firm's total assets
2. Overall performance and success of your firm
3. Sales volume for your firm
4. Cash flow in your firm
5. Market share for your firm

References

Abernathy, W. J., & Clark, K. B. (1985). Innovation: Mapping the winds of creative destruction. *Research Policy, 14*, 3–22.

Amburgey, T., Dawn, K., & Barnett, W. P. (1993). Resetting the clock: the dynamics of organizational change and failure. *Administrative Science Quarterly, 38*(1), 51–73.

Anderson. P., & Tushman, M. L. (1990). Technological discontinuities and dominant designs: A cyclical model of technological change. *Administrative Science Quarterly, 35*, 604–633.

Baron, D. P. (1995). Integrated strategy: Market and nonmarket components. *California Management Review, 37*(2), 47–65.

Baron, R. (2006). Opportunity recognition as pattern recognition: How entrepreneurs "connect the dots" to identify new business opportunities, *Academy of Management Perspectives, 20*(1), Feb., 104–119.

Bartel, A. P., & Thomas, L. G. (1985). Direct and indirect effects of regulation: A look at OSHA's impact. *Journal of Law and Economics, 28*, 1–27.

Baysinger, B. D. (1984). Domain maintenance as an objective of business political activity: An expanded typology. *Academy of Management Review, 9*, 248–258.

Becker, G. S. (1983). A theory of competition among pressure groups for political influence. *The Quarterly Journal of Economics, 98*, 371–400.

Birnbaum, P. H. (1984). The choice of strategic alternatives under increasing regulation in high technology companies. *Academy of Management Review, 27*, 489–510.

Blumentritt, T. (2003). Foreign subsidiaries' government affairs activities: The influence of managers and resources. *Business & Society, 42*, 202–233.

Boddewyn, J. J., & Brewer, T. L. (1994). International-business political behavior: New theoretical directions. *Academy of Management Review, 19*(1), 119–143.

Bonardi, J.-P., Hillman, A. J., & Keim, G. D. (2005). The attractiveness of political markets: Implications for firm strategy. *Academy of Management Review, 30*, 397–413.

Bourgeois, L. J. (1985). Strategic goals, perceived uncertainty, and economic performance in volatile environments. *Academy of Management Journal, 28*(3), 548–573.

Burer, M. J., & Wüstenhagen, R. (2008). Cleantech venture investors and energy policy risk: an exploratory analysis of regulatory risk management strategies. In R. Wüstenhagen, J. Hammischmidt, S. Sharma, and M. Starik (Eds.), *Sustainable innovation and entrepreneurship*, pp. 290–309. Northampton, MA: Edgar Elgar.

Burris, V. (2001). The two faces of capital: Corporations and individual capitalists as political actors. *American Sociological Review, 66*, 361–381.

Capelos, T. (2010). Feeling the issue: how citizens' affective reactions and leadership perceptions shape policy evaluations. *Journal of Political Marketing, 9*, 9–33.

Carter, N. (1990). Small firm adaptation: Responses of physicians' organizations to regulatory and competitive uncertainty. *Academy of Management Journal, 33*, 307–333.

Cook, R., & Barry, D. (1995). Shaping the external environment: A study of small firms' attempts to influence public policy. *Business & Society, 34*, 317–344.

Cyert, R. M., & March, J. G. (1963). *A behavioral theory of the firm*. Englewood Cliffs, NJ: Prentice-Hall.

David, P., Bloom, M., & Hillman, A. J. (2007). Investor activism, managerial responsiveness, and corporate social performance. *Strategic Management Journal, 28*(1), 91–100.

Delmas, M. A,. & Montes-Sancho, M. J. (2010). Voluntary agreements to improve environmental quality: symbolic and substantive cooperation. Strategic Management Journal, *31*(6), 575–601.

Dess, G. G., & Beard, D. W. (1984). Dimensions of organizational task environments. *Administrative Science Quarterly, 29*, 52–73.

Di Maggio, P. J., & Powell, W. W. (1983). The iron cage revisited: Institutional isomorphism and collective rationality in organizational fields. *American Sociological Review, 48*, 147–160.

Dutton, J., & Jackson, S. (1987). Categorizing strategic issues: Links to organizational action. *Academy of Management Review, 12*(1), 76–90.

Eisenhardt, K., & Martin, J. (2000). Dynamic capabilities: What are they? *Strategic Management Journal, 21*, 1105–22.

Finegold, D., Brendley, K. W., Lempert, R., Henry, D., Cannon, P., Boultinghouse, B., & Nelson, M. (1994). *The decline of the U.S. machine tool industry and prospects for its sustainable recovery* (Vol. 1). Santa Monica, CA: RAND.

Frynas, J. G., Mellahi, K., & Pigman, G. A. (2006*).* First mover advantages in international business and firm-specific political resources. *Strategic Management Journal, 27*(4), 321–345.

Gale, J., & Buchholz, R. (1987). The pursuit of competitive advantage: What business can gain from government. In. A. Marcus, A. Kaufman, & D. Beam (Eds.), *Business strategy and public policy* (pp. 31–41). New York: Quorum Books.

Gavetti, G., & Levinthal, D. (2000). Looking forward and looking backward: Cognitive and experiential search. *Administrative Science Quarterly, 45*, 113–137.

Getz, K. (1997). Research in corporate political action: Integration and assessment. *Business & Society, 36*, 32–77.

Glick, W., Huber, G., Miller, C., Doty, C., & Sutcliffe, K. (1990). Studying changes in organizational design and effectiveness. *Organization Science, 1*, 293–312.

Glick, W., Jenkins, G., & Gupta, N. (1986). Method vs. substance. *Academy of Management Journal, 29*, 441–464.

Godwin R., & Seldon B. (2002). What corporations really want from government: the public provision of private goods. In A. Cigler & B. Loomis (Eds.), *Companies in national politics*, 6th ed. (pp. 205–224). Washington: CQ Press.

Hansen, W. L., & Mitchell, N. J. (2000). Disaggregating and explaining corporate political activity: Domestic and foreign corporations in national politics. *American Political Science Review, 94*, 891–903.

Hart, D. (2004). Business is not an interest group. *Annual Review of Political Science, 7*, 47–69.

Henderson, A. D., & Stern, I. (2004). Selection-based learning: The coevolution of internal and external selection in high-velocity environments. *Administrative Science Quarterly, 49*(1), 139–175.

Henisz, W. (2000). The institutional environment for multinational investment. *Journal of Law, Economics and Organization, 16*, 334–364.

Hillman, A. J. (2005). Politicians on the board of directors: Do connections affect the bottom line? *Journal of Management, 31,* 464–481.

Hillman, A. J., & Hitt, M. A. (1999). Corporate political strategy formulation: A model of approach, participation, and strategy decisions. *Academy of Management Review, 24,* 825–842.

Hillman, A. J., Keim, G. D., & Schuler, D. (2004). Corporate political activity: A review and research agenda. *Journal of Management, 30,* 837–857.

Hillman, A. J., & Zardkoohi, A., & Bierman, L. (1999). Corporate political strategies and firm performance: Indications of firm-specific benefits from personal service in the US government. *Strategic Management Journal, 20*(1), 67–81.

Holburn, G. L. F., & Zelner, B. (2010). Political capabilities, policy risk, and international investment strategy: evidence from the global electric power generation industry. *Strategic Management Journal, 31* (12), 1290–1315.

Holburn, G., & Vanden Bergh, R. (2008). Making friends in hostile environments. *Academy of Management Review, 33,* 521–540.

Jackson, S., & Dutton, J. (1988). Discerning threats and opportunities. *Administrative Science Quarterly, 33*(3), 370–387.

Kaufman, A., Englander, E., & Marcus, A. (1993). Selecting and organizational structure for implementing issues management: A transaction costs and agency theory perspective. In B. Mitnick (Ed.), *Corporate political agency* (pp. 148–168). London: Sage.

Lau, R., & Redlawsk, D. (1997). Voting correctly. *American Political Science Review, 91,* 585–599.

Langenfeld, J. L., & Silvia, L. (1993). Federal Trade Commission horizontal restraint cases: An economic perspective. *Antitrust Law Journal, 61*(3), 653–697.

Leone, R. (1986). *Who profits: Winners, losers, and government regulation.* New York: Basic Books.

Lester, R. H., Hillman, A., Zardkoohi, A., & Cannella, A. A. (2008). Former government officials as outside directors: The role of human and social capital. *Academy of Management Journal, 51,* 999–1013.

Levinthal, D. A., & March, J. G. (1993). The myopia of learning. [Special issue, winter.] *Strategic Management Journal, 14,* 95–112.

Levitt, B., & March, J. G. (1988). Organizational learning. *Annual Review of Sociology, 14,* 319–340.

March, J. G., & Simon, H. A. (1958). *Organizations.* Wiley: New York.

Marcus, A. (1984). *The adversary economy: Business responses to changing government requirements.* Westport, CT: Quorum Books.

Marcus, A. (1992). *Controversial issues in energy policy.* Beverly Hills, CA: Sage Press.

Marcus, A., Anderson, M., Cohen, S., & Sutcliffe, K. (2010). Prolonged gestation and commitment to an emerging organizational field. In R. Wüstenhagen, and Wuebker (Eds.), *Handbook of research on energy entrepreneurship,* Northampton, MA: Edward Elgar Publishing.

Marcus, A., & Geffen, D. (1998). The dialectics of competency acquisition. *Strategic Management Journal, 19,* 1145–1168.

Marcus, A., Kaufman, A., & Beam, D. (1987). *Business strategy and public policy.* Waterport, CT: Quorum Books.

Martin, C. (2000). *Stuck in neutral: business and the politics of human capital investment policy.* Princeton, NJ: Princeton Univ. Press.

Masters, M. S., & Keim, G. D. (1986). Variation in corporate PAC and lobbying activity: An organizational and environmental analysis. *Research in Corporate Social Performance and Policy, 8,* 249–271. JAI Press.

Meznar, M., & Nigh, D. (1995). Buffer or bridge? Environmental and organizational determinants of public affairs activities in American firms. *Academy of Management Journal, 38,* 975–996.

Miler, K. (2009). The limitations of heuristics for political elites. *Political Psychology, 30,* 863–894.

Mitnick, B. M. (1981). The strategic uses of regulation and deregulation. *Business Horizons, 24,* 71–83.

Ocasio, W. (1997). Toward an attention-based view of the firm. *Strategic Management Journal, 18,* 187–206.

Oliver, C., & Holzinger, I. (2008). The effectiveness of strategic political management. *Academy of Management Review, 33,* 496–520.

Olson, M. (1965). *The logic of collective action: Public goods and a theory of groups.* Cambridge, MA: Harvard University Press.

Ozcan, P., Eisenhardt, K. M. (2009). Origin of alliance portfolios: Entrepreneurs, network strategies, and firm performance. *Academy of Management Journal, 52*(2), 246–279.

Pearce, J., De Castro, J., & Guillen, M. (2008). Influencing politics and political systems. *Academy of Management Review, 33,* 493–495.

Peltzman, S. (1976). Toward a more general theory of regulation. *Journal of Law and Economics, 19,* 211–240.

Pfeffer, J., & Salancik, G. (1978). *The external control of organizations: A resource-based perspective.* New York: Harper & Row.

Podolny, J. M., & Stuart, T. E. (1995). A role-based ecology of technological change. *American Journal of Sociology, 100,* 1224–1260

Porter, M. (1980). *Competitive strategy: Techniques for analyzing industries and competitors.* New York: Free Press.

Posner, R. A. (1974). Theories of economic regulation. *Bell Journal of Economics and Management Science, 5,* 335–358.

Rudy, R. (2010). *Substitute or complement? An examination of the relationship between technological innovation and investment in strategic political management by firms.* Paper presented at the Academy of Management annual meetings, Montreal, Canada.

Russo, M. (2001). Institutions, exchange relations, and the emergence of new fields: Regulatory policies and independent power production in America, 1978–1992. *Administrative Science Quarterly, 46*(1), 57–86.

Russo, M. (2003). The emergence of sustainable industries: Building on natural capital. *Strategic Management Journal, 24*(4), 317–331.

Salamon, L. M., & Siegfried, J. J. (1977). Economic power and political influence: The impact of industry structure on public policy. *American Political Science Review, 71*(3), 1026–1043.

Salop, S. C., Scheffman, D. T., & Schwartz, W. (1984). A bidding analysis of special interest in regulation: raising rivals' costs in a rent seeking society. In *The political economy of regulation: Private interests in the regulatory process,* 102–127.

Schuler, D. (1996). Corporate political strategy and foreign competition: The case of the steel industry. *Academy of Management Journal, 39,* 720–737.

Schuler, D. (1999). Corporate political action: Rethinking the economic and organizational influences. *Business and Politics, 1,* 83–97.

Schuler, D., & Rehbein, K. (1997). The filtering role of the firm in corporate political involvement. *Business & Society, 36,* 116–139.

Schuler, D. A., Rehbein, K., & Cramer, R. D. (2002). Pursuing strategic advantage through political means: A multivariate approach. *Academy of Management Journal, 45,* 659–672.

Shaffer, B. (1995). Firm-level responses to government regulation: Theoretical and research approaches. *Journal of Management, 21,* 495–514.

Simon, H. A. (1970) *Administrative behavior: A study of decision-making process in administrative organization.* (2nd ed.) New York: Macmillan.

Sine, W., Haveman, H., & Tolbert, P. (2005). Risky business? Entrepreneurship in the new independent-power sector. *Administrative Science Quarterly, 50*(2): 200–232.

Smith, M. (2000). *American business and political power: Public opinion, elections, and democracy.* Chicago: Univ. of Chicago Press.

Sorenson, O. (2000). Letting the market work for you: An evolutionary perspective on product strategy. *Strategic Management Journal, 21*(5), 577–592.

Stigler, G. J. (1971). The theory of economic regulation. *Bell Journal of Economics and Management Science, 2*(1), 3–21.

Suarez, S. (2000). *Does business learn? Tax breaks, uncertainty, and political behavior.* Ann Arbor: University of Michigan Press.

Teece, D., Pisano, G., & Shuen, A. (1997). Dynamic capabilities and strategic management. *Strategic Management Journal, 18,* 509–533.

Tushman, M. L., & O'Reilly, C. (1997). *Winning through innovation.* Boston, MA: Harvard Business School Press.

Ungson, G. R., James, C., & Spicer, B. H. (1985). The effects of regulatory agencies on organizations in wood products and high technology electronics industries. *Academy of Management Journal, 28,* 426–445.

Vogel, D. (1978). Why American businessmen distrust their state: The political consciousness of American corporate executives. *British Journal of Political Science, 8,* 45–78.

Van de Ven, A. H., & Garud, R. (1989). A framework for understanding the emergence of new industries. *Research on Technological Innovation, Management and Policy, 4,* 195–225.

Yoffie, D. (1987). Corporate strategies for political action: A rational model. In A. Marcus, A. Kaufman, & D. Beam (Eds.), *Business strategy and public policy* (pp. 43–60). New York: Quorum Books.

Yoffie, D. (1988). How an industry builds political advantage. *Harvard Business Review, 66*(3), May–June, 82–89.

Yoffie, D., & Bergenstein, S. (1985). Creating political advantage: The rise of the corporate political entrepreneur. *California Management Review, 28*(1), 124–139.

York, J., & Lenox, M. (2009). It's not easy building green: The interaction of private and public institutions in the adoption of voluntary standards. Working chapter, University of Virginia.

Zardkoohi, A. (1985). On the political participation of the firm in the electoral process. *Southern Economic Journal, 51,* 804–817.

Zollo, M., & Winter, S. G. (2002). Deliberate learning and the evolution of dynamic capabilities. *Organization Science, 13*(3), 339–351.

PART II

Innovation Matters

The chapters in Part II explore innovation as a key driver of the green economy. Technological innovation can help in the creation of eco-efficiencies, conservation of energy and materials, and waste reduction. Business model innovations can help bring new sustainable products and services to market and shape profitability and structures of industries. Innovation in entrepreneurship is necessary for the creation of mission-driven organizations, the type of social economy businesses that contribute to the formation of the green economy. Innovation may also mediate the financial and social/ecological performance of firms. In the long run it may drive the social and financial performance of firms. Innovative technologies also may be attractive venues for investment capital to flow into. Venture capital is critical for commercialization of innovations. Venture investment paths vary between following past established paths or deviating from them, depending on the maturity of the innovations.

Innovations in the green economy are mediated by numerous nonorganizational level variables—both at the industry level (clusters) and at the individual level (gender of entrepreneurs). Although all the external influences are not known, we are beginning to gain an understanding of some of the critical ones. The role of location within already existing industry clusters and the ability to leverage the connections and resources within clusters play a role in fostering innovative industries. The green economy can be catalyzed by building upon existing innovative clusters, to usher in new green firms that build on an existing industrial ecosystem. Another critical external variable is the gender of entrepreneurs. Individuals practicing sustainability in their own lives and work environments can be a great force. Women entrepreneurs in the fashion industries (labeled *shecopreneurs*) provide a case in point. They reconceptualize ecological and social constraints to experiment with sustainable practices based on the duty to care. Experiments among a few can bring about large-scale social changes through invention, adoption, and reimagination of the real.

CHAPTER FIVE

Rethinking Sustainability, Innovation, and Financial Performance

TIMO BUSCH, BRYAN T. STINCHFIELD, AND
MATTHEW S. WOOD

In light of the recent financial crises, many economists and politicians claim that a paradigm change in modern capitalism is needed, from short-term profit maximization to a long-term value value-creating and value-maintaining strategy. In this context, scholars have emphasized stakeholder claims, institutional change, corporate responsibilities, and the role of ecological conditions on the competitive environment (Buysse & Verbeke, 2003; Delmas & Toffel, 2004; Henriques & Sadorsky, 1999; Hoffman, 1999; Kassinis & Vafeas, 2006; Aragon-Correa & Sharma, 2003; Darnall & Edwards, 2006; Sharma & Vredenburg, 1998; Husted & Allen, 2007; Matten & Crane, 2005; Scherer & Palazzo, 2007). For managers this entails investing in resources that enhance the firm's environmental and social performance while continuing to pursue economic growth. The goals are to minimize the firm's negative effects on the natural environment and society without compromising profits. Are these goals mutually exclusive? We find that they are compatible in the long run. They are different sides of the same coin, with innovation being the missing link between them.

Proceeding from early investigations (Bowman & Haire, 1975; Bragdon & Marlin, 1972), management researchers have looked at the relationship between a firm's environmental and social performance (ESP) and corporate financial performance (CFP) from different angles. Some studies examine why firms should address environmental and/or social issues (Gladwin, Kennelly, & Krause, 1995; Hart, 1995; Shrivastava, 1995); others look at why firms pursue high levels of ESP (Bansal, 2005; Sharma & Henriques, 2005); and still others take an instrumental perspective by examining the links between ESP and CFP (King & Lenox, 2002; Klassen & Whybark, 1999). A few scholars have attempted to generalize the findings of studies done thus far; they suggest that research has yielded mixed results (e.g., Salzmann, Ionescu-Somers, & Steger, 2005). Other analyses indicate that corporate virtue in form of sustainability efforts is likely to pay off (e.g., Margolis & Walsh, 2003; Orlitzky, Schmidt, & Rynes, 2003). However, currently there is much confusion regarding the terminology, performance measurements, and the generalizability of these results (Griffin & Mahon, 1997; Peloza, 2009).

We heed calls in the literature to incorporate a contingency perspective when investigating the ESP-CFP relationship (Barnett, 2007; Berchicci & King, 2007; Rowley & Berman, 2000). We do this by introducing a short-term and long-term analysis of the ESP–CFP relationship. Including this contingency perspective enables generalizations within the debate by emphasizing *when* ESP affects CFP. We also build upon recent work (Hull & Rothenberg, 2008) and theorize under what conditions the ESP-CFP relationship should be positive, that is, when a firm is innovative. We investigate the interaction effect of innovation and ESP.

A Balanced Environmental and Social Performance Construct

Contained within the influential Brundtland Report of the World Commission on Environment and Development (WCED), 1987, three central dimensions are discussed as to how firms can address the challenge of global sustainable development: environmental integrity, social equity, and economic well-being (Bansal, 2005). Environmental integrity requires organizations to first understand their negative impacts on global ecosystems and natural resources and then take actions to mitigate those impacts (Whiteman & Cooper, 2000). Social equity is the understanding that corporations have not only a fiduciary responsibility to their shareholders, but also the responsibility in terms of achieving social equity among a diverse group of stakeholders such as customers, employees, and community residents (Donaldson & Preston, 1995). Economic well-being is commonly understood as the third leg of the sustainability triangle, for corporations also must generate profits and maintain their competitiveness (Ambec & Lanoie, 2008). For management research, the main questions stemming from this triple bottom line approach have been: What is the relationship between the first two central dimensions and the latter, and how can firms formulate strategies to meet all three goals?

Previous studies (e.g., McWilliams & Siegel, 2000; Waddock & Graves, 1997) utilized a weighting scheme for the different social and environmental categories in order to construct a score that measures corporate social performance (or corporate social responsibility). Such schemes represented socially and ethically oriented performance metrics and marginally included ecological considerations. Starik and Rands (1995) argue that achieving progress towards sustainable development requires an effective integration of the different dimensions. Other scholars have emphasized that a challenge for corporate strategy is balancing ecological and social considerations while achieving attractive financial returns (Ambec & Lanoie, 2008; McWilliams & Siegel, 2001). In a competitive landscape increasingly concerned with sustainability, it is not clear why a specific environmental or social issue should be emphasized. Thus, biased weighting schemes are unable to reflect a balanced picture of how firms address social *and* environmental issues. Corporate attention to these dimensions is not mutually exclusive and linkages exist across them. For example, the global environmental issue of climate change may create water scarcity in many regions, which in turn can cause negative social implications such as conflict and poverty (Barnett & Adger, 2007). Thus, it is difficult to judge which firms are doing better in terms of corporate sustainability: the ones that prioritize curbing emissions in order to mitigate climate change or the ones that prioritize improving living conditions in the poorest regions of developing countries. We suggest a balanced ESP construct that adequately reflects both dimensions by equally weighting environmental and social aspects. We use the term *corporate environmental and social performance* to refer to a variety of voluntary and/or coercive activities undertaken by a firm in order to improve its performance with regard to the natural environment and in response to social and ethical issues. This ESP construct by itself does not include a financial component and is therefore distinct from Bansal's (2005) corporate sustainable development construct, which includes environmental, social, and financial considerations.

The resource-based view of the firm argues that rent-earning resources and capabilities determine the competitive advantage of firms (Barney, 1991). A firm's resources are defined as "those (tangible and intangible) assets which are tied semipermanently to the firm" (Wernerfelt, 1984, p. 172). From this, Hart (1995) advocates a theory of the "natural resource-based view." Under this framework, firms can improve ESP and simultaneously secure a competitive advantage by: 1) achieving lower costs through continuous improvement of pollution reduction technologies and processes, 2) preempting competitors by integrating a variety of stakeholders into creating more ecologically friendly products, and 3) securing a favorable future position through "minimizing [the] environmental burden of firm growth and development" (Hart, 1995, p. 992). Many studies empirically have demonstrated a positive linkage between ESP and CFP (e.g., Hart & Ahuja, 1996; King & Lenox, 2002).

Alternatively, there may be situations where firms would not invest in resources that enhance a firms' ESP if they can "gain little by providing public goods" and market pressure drives them to profit-maximizing choices (Berchicci & King, 2007, p. 515). For example, in the ecological context authors have argued that a high level of environmental performance might be disadvantageous for CFP (Filbeck & Gorman, 2004; Walley & Whitehead, 1994) and still others find a neutral relationship (e.g., Elsayed & Paton, 2005). Similar mixed results can be found regarding empirical studies in the social context (cf. Ullmann, 1985). In sum, there are analyses proposing that existing studies are inconclusive (McGuire, Sundgren & Schneeweis, 1988; Salzmann et al., 2005; Ullmann, 1985), although others claim that there is a positive—or at least no negative—relationship (Margolis, Elfenbein, & Walsh, 2009; Margolis & Walsh, 2003; Orlitzky et al., 2003).

As such, investments in ESP are detrimental in some cases and advantageous in others. We suggest that this differentiation can be explained by including a contingency perspective and considering the specific time horizon under analysis. For example, firms focusing on the introduction of environmentally friendly products and services are often faced with immediate higher production costs, which can result in higher consumer costs (Marcus & Fremeth, 2009). These higher costs may not be well-received by the market as the majority of consumers tend to stick with the less ecologically sustainable but cheaper products (Marcus, 2005). Based on the premise that it takes time to develop environmental and social service markets, and that it takes time for the costs of such products and services to drop to a level that average consumers are willing to afford, it appears that investments in resources to develop such products and services may negatively influence CFP—at least over the short-term (Marcus, 2005; Marcus & Fremeth, 2009). Furthermore, firms may not realize cost savings of certain ESP investments if they lack the required capabilities (Christmann, 2000), which usually cannot be obtained in the short-term. We reflect these arguments in the following hypothesis:

H 1a: The relationship between ESP and short-term CFP is negative.

Going beyond this consideration of immediate financial effects, recent research suggests that a firm's social performance positively affects its long-term CFP (Brammer & Millington, 2008). Similar arguments can be made regarding environmental performance: the development of a proactive environmental strategy designed to increase environmental performance can be a source for unique competitively valuable organizational capabilities, which can in turn have implications on competitiveness (Hart, 1995; Sharma & Vredenburg, 1998). These arguments are consistent with Porter's (1980) analysis of firms' competitive advantage, which said that successful differentiation is expected to lead to superior industry returns. In this sense, superior ESP management activities are strategic moves intended to differentiate the firm from competitors (Orsato, 2006). Furthermore, previous literature has discussed the benefits of ESP in terms of achieving increased efficiency, reduction of raw material and energy inputs, fewer fines and lawsuits, enhanced legitimacy, and greater employee morale and organizational commitment (Ambec & Lanoie, 2008; Carroll, 1999; Kassinis & Vafeas, 2006; King & Lenox, 2002; Klassen & Whybark, 1999; Russo & Fouts, 1997; Shrivastava, 1995). We consider the resulting financial benefits as long-term outcomes after initial investments for required resources have been amortized and corresponding ESP efforts have been acknowledged by stakeholders. Therefore, our second hypothesis explicitly focuses on the long-term payoff of investments in ESP-related resources.

H 1b: *The relationship between ESP and long-term CFP is positive.*

Innovation, Corporate Financial and Environmental and Social Performance

Schumpeter (1934) is often credited with the initial idea that innovations can lead to competitive advantage that can be exploited by innovative firms. According to Larsen (1993), one of the most common definitions of innovation includes the "development and implementation of new ideas by people who over time engage in transactions with others within an institutional order" (Van de Ven, 1986, p. 590). These "new ideas" include technical innovations, such as new products and services, and administrative innovations, such as new policies, strategies, and organizational structures as well as a recombination of old ideas. Similarly, Damanpour (1991, p. 556) describes an innovation as a "new product or service, a new production process technology, a new structure or administrative system, or a new plan or program pertaining to organizational members." As overlap between these definitions, we use the term *innovation* to refer to any invention, new technology, idea, product, or process that has been introduced by the focal firm (Damanpour & Gopalakrishnan, 2001; Wood, 2009).

A substantial body of research suggests that the relationship between a firm's level of innovation and CFP should be positive (Christensen & Bower, 1996; McWilliams & Siegel, 2000; O'Reilly & Tushman, 2004; Schumpeter, 1934; Zahra & Covin, 1995). For example, theoretical and empirical research investigating the connection between innovation and CFP shows that innovation provides firms with commercially superior products (Cooper & Kleinschmidt, 1987), better mechanisms to cope with environmental uncertainties (Damanpour & Evan, 1984), and an increased ability to create new resource configurations (Yiu & Chung-Ming, 2008). Specifically in the short-term, innovative firms can capture early mover advantages such as securing relationships with key suppliers (Doz, 1996), carving out attractive market share (Robinson, 1988), and forging customer loyalty (Parry & Bass, 1989). In the longer term, innovative firms can influence regulatory regimes (Frynas, Mellahi, & Pigman, 2006), forge favorable product standards (Rumelt, 1987), and create a self-reinforcing culture of attracting innovative employees (Ireland & Webb, 2007). As such, one can expect to find a positive relationship between innovation and CFP both in the short- and long-term:

H2a: *The relationship between innovation and short-term CFP is positive.*

H2b: *The relationship between innovation and long-term CFP is positive.*

We argue for ESP having both a positive and a negative effect on CFP, depending on the underlying timeframe used in the analysis. Furthermore, we hypothesize that innovation has a positive effect on CFP regardless of the time horizon. Considering this triptych inquiry of ESP, innovation, and CFP, the question arises as to how these three variables interact with each other. A starting point in this debate is research conducted by Waddock & Graves (1997), who find a positive relation between corporate social responsibility (CSR) and *past* CFP as well as a positive relation between CSR and *future* CFP. Further, McWilliams & Siegel (2000) propose that many such analyses are mis-specified as they leave out important control variables and so they conduct a similar analysis but include research and development (R&D) as a measure of innovation. They find that CSR and innovation are highly correlated and suggest the effect of CSR on CFP is neutral when innovation is taken into account. Extending McWilliams & Siegel (2000), Hull & Rothenberg (2008) use innovation as a moderator for the relationship of corporate social performance (CSP) and CFP. As result, they find a moderating relationship and support the initial argument that a positive relationship exists between CSP and CFP, but only in the context of low levels of innovation.

The Substitution Hypothesis. Hull and Rothenberg (2008) consider the CSP and innovation to be interchangeable. Following their substitution hypothesis, less innovative firms might chose to differentiate themselves from their competitors in order to improve their firm performance by improving CSP. In competitive environments requiring a high level of innovation, the authors suggest that CSP has a smaller effect on firm performance. They find that CSP has a greater impact on performance of those firms with low levels of innovation. If this observation holds true, then we could extend this argument to the broader concept of ESP where environmental criteria are equally weighted with social factors. Thus, the substitution hypothesis proposes that managers have to decide between investments in ESP and increasing their innovativeness in order to increase CFP. We can analyze this tradeoff affecting CFP both in the short-term and in the long-term. Therefore, the hypothesized moderated relationships based on the substitution argument are as follows:

> *H 3a: The negative relationship between ESP and short-term CFP is moderated by the level of firm's innovation, such that the relationship becomes stronger in the presence of a high level of innovation.*
>
> *H 3b: The positive relationship between ESP and long-term CFP is moderated by the level of firms' innovation, such that the relationship becomes stronger in the presence of a low level of innovation.*

Figure 5.1 summarizes our research model for the substitution hypothesis. Following Aiken and West (1991), we illustrate the moderation effect displayed in hypotheses 3a and 3b for values (*b*) for the level of innovation at one standard deviation below the mean (low level of innovation; $b = \mu - \delta$),

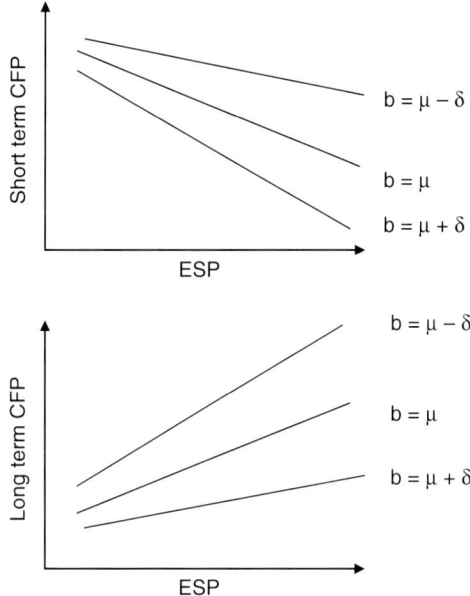

Figure 5.1 The moderating effect of innovation on ESP and CFP.

at the mean (there is no innovation effect; $b = \mu$), and one standard deviation above the mean (high level of innovation; $b = \mu + \delta$).

The Complementary Hypothesis. Counter to the substitution hypothesis above, scholars have claimed that successful firms require a "strategy that integrates the goals of innovation and sustainable development" (Hall & Vredenburg, 2003, p. 61). As such, firms require a complementary consideration of both ESP *and* innovation in order to differentiate themselves from their competitors (Reinhardt, 1998). In this context, some researchers have argued that the relationship between ESP and innovation is sequential while others view ESP as a precursor to innovation (Fowler & Hope, 2007; Hart, 1997; Larson, 2000). Hart (1997), for example, argues that the vision of corporate sustainability acts as a roadmap that guides innovation. This view has received empirical support from Fowler and Hope (2007) and Larson (2000), who analyzed the vision, organizational formation, and product development of entrepreneurial firms that remained committed to their corporate vision of sustainable development, which then dictated the types of organizational and technological innovations they deployed. Still other researchers have proposed that a high level of innovativeness is required for improved ESP. For example, Porter and van der Linde (1995) consider innovation, triggered by environmental regulation, as a precondition for improved ESP and competitiveness. As a result, a firm's commitment to new technological, administrative, and strategic innovations can be seen as a driver for ESP. In this way, innovation is likely to become the key mechanism by which ESP influences CFP. Following this latter logic we hypothesize:

H4a: The negative relationship between ESP and short-term CFP will be mediated by the level of firms' innovation, such that ESP impacts performance through innovation.

H4b: The positive relationship between ESP and long-term CFP will be mediated by the level of firms' innovation, such that ESP impacts performance through innovation.

Sample and Data Collection

We base our analysis on a data set obtained from KLD Research and Analytics, Inc. Prior research has used subsets of the KLD databases to construct an index of corporate social performance (Waddock & Graves, 1997; McWilliams & Siegel, 2000; Hull & Rothenberg, 2008); however, our data set diverges from these previous studies in two important ways. First, we obtained a KLD data set for the years 2001 through 2003 and our data set included a larger set of companies ($N = 252$) than what has been used in recent studies (e.g., Hull & Rothenberg, 2008, $N = 69$). Second, one of the primary constructs under investigation is ESP, which is a balanced combination of firms' social and environmental performance. Thus, our ESP variable reflects an equal weighting of KLD's social and environmental ratings, whereas in previous research the environmental ratings were dominated by social ratings.

As starting point we used the complete KLD data set for 2001, 2002, and 2003, which provided data for 900 North American based firms. Using these firms as a reference, we then used Compustat to gather data on financial performance, innovation, and the control variables. However, a number of firms that were listed in the KLD index did not have complete data within the Compustat database for the required years. After eliminating those firms where complete information was not available, we obtained a final sample of 252 firms for all three years.

In our analysis, we consider ESP as an independent variable that accounts for both the environmental and social performance of a firm. For the corporate social and environmental ratings, KLD evaluates over 125 corporate social and environmental aspects and groups them into 13 broad categories. Seven of these thirteen categories were deemed relevant for the development of our ESP measure and are included in our analysis. These categories are:

1. community (e.g., charitable donations and support for employee volunteer programs);
2. corporate governance (e.g., firm has not been engaged in controversial governance practices and places limits on executive and board member compensation);

3. diversity (e.g., firm promotes hiring of women, minorities, and the disabled);
4. employee relationships (e.g., firm has good relations with its unions and has a strong record in promoting the health and safety of its workers);
5. human rights (e.g., firm is recognized for its open, respectful, and transparent relationships with indigenous peoples and overseas factory workers);
6. product (e.g., quality of firm's products and avoidance of antitrust and product safety concerns);
7. environmental, which includes such items as energy efficiency, pollution prevention, recycling, clean energy, environmental regulatory problems, and the degree to which the firm generates revenues from industries that put forth large amounts of carbon dioxide emissions.

Those categories excluded from our analyses include KLD evaluations about controversial business issues, namely alcohol, gambling, tobacco, firearms, military, and nuclear power. Although many sustainability-oriented rating concepts for financial markets' indices and funds, as well as previous studies, include such "exclusion-criteria," we decided not to take them into account when developing our ESP score for two reasons. First, some of the items may actually have debatable or even positive sustainability effects as compared to other options. For example, many policymakers consider nuclear power an important short-term solution for curbing CO_2 emissions. Similarly, it can be argued that military equipment is needed to obtain or maintain peace in certain areas of the world. Second, many of the other categories do not have an immediate effect on social developments or the natural environment. For example, human action—or, more precisely, human irresponsible action or abuse—is required in order for some of these categories to have negative effects on human or society. One example for this would be the consumption of alcohol. Our goal is not to expand the concept of corporate social and environmental responsibility to include such "third-party behavior-dependent" items.

For each of the KLD categories, KLD provides several items that are labeled "strengths" and several that are labeled "concerns." Each item is coded with a "1" if the firm has a strength / concern and otherwise "0." For each firm we then added the KLD social "strength" scores and afterwards subtracted the social "concerns." From this overall social score we calculated z-scores to arrive at a standardized social score (Choi & Wang, 2009). Next we did the same for the environmental scores; we subtracted the sum of "concerns" from the sum of "strengths" and then calculated z-scores for the environmental dimension. Finally, we averaged the standardized social scores with the standardized environmental scores with equal weight to arrive at an ESP score. As such, we obtained a balanced score that reflects firms' corporate sustainability (environmental and social) efforts.

The second independent variable, innovation, was operationalized as R&D intensity, as it is commonly done in the ESP–CFP literature (Choi & Wang, 2009; McWilliams & Siegel, 2000). This measure was constructed by taking each firm's R&D spending and dividing it by the firm's sales and then averaging these values across the three-year period from 2001 through 2003. By using the three-year average of R&D intensity, we control for the influence of single-year fluctuations in investments in innovative activities.

The dependent variable, corporate financial performance (CFP), was measured using Tobin's q. This measure is a dynamic performance indicator that reflects the stock market's expectations about the profitability and growth potential of the firm as well as internal efficiency metrics, such as equity and assets (Kor & Mahoney, 2005). In this case, Tobin's q is an appropriate measure, because we seek to understand the influence of ESP and innovation on the economic value generated by the firm, in both the long- and short-term (Wernerfelt & Montgomery, 1988). This measure is more appropriate than return on assets (ROA) or other accounting-based measurements, which are often used in studies investigating the ESP-CFP relationship. The payoff of investments in ESP-related resources and innovation may or may not be reflected in the balance sheet and in the firm's market value, both of which constitute important parameters of economic value creation. As such, we calculated Tobin's q by dividing the sum of the firm's equity (market value), book value of long-term debt, and net current liabilities by the firm's total assets (Chung & Pruitt, 1994; King & Lenox, 2002).

Because of the temporal nature of our research question and hypotheses, it was necessary to construct both a short- and long-term performance measure. The short-term measure was operationalized by using Tobin's q for the year 2004, which is the first year following the 2001–2003 time periods that were used to calculate our ESP and innovation measures. Among studies that conceptualize the concept of long-term CFP, Eisenmann (2006) operationalizes long-term CFP of Internet companies as roughly two years after their IPO. Prashant, Dyer, & Singh (2002) use the period 1993-1997 to assess the long-term CFP of alliances. Similarly, Combs, Crook and Shook (2004) and Tosi, Werner, Katz, and Gomez-Mejia (2000) consider the five-year average for measuring long-term return on equity. We derive our long-term CFP measure by averaging Tobin's q's for the second year (2005) through the fourth year (2007) following the investments in ESP and innovation (2001–2003).

A number of control variables are used as previous research has identified important factors affecting CFP. We used *firm size* since it can affect firm performance through economies of scale, monopoly power, and bargaining power. In this study, firm size is operationalized as the three year average of firm sales from 2001–2003. Furthermore, research has shown that a firm's *risk* is an important factor to be controlled (e.g., Choi & Wang, 2009; Waddock & Graves, 1997). Therefore, we used the three-year average of long-term debt to total assets as proxy for the riskiness of the firm. Finally, *industry membership* has been cited as an influential factor on both ESP and CFP (e.g., Derwall, Guenster, Bauer & Koedijk, 2005; Ullmann, 1985). As such, we included dummy variables for each of the nine industries in our sample, as identified by the two-digit GICS code. The industries are energy, materials, industrials, consumer-discretionary, consumer-staples, health care, financials, telecom, and utilities.

Analysis and Results

Our hypotheses were tested using hierarchical, mediated, and moderated regression analysis. These statistical tools allowed us to determine the effects of each variable separately and the interaction effects between the independent variables (Howell, 2007). More specifically, hierarchical regression analyses were used to identify main and interaction effects. In this type of analysis, the interaction effects are found to be significant only if they explain a significantly greater portion of the variance in the dependent variable. Thus, moderated regression analysis helps test the significance of interaction effects by regressing the dependent variable onto two or more main variables (one independent and one moderator) and the cross-product of those main variables (Sharma, Durand, and Gur-Arie, 1981). If the addition of the interaction term significantly increases the power of the regression equation to explain the variance in the dependent variable, then the contingency relationship can be said to exist. Of course, moderation is only possible if it has been shown that strategic choice is not acting as a mediating variable.

Barron and Kenny (1986) provide a widely accepted technique for testing mediation and they recommend that four conditions be satisfied for a researcher to claim mediation. The first requirement is that there must be a relationship between the independent variable of ESP and the dependent variable of CFP. Second, there must be a significant relationship between the mediating variable of innovation and the independent variable; if this relationship does not exist, then the variable cannot mediate anything. Third, the mediating variable must be related to the dependent variable. Finally, the effect of the independent variable on the dependent variable must be significantly weakened in the presence of the moderator variable.

Because we selected regression as the analytic technique, we first explored graphical representations of the data in order to ensure that our data satisfied the assumptions required to accurately apply regression techniques. The assumptions analysis revealed that the relationships between the variables were in fact linear, and it also revealed that all data points were viable—indicating that there were no outliers in our sample. Examination of the graphical representations of the CSP variable indicate that the data were normally distributed and thus appropriate for use in regression analysis. However, the graphical representation of the innovation variable revealed that the data were suffering from positive skewness and kurtosis. As such, we conducted a data transformation (Hair, Black, Babin, Anderson, & Tatham, 2006) by taking the natural log of the innovation measure (R&D/sales). The graphical representation of the transformed data indicated that innovation measure was indeed normally distributed, therefore appropriate for use in the regression analysis. Further data exploration was conducted to ensure that the assumptions of normality and linearity had all been adequately satisfied (Hair et al., 2006). Once we were sure that the assumptions for linear regression had been satisfied, we continued with our regression-based hypothesis testing.

The descriptive statics and correlations for our variables are reported in Table 5.1 and the standardized regression coefficients are reported in for short-term CFP in Table 5.2 and for long-term CFP in Table 5.3. We report standardized coefficients so that differences in the strength of the relationships can over time be evaluated (Hair, et al., 2006). We tested our hypotheses using four separate regression models (see Table 5.2). Model 1 is the control model, Model 2 tests the ESP and innovation hypotheses, Model 3 tests the moderation hypotheses and Model 4 tests the mediation hypotheses.

Short-Term Financial Performance. We first tested hypothesis 1a which explored the possibility that there is a negative relationship between ESP and short-term financial performance. Results from Model 2 (see Table 5.2) indicate that there is a positive but non-significant relationship between ESP and short-term CFP ($\beta = .01$, $p \geq .10$). Thus, hypothesis H1a is not supported. Next, we tested hypothesis H2a, which argues for a positive relationship between innovation and short-term performance. The regression in Model 2 revealed that the coefficient for innovation was positive and significant ($\beta = .38$, $p < .01$) thereby supporting hypothesis H2a. Next, we tested hypothesis H3a, which argues that innovation would moderate the relationship between ESP and short-term CFP. Model 3 indicates that there is a negative and marginally significant effect for the addition of the interaction term ($\beta = -.12$, $p < .10$). As such, H3a is marginally supported. This finding seems to indicate that high levels

Table 5.1 Descriptive statistics and correlations

	Mean	SD	1	2	3	4	5	6	7
1. Size	4873.23	1.46	—						
2. Risk	1.14	.16	.01	—					
3. ESP	.07	.99	−.34**	−.05	—				
4. Innovation	254.35	702.88	−.21**	−.16**	.14*	—			
5. ESP ⋆ Innovation	1	0	.43**	−.11	.38**	−.33	—		
6. Tobin's q (2004)	2.17	1.34	−.10	−.02	.08	.39**	−.13*	—	
7. Tobin's q (2005 through 2007)	2.07	1.06	−.08	−.10	.15**	.32**	−.02	.81**	—

$N = 252$. *Correlation is significant at the .05 level; **correlation is significant at the .01 level.

Table 5.2 Regression results for short-term CFP (Tobin's q for 2004)[a]

Variable	Model 1 (Control)	Model 2 (Independent variables)	Model 3 (Moderation)	Model 4 (Mediation)
Industry:				
Energy	−.08 (.43)	−.03 (.42)	−.01(.42)	
Materials	.01 (.33)	.05 (.31)	.04 (.32)	
Industrials	−.15 (.28)**	−.04 (.27)	−.06 (.27)	
Consumer discres.	−.10 (.27)	−.03 (.25)	−.03 (.25)	
Consumer staple	−.01 (.45)	.05 (.43)	.04 (.43)	
Health care	.12 (.38)*	.14 (.27)**	.13 (.27)*	
Financials	−.08 (.33)	−.02 (.31)	−.04 (.31)	
Telecom	−.07 (.79)	−.06 (.74)	−.06 (.74)	
Utilities	−.09 (.44)	−.04 (.41)	−.04 (.41)	
Firm size	−.08 (.01)	−.01 (.00)	.03 (.01)	
Risk	.01 (.52)	.06 (.50)	.05 (.50)	
ESP		.01 (.12)	−.02 (.12)	
Innovation		.38 (.14)***	.38 (.13)***	
ESP × Innovation			−.12 (.19)*	
Mediation:				
Path A				.14**
Path B				.39***
Path C				.08 (ns)
Path C'				n/a
F Change		17.57***	2.74*	
R^2 Change		11.90%	1.01%	
R^2 Total	7.5%	19.40%	20.41%	

Dependent Variable (DV) = Corporate Financial Performance (CFP) measured via Tobin's q for FY 2004.
[a]Reporting standardized beta coefficients with standard error in parentheses.
*** Significant at .01 level.
** Significant at .05 level.
* Significant at .10 level.

of investment in innovation negatively impact performance in companies that are pursuing a high level of ESP, at least in the short-term.

We then tested hypothesis H4a, which argued that innovation would mediate the ESP–CSF relationship in the short term. As previously discussed, Barron and Kenny (1986) established four conditions that must be satisfied for a mediation type relationship to exist. We closely followed the Baron and Kenny (1986) technique and Model 4 (Table 5.2) reports the results of our test for mediation. In an attempt to provide a clear conceptual link between our test and the Baron and Kenny (1986) approach, we also diagram our results in Figure 5.2. The figure illustrates the strength of the various relationships among the variables via regression coefficients. The key element here is the significance of the paths and ultimately the change in the strength of the relationships between ESP and CFP (paths C and C') in the presence of the innovation variable. Mediation exists if there is a significant reduction in the strength of the ESP-CFP relationship, as measured by the Sobel test, when innovation is present.

Barron and Kenny (1986) suggest that the first step to test for mediation is to examine the direct relationship between ESP and innovation (Path A in Figure 5.2); we found it to be positive and significant ($\beta = .14$, $p < .05$); see Table 5.2. Next we tested the direct relationship between innovation and CFP (Path B) and found it to be positive and significant ($\beta = .39$, $p < .01$). We then tested the direct relationship between ESP and CFP (Path C); our analysis revealed that this relationship was positive but not significant ($\beta = .08$, $p > .10$. Because there was not a significant direct relationship between ESP and short-term CFP, it is not possible for innovation to mediate the ESP and short-term CFP relationship; there simply is not a significant relationship to mediate. Thus there is no evidence that innovation mediates the ESP–CFP relationship in the short-term situation and H4a is not supported.

Long-Term Financial Performance. In order to test our long-term performance hypotheses, we again used four different regression models; these results are reported in Table 5.3. Hypothesis H1b proposes a positive relationship between ESP and long-term CFP. Model 2 revealed that there was a positive and marginally significant relationship between ESP and CFP ($\beta = .11$, $p < .10$), providing marginal support for hypothesis H1b. Next, we tested hypothesis H2b, which predicts a positive relationship between innovation and long-term performance. Model 2 provides support for H2b by indicating a positive and significant relationship between innovation and CFP ($\beta = .29$, $p < .01$). We then used Model 3 to test hypothesis H3b, which suggests that innovation would moderate the relationship between ESP and long-term CFP. Regression results indicated that there is a positive but nonsignificant effect for the addition of the interaction term ($\beta = .03$, $p > .10$). Thus, innovation does not moderate the relationship between innovation and long-term performance, and so hypothesis H3b was not supported.

To test for the idea that innovation mediates the ESP and long-term CFP relationship, we again utilized the Barron and Kenny (1986) technique and report our results in Model 4 (Table 5.3); we illustrate them in Figure 5.3. Results show that the direct relationship between ESP

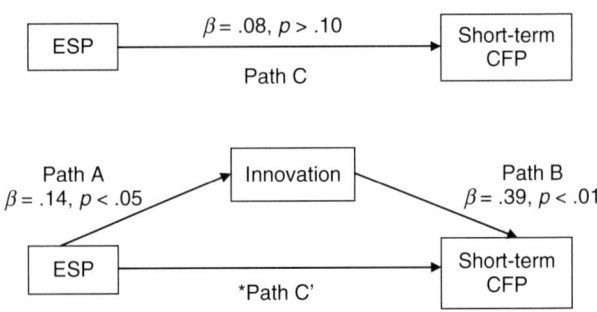

*Since Path C is non-significant, Path C' cannot exist—indicating no mediation (Barron and Kenney,1986).

Figure 5.2 Mediated model of ESP, innovation, and short-term CFP.

Table 5.3 Regression results for long-term CFP (Tobin's q scores, average of years 2005 through 2007)[a]

Variable	Model 1 (Control)	Model 2 (Independent variables)	Model 3 (Moderation)	Model 4 (Mediation)
Industry:				
Energy	.01 (.34)	.08 (.34)	.07 (.34)	
Materials	.01 (.26)	.06 (.25)	.06 (.25)	
Industrials	−.05 (.22)	−.04 (.22)	.05 (.22)	
Consumer discres.	−.01 (.21)	−.04 (.20)	.04 (.20)	
Consumer staple	.23 (.35)**	.08 (.35)	.08 (.34)	
Health care	.25 (.22)**	.25 (.22)**	.26 (.22)**	
Financials	−.07 (.26)	−.02 (.25)	−.23 (.25)**	
Telecom	−.01 (.62)	.01 (.59)	.01 (.59)	
Utilities	−.07 (.34)	−.02 (.33)	−.02 (.33)	
Firm Size	−.08 (.01)	−.01 (.00)	−.01 (.00)	
Risk	−.09 (.42)	−.05 (.50)	−.05 (.40)	
ESP		.11 (.09)*	.12 (.09)*	
Innovation		.29 (.10)***	.29 (.10)***	
ESP × Innovation			.03 (.16)	
Mediation:				
Path A				.14**
Path B				.32***
Path C				.16***
Path C'				.11*
Sobel Test				$Z = .02, p < .05$
F Change		12.19***	.15	
R^2 Change		8.40%	.01%	
R^2 Total	10.10%	18.50%	18.51%	

DV = Corporate Financial Performance (CFP) measured via Tobin's q for FY 2005 through 2007.
[a] Reporting standardized beta coefficients with standard error in parentheses.
*** Significant at $p < .01$ level.
** Significant at $p < .05$ level.
* Marginally significant at $p < .10$ level.

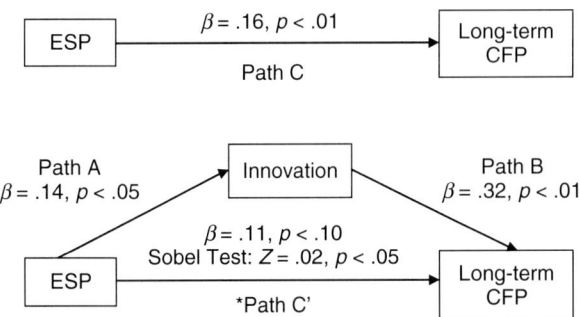

*Path C goes from highly significant to less significant in Path C'. This difference is statistically significant–indicating mediation exists (Barron and Kenny, 1986).

Figure 5.3 Mediated model of ESP, innovation, and long-term CFP.

and innovation (Path A) is positive and significant ($\beta = .14$, $p < .05$). The direct relationship between innovation and CFP (Path B) is also positive and significant ($\beta = .32$, $p < .01$). The relationship between ESP and CFP (Path C) is positive and significant ($\beta = .16$, $p < .01$). Finally, the strength of the relationship between ESP and CFP (Path C') is reduced in the presence of innovation ($\beta = .11$, $p < .10$). In order to claim mediation, the reduction in the strength of the relationship between ESP and CFP (Path C') must be statistically significant. We tested this difference between path C and Path C' (Figure 5.3) using a Sobel test and found that the reduction was indeed significant, $Z = .02$, $p < .05$. Therefore, we find that innovation does in fact mediate the ESP–CFP relationship in the long-term and hypothesis H4b is supported.

Contrasting Short-term Versus Long-term Results. One of the goals of our research was to look at the interplay between ESP, innovation, and CFP over time. To that end, we now compare the differences in results in our short-term and long-term analyses. For the effect of ESP on CFP we found that the relationship was not significant in the short term ($\beta = .01$, $p > .10$) but in the long-term it was significant at the .10 level ($\beta = .11$). Since we are using standardized coefficients we can directly compare the coefficients and this process reveals that there is a positive difference of .10 in the coefficients. This gives some indication that the impact of ESP on financial performance is becoming stronger over time. Next we compared the coefficients for the relationship between innovation and CFP. Here we find that this relationship was positive and significant in both the short-term ($\beta = .38$, $p < .01$) and the long-term ($\beta = .29$, $p < .01$) with a coefficient difference of .08. However, this difference represents a reduction in the strength of the relationship, indicating that the effect of innovation on CFP is weakening over time during the considered time frame.

Next we compared the moderated relationships. For short-term CFP, the moderated relationship was negative and marginally significant ($\beta = -.12$, $p < .10$) but in the long-term it was positive and not significant ($\beta = .03$, $p > .10$). What is interesting here is that the sign flipped from negative to positive, indicating that firms that invest a great deal of resources into ESP and innovation may suffer negative CFP effects in the short-term, but as time goes by that negative effect may turn positive.

Finally, we compared the mediated relationships (Model 4). For short-term performance, we found no support for the existence of a mediated relationship. However, the relationship did exist in the long term. This outcome seems to indicate that innovation becomes the mechanism by which ESP impacts CFP, but that this relationship takes time to emerge. When coupled with our other findings, it appears that collective investments in ESP and innovation may hurt short-term CFP by way of a moderated effect, but in the longer term, innovation helps to improve the ESP-CFP relationship by way of mediation.

Implications

This chapter extends the academic debate regarding firm performance within the context of ESP by moving beyond the questions of whether or not it pays to improve both environmental and social performance, and instead asks a more specific question of "When does it pay?" (King & Lenox, 2001; Orsato, 2006). With respect to this question, researchers have argued that is important to take a contingency perspective (Barnett, 2007; Berchicci & King, 2007; Rowley & Berman, 2000) and that there may be a non-linear relationship between environmental and social performance and CFP (Peloza, 2009). In fact, Brammer & Millington (2008) find in a recent study that firms with unusually high, as well as with unusually low, levels of social performance also have higher levels of CFP than other firms. This U-shaped curvilinear relationship is akin to Porter's (1985) "stuck in the middle" phenomenon. Drawing on the resource-based view of the firm, our empirical results show that this contingency in the performance debate can be explained by different time horizons. We now turn our attention to elaborating on these findings in the following paragraphs.

Our result of hypothesis H1a is consistent with scholars who found ambiguous findings concerning the ESP–CFP relationship (e.g., Salzmann, Ionescu-Somers, & Steger, 2005). When the focus is on the single relationship between ESP and short-term CFP we too did not find a clear relationship. However, the relationship between ESP and financial performance becomes significant when the focus is on the long-term (Hypothesis H1b). This result supports our initial assumption that is important to include a temporal perspective when investigating the ESP–CFP relationship. We conclude that the relationship between ESP and financial performance is a time-dependent inverse-U-shaped relationship. As illustrated in Figure 5.4, CFP varies for a given level of investments in ESP-related resources, depending on the time frame under consideration. In the short-term (within the one-year time-frame 0-t_1), there is no distinct result between ESP and CFP, and the *type* of ESP activity may have differential effects on CFP. This means the realization of low-hanging fruits through increasing eco-efficiency has a positive effect in CFP (upper curve) while investments into expensive resources required for developing new environmentally sound products may result in a negative CFP (lower curve).

The short-term CFP focus of many studies may explain why the generalizability of the results appears to be difficult. In contrast, in the long-term (within the time frame t_1 through t_b) investments in ESP-related resources indeed seem to pay off. Research found that the development of a proactive environmental strategy can be the source for unique competitively valuable organizational capabilities (Russo & Fouts, 1997; Sharma & Vredenburg, 1998). The acquisition and development of these capabilities takes time and thus the positive effect on CFP can be expected to take a long time. However, when only considering the ESP–CFP relationship,

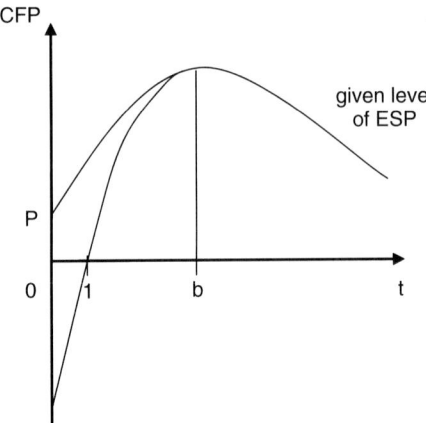

Figure 5.4 Time-dependent inverse-U-shaped relationship between a given level of investments in ESP-related resources and CFP.

and not including innovation in the analysis, the literature has discussed that this positive effect might diminish again after a certain time (after t_b): ongoing investments in resources in order to maintain a high level of ESP may exceed cost savings generated from such activities (Sharma & Vredenburg, 1998) or competing firms may be able to imitate strong stakeholder relationships that initially provided some firms with a competitive advantage (Choi & Wang, 2009; Hillman & Keim, 2001). In sum, an important implication of our results is that the generalizability of results may be significantly increased by incorporating a temporal perspective when investigating the ESP–CFP relationship.

Regarding innovation, our results are consistent with current research in the domain of innovation management: a positive relationship exists between innovation and CFP in both the short-term and long-term. However, the effect seems to weaken over time such that we observe a statistically significant change in standardized betas ($T = 1.34$) between short-term and long-term performance. While there are many possible explanations for this, it is likely that the decreasing intensity of the *effect* of innovation is due to the weakening *nature* of innovation. To elaborate, previous research has shown that innovations are often copied by competitors (VanderWerf & Mahon, 1997) and that knowledge spillovers allow copycat firms to erode innovators' first mover advantage (e.g., Acs, Braunerhjelm, Audretsch, & Carlsson, 2009). Our empirical results are consistent with these ideas and suggest that the effects of innovative behavior are generally positive, but have a greater impact on short-term performance compared to the long-term. This lends support to the dominant management thinking that for innovation to benefit the firm it should not be a one-time or ad hoc exercise, but rather a continuous effort (e.g., Barringer & Ireland, 2008).

The results of testing the substitution and complementary hypotheses contribute to a better understanding of the interplay between ESP, innovation, and CFP. Our results with respect to ESP, innovation, and short-term CFP are similar to those of Hull and Rothenberg (2008), who found a negative moderating effect of innovation on corporate social performance and short-term CFP. We conclude that under a short-term horizon, firms with limited resources are faced with a choice between investments in innovation or in ESP. Simultaneous investments in resources that allow for both activities generate higher costs to include management's time and attention and, thus, harm CFP in the short run. We deduce that the full benefit of ESP does not accrue immediately but takes time to pay off through high levels of innovation. Notably, the moderating effect occurs in the short-term, but it is not a time-consistent effect. In the long-term, innovation has a mediating effect. As such, efforts to increase ESP without any simultaneous investments in resources that trigger a firm's innovativeness will have little effect on CFP and might even result in diminishing CFP, as postulated under the time-dependent inverse-U-shaped relationship proposition. However, the combination of investing in innovation and enhancing ESP results in superior financial performance and may lead to a sustained competitive advantage. In sum, innovation acts as the organizational action through which ESP can contribute to achieving substantial and sustainable improvements of CFP.

In the academic debate regarding whether it pays to be "good" and/or "green," there is much confusion with respect to the utilized terminology and applied performance measurements. We suggest that using precise and consistent terminologies within this debate could significantly reduce this confusion. Some authors implicitly follow this line by focusing their investigation on corporate eco-efficiency (Derwall, et al., 2005) or corporate charitable giving (Brammer & Millington, 2008). When generalizing the results, such studies are limited by their theoretical and managerial implication to the specific focus of the study. For example, the eco-efficiency literature proposes that corporate efforts to enhance a firm's eco-efficiency should—if successfully implemented—optimize a firm's production processes by a reduction of the material and energy flows and simultaneously achieve cost benefits (DeSimone & Popoff, 1997). As such, research in this area can empirically test the cost-effectiveness of specific corporate activities. However, general statements as to whether corporate environmental performance, or even corporate social responsibility, pays off cannot be precisely derived. For studies investigating general questions as whether the environmental, social, or ethical efforts of a firm result in better CFP, we suggest using a clearly defined construct. Using the widely accepted Brundtland Report (World Commission on Environment and Development, 1987) to anchor our definition of "sustainability,' we then separated the financial component and defined ESP as an equally balanced construct of a firm's environmental and social performance. We hope that this new construct can reduce further confusion regarding the terminology and performance measurements in future research.

Managerial Implications. Our results reconfirm the important role of innovation: in the short-term, managers may be forced to choose between investments in resources that enhance the firm's ESP or innovation. As such, from a short-term profit maximizing strategy, managers have to decide in which areas it is better to invest, innovation or ESP. Two basic situations are possible. On the one hand, the essential resources to identify and realize eco-efficiency potentials are likely to be acquired at a low cost and with minimal effort. For example, the World Business Council for Sustainable Development (see http://www.wbcsd.org) offers a great deal of publicly available information and easily implementable tools as to how firms can reap such low-hanging fruits. In such cases, it is a firm-specific tradeoff situation whether to invest in resources that enhance its ESP or innovativeness. On the other hand, in order to develop and implement a highly sophisticated and supply-chain-wide ESP strategy usually requires significant managerial effort and costs. In such cases, a purely short-term CFP-driven management strategy would suggest investing in resources that enhance the firm's innovativeness.

However, in a more strategic (i.e., long-term) perspective, the complementary hypothesis holds, which is that innovation is a key mechanism required for realizing and maximizing the effects of ESP initiatives on financial performance. Our findings of a mediated relationship suggests that without innovation, ESP efforts will fall short of expectations because innovation has to be present for ESP to influence long-term CFP. Thus, managers who focus on long-term value creation may be well advised to direct resources towards increasing both ESP and innovativeness. In practical terms, this means that companies who simply focus on increasing ESP will have difficulty recognizing long-term performance benefits. Rather, our findings suggest that these firms must also introduce innovative products, services, and processes, because it is through these innovations that ESP improves long-term financial performance. In sum, it is important for corporate managers to realize that a competitive advantage requires focusing on both ESP and innovation, and this requires a long-term investment horizon versus quick fixes. In fact, our comparison of short-term versus long-term performance indicates that a strategy reflecting the insights of our triptych inquiry cannot be realized overnight. For firms looking for predictable long-term growth and performance, resources should simultaneously be devoted to ESP and innovation-focused programs.

Limitations and Future Research. Regarding the ESP–CFP relationship, our results suggest that scholars should look at the different impacts on short-term versus long-term CFP while recognizing the interactive role of innovation. Although our intention was to construct an ESP score that equally weights environmental and social dimensions, investigating specific constructs within these dimensions and their effect on CFP may further shed light on this triptych inquiry. For example, it could be investigated whether there is a difference when just considering output-based environmental performance data (e.g., a firm's level of greenhouse gas emissions) and process-based management indicators for ESP (e.g., the sophistication of a firm's carbon management) (cf., Ginsberg, 1988). It could be the case that more symbolic actions differ from substantive efforts in terms of a firm's short- and, notably, long-term CFP (cf., Berrone, Gelabert & Fosfuri, 2009). Furthermore, we followed the literature on measuring innovation as R&D expenses (McWilliams & Siegel, 2000; Hull & Rothenberg, 2008). R&D expenses usually reflect more technological innovations (e.g., Choi & Wang, 2009), but more "soft" innovation decisions relating to organizational practices may be less dependent on R&D expenditures. Future research could address this by using more fine-grained measures for corporate innovation activities.

Moreover, we introduced a time-dependent inverse-U-shaped relationship between ESP and CFP. As a limitation, we did not test for diminishing CFP at the end of the curve. Future research could empirically investigate whether this possibility actually occurs. Furthermore, since our data suggests a mediating relationship in the long run, more studies investigating how innovation acts as mediator between ESP and CFP seem especially relevant. Lastly, researchers should investigate whether these results hold during a time of global economic cycles as this data was gathered prior to entering the worst of the recent recession and scholars argue that specific value-creating resources may also be the sources of losses in times of financial turmoil (Choi & Wang, 2009; Leonard-Barton, 1992). Although many publicly traded firms have experienced a sharp decline in their performance and equity, researchers could investigate the degree to which investments in ESP and innovation either hinders or buffers (Thompson, 1967) firm performance during periods of heightened uncertainty.

The purpose of the triptych inquiry of ESP, innovation, and corporate financial performance was to empirically explore when it pays for firms to address the first two challenges of sustainability – environmental integrity and social equity. We have expanded upon previous work in this area by investigating two time-related performance periods and emphasized the interaction effect of innovation. The results suggest that it pays to increase a firm's level of ESP when firms have the ability to innovate *and* when the financial goals are not limited to short-term planning horizons. We conclude that for innovative firms, there is no mutual exclusivity among corporate environmental, social, and financial performance. These dimensions are collectively reinforcing and can contribute to the long-term survival of the firm in competitive markets and within its natural and social environment.

References

Acs, Z., Braunerhjelm, P., Audretsch, D., & Carlsson, B. (2009). The knowledge spillover theory of entrepreneurship. *Small Business Economics, 32*(1), 15–30.
Aiken, L. S., & West, S. G. (1991). *Multiple regression: Testing and interpreting interactions*. Newbury Park: Sage Publications.
Ambec, S., & Lanoie, P. (2008). Does it pay to be green? A systematic overview. *Academy of Management Perspectives, 22*(4), 45–62.
Aragon-Correa, J. A., & Sharma, S. (2003). A contingent resource-based view of proactive corporate environmental strategy. *Academy of Management Review, 28*(1), 71–88.
Bansal, P. (2005). Evolving sustainably: A longitudinal study of corporate sustainable development. *Strategic Management Journal, 26*(3), 197 218.
Barnett, J., & Adger, W. N. (2007). Climate change, human security and violent conflict. *Political Geography, 26*, 639–655.
Barnett, M. L. (2007). Stakeholder influence capacity and the variability of financial returns to corporate social responsibility. *Academy of Management Review, 32*(3), 794–816.
Barney, J. B. (1991). Firm resources and sustained competitive advantage. *Journal of Management, 17*(1), 99–120.
Barron, R. M., & Kenny, D. A. (1986). The moderator mediator variable distinction in social psychological research—Conceptual, strategic, and statistical considerations. *Journal of Personality and Social Psychology 51*(6), 1173–1182.
Barringer, B., & Ireland, D. (2008). *Entrepreneurship: Successfully launching new ventures*. New York: Prentice Hall.
Berchicci, L., & King, A. (2007). Chapter 11: Postcards from the edge. *The Academy of Management Annals, 1*(1), 513–547.
Berrone, P., Gelabert, L., & Fosfuri, A. (2009). The impact of symbolic and substantive actions on environmental legitimacy. Barcelona: IESE Business School.
Bowman, E. H., & Haire, M. (1975). A strategic posture toward corporate social responsibility. *California Management Review, 18*(2), 49–58.
Bragdon, J. H., & Marlin, J. A. T. (1972). Is pollution profitable? *Risk Management, 19*(4), 9–18.
Brammer, S., & Millington, A. (2008). Does it pay to be different? An analysis of the relationship between corporate social and financial performance. *Strategic Management Journal, 29*, 1325–1343.
Buysse, K., & Verbeke, A. (2003). Proactive environmental strategies: A stakeholder management perspective. *Strategic Management Journal, 24*(5), 453–470.
Carroll, A. (1999). Corporate social responsibility. *Business & Society, 38*(3), 268–295.
Choi, J., & Wang, H. (2009). Stakeholder relations and the persistence of corporate financial performance. *Strategic Management Journal, 30*, 895–907.
Christensen, C. M., & Bower, J. L. (1996). Customer power, strategic investment, and the failure of leading firms. *Strategic Management Journal, 17*(3), 197–218.
Christmann, P. (2000). Effects of "best practices" of environmental management on cost advantage: The role of complementary assets. *Academy of Management Journal, 43*(4), 663–680.
Chung, K. H., & Pruitt, S. W. (1994). A simple approximation of Tobin's-Q. *Financial Management, 23*(3), 70–74.
Combs, J. G., Crook, T. R., & Shook, C. L. (2004). The dimensionality of organizational performance and its implications for strategic management research. In D. D. Bergh (Ed.), *Research methodology in strategy and management* (pp. 259–286). Greenwich, CT: JAI Press.
Cooper, R., & Kleinschmidt, E. (1987). New products: What separates winners from losers? *Journal of Product Innovation Management, 4*, 169–184.
Damanpour, F. (1991). Organizational innovation —A metaanalysis of effects of determinants and moderators. *Academy of Management Journal, 34*(3), 555–590.
Damanpour, F., & Evan, W. M. (1984). Organizational innovation and performance—The problem of organizational lag. *Administrative Science Quarterly, 29*(3), 392–409.
Damanpour, F., & Gopalakrishnan, S. (2001). The dynamics of the adoption of product and process innovations in organizations. *Journal of Management Studies, 38*(1), 45–65.

Darnall, N., & Edwards, D. (2006). Predicting the cost of environmental management system adoption: The role of capabilities, resources and ownership structure. *Strategic Management Journal, 27*(4), 301–320.

Delmas, M., & Toffel, M. W. (2004). Stakeholders and environmental management practices: An institutional framework. *Business Strategy and the Environment, 13*(4), 209–222.

Derwall, J., Guenster, N., Bauer, R., & Koedijk, K. (2005). The eco-efficiency premium puzzle. *Financial Analysts Journal, 61*(2), 51–63.

DeSimone, L. D., & Popoff, F. (1997). *Eco-Efficiency, The business link to sustainable development*. Cambridge, MA and London: MIT Press.

Donaldson, T., & Preston, L. E. (1995). The stakeholder theory of the corporation—Concepts, evidence, and implications. *Academy of Management Review, 20*(1), 65–91.

Doz, Y. L. (1996). The evolution of cooperation in strategic alliances: Initial conditions or learning processes? *Strategic Management Journal, 17* (Summer special issue), 55–83.

Eisenmann, T. (2006). Internet companies' growth strategies: Determinants of investment intensity and long-term performance. *Strategic Management Journal, 27*(12), 1183–1204.

Elsayed, K., & Paton, D. (2005). The impact of environmental performance on firm performance: Static and dynamic panel data evidence. *Structural Change and Economic Dynamics, 16,395–412*.

Filbeck, G., & Gorman, R. F. (2004). The relationship between the environmental and financial performance of public utilities. *Environmental & Resource Economics, 29*(2), 137–157.

Fowler, S. J., & Hope, C. (2007). A critical review of sustainable business indices and their impact. *Journal of Business Ethics, 76*(3), 243–252.

Frynas, J., Mellahi, K., & Pigman, G. (2006). First mover advantages in international business and firm-specific political resources. *Strategic Management Journal, 27*(4), 321–345.

Ginsberg, A. (1988). Measuring and modeling changes in strategy—Theoretical foundations and empirical directions. *Strategic Management Journal, 9*(6), 559–575.

Gladwin, T. N., Kennelly, J. J., & Krause, T. S. (1995). Shifting paradigms for sustainable development—Implications for management theory and research. *Academy of Management Review, 20*(4), 874–907.

Griffin, J. J., & Mahon, J. F. (1997). The corporate social performance and corporate financial performance debate: Twenty-five years of incomparable research. *Business & Society, 36*(1), 5–31.

Hair, J., Black, W., Babin, B., Anderson, R., & Tatham, R. (2006). *Multivariate data analysis*. Upper Saddle River, NJ: Pearson Prentice Hall.

Hall, J., & Vredenburg, H. (2003). The challenges of innovating for sustainable development. *MIT Sloan Management Review, 45*(1), 61–68.

Hart, S. L. (1995). A natural-resource-based view of the firm. *Academy of Management Review, 20*(4), 986–1014.

Hart, S. L. (1997). Beyond greening: Strategies for a sustainable world. *Harvard Business Review, 75*(1), 66–76.

Hart, S. L., & Ahuja, G. (1996). Does it pay to be green? An empirical examination of the relationship between emission reduction and firm performance. *Business Strategy and the Environment, 5*(1), 30–37.

Henriques, I., & Sadorsky, P. (1999). The relationship between environmental commitment and managerial perceptions of stakeholder importance. *Academy of Management Journal, 42*(1), 87–99.

Hillman, A. J., & Keim, G. D. (2001). Shareholder value, stakeholder management, and social issues: What's the bottom line? *Strategic Management Journal, 22*(2), 125–139.

Hoffman, A. J. (1999). Institutional evolution and change: Environmentalism and the US chemical industry. *Academy of Management Journal, 42*(4), 351–371.

Howell, D. (2007). *Statistical Methods for Psychology* (6th ed). Belmont, CA: Thompson.

Hull, C. E., & Rothenberg, S. (2008). Firm performance: The interactions of corporate social performance with innovation and industry differentiation. *Strategic Management Journal, 29*(7), 781–789.

Husted, B. W., & Allen, D. B. (2007). Corporate social strategy in multinational enterprises: Antecedents and value creation. *Journal Of Business Ethics, 74*(4), 345–361.

Ireland, R. D., & Webb, J. W. (2007). A cross-disciplinary exploration of entrepreneurship research. *Journal of Management, 33*(6), 891–927.

Kassinis, G., & Vafeas, N. (2006). Stakeholder pressures and environmental performance. *Academy of Management Journal, 49*(1), 145–159.

King, A., & Lenox, M. (2001). Does it really pay to be green? An empirical study of firm environmental and financial performance. *Journal of Industrial Ecology, 5*(1), 105–115.

King, A., & Lenox, M. (2002). Exploring the locus of profitable pollution reduction. *Management Science, 48*(2), 289–299.

Klassen, R. D., & Whybark, D. C. (1999). The impact of environmental technologies on manufacturing performance. *Academy of Management Journal, 42*(6), 599–615.

Kor, Y., & Mahoney, J. (2005). How dynamics, management, and governance of resource deployments influence firm-level performance. *Strategic Management Journal, 25*(5), 489–496.

Larsen, T. (1993). Middle managers' contribution to implemented information technology innovation. *Journal of Management Information Systems, 10*(2), 155–176.

Larson, A. (2000). Sustainable innovation through an entrepreneurship lens. *Business Strategy and the Environment, 9*(5), 304–317.

Leonard-Barton, D. (1992). Core capabilities and core rigidities: A paradox in managing new product development. *Strategic Management Journal, 13*, 111–125.

Marcus, A. (2005). *Management Strategy.* Burr Ridge, IL: McGraw Hill /Irwin.

Marcus, A., & Fremeth, A. (2009). Green management matters regardless. *Academy of Management Perspectives, 23*(4), 17–26.

Margolis, J. D., Elfenbein, H. A., & Walsh, J. P. (2009). Does it pay to be good . . . and does it matter? A meta-analysis of the relationship between corporate social and financial performance. *Working chapter.*

Margolis, J. D., & Walsh, J. P. (2003). Misery loves companies: Rethinking social initiatives by business. *Administrative Science Quarterly, 48*(2), 268–305.

Matten, D., & Crane, A. (2005). Corporate citizenship: Toward an extended theoretical conceptualization. *Academy of Management Review, 30*(1), 166–179.

McGuire, J. B., Sundgren, A., & Schneeweis, T. (1988). Corporate social responsibility and firm financial performance. *Academy of Management Journal, 31*(4), 854–872.

McWilliams, A., & Siegel, D. (2000). Corporate social responsibility and financial performance: Correlation or misspecification? *Strategic Management Journal, 21*(5), 603–609.

McWilliams, A., & Siegel, D. (2001). Corporate social responsibility: A theory of the firm perspective. *Academy of Management Review, 26*(1), 117–127.

Murphy, C. J. (2002). The O'Reilly, C. A., & Tushman, M. L. (2004). The ambidextrous organization. *Harvard Business Review, 82*(4), 74–81.

Orlitzky, M., Schmidt, F. L., & Rynes, S. L. (2003). Corporate social and financial performance: A meta-analysis. *Organization Studies, 24*(3), 403–441.

Orsato, R. (2006). Competitive environmental strategies: When does it pay to be green? *California Management Review, 48*(2), 127–143.

Parry, M., & Bass, F. (1989). When to lead or follow? It depends. *Marketing Letters, 1*(3), 187–198.

Peloza, J. (2009). The challenge of measuring financial impacts from investments in corporate social performance. *Journal of Management, 35*(6), 1518–1541.

Porter, M. E. (1980). *Competitive strategy: Techniques for analyzing industries and competitors.* New York: Free Press.

Porter, M. E. (1985). *Competitive advantage: creating and sustaining, superior performance.* New York: The Free Press.

Porter, M. E., & Van der Linde, C. (1995). Toward a new conception of the environment—Competitiveness relationship. *Journal of Economic Perspectives, 9*(4), 97–118.

Prashant, K., Dyer, J., & Singh, J. (2002). Value creation and success in strategic alliances: Alliancing skills and the role of alliance structure systems. *European Management Journal, 19*(5), 463–472.

Reinhardt, F. L. (1998). Environmental product differentiation: Implications for corporate strategy. *California Management Review, 40*(4), 43–73.

Robinson, W. (1988). Sources of market pioneer advantages: The case of industrial goods industries. *Journal of Marketing Research, 25*, 87–94.

Rowley, T., & Berman, S. (2000). A brand new brand of corporate social performance. *Business & Society, 39*(4), 397–418.

Rumelt, R. (1987). Theory, strategy and entrepreneurship. In D. J. Teece (Ed.), *The competitive challenge: Strategies for industrial innovation and renewal* (pp. 137–158). Cambridge, MA: Ballinger.

Russo, M. V., & Fouts, P. A. (1997). A resource-based perspective on corporate environmental performance and profitability. *Academy of Management Journal, 40*(3), 534–559.

Salzmann, O., Ionescu-Somers, A., & Steger, U. (2005). The business case for corporate sustainability: Literature review and research options. *European Management Journal, 23*(1), 27–36.

Scherer, A. G., & Palazzo, G. (2007). Toward a political conception of corporate responsibility: Business and society seen from a Habermasian perspective. *Academy of Management Review, 32*(4), 1096–1120.

Schumpeter, J. A. (1934). *The theory of economic development*. Cambridge, MA: Harvard University Press.

Sharma, S., Durand, R. M. and Gur-Arie, O. (1981) Identification and analysis of moderator variables. *Journal of Marketing Research, 18*(3): 291–300.

Sharma, S., & Henriques, I. (2005). Stakeholder influences on sustainability practices in the Canadian forest products industry. *Strategic Management Journal, 26*(2), 159–180.

Sharma, S., & Vredenburg, H. (1998). Proactive corporate environmental strategy and the development of competitively valuable organizational capabilities. *Strategic Management Journal, 19*(8), 729–753.

Shrivastava, P. (1995). Environmental technologies and competitive advantage. *Strategic Management Journal, 16*, 183–200.

Starik, M., & Rands, G. P. (1995). Weaving an integrated Web—Multilevel and multisystem perspectives of ecologically sustainable organizations. *Academy of Management Review, 20*(4), 908–935.

Thompson, J. D. (1967). *Organizations in action*. New York: McGraw-Hill.

Tosi, H. L., Werner, S., Katz, J. P., & Gomez-Mejia, L. R. (2000). How much does performance matter? A meta-analysis of CEO pay studies. *Journal of Management, 26*(2), 301–339.

Ullmann, A. A. (1985). Data in search of a theory: A critical examination of the relationships among social performance, social disclosure, and economic performance of U. S. firms. *Academy of Management Review, 10*(3), 540–557.

Van de Ven, A. H. (1986). Central problems in the management of innovation. *Management Science, 32*(5), 590–607.

VanderWerf, P. A., & Mahon, J. F. (1997). Meta-analysis of the impact of research methods on findings of first mover advantage. *Management Science, 43(11)*, 1510–1519.

Waddock, S. A., & Graves, S. B. (1997). The corporate social performance—Financial performance link. *Strategic Management Journal, 18*(4), 303–319.

Walley, N., & Whitehead, B. (1994). It's not easy being green. *Harvard Business Review, 72*(3), 46–&.

Wernerfelt, B. (1984). A resource-based view of the firm. *Strategic Management Journal, 5*(2), 171.

Wernerfelt, B., & Montgomery, C. A. (1988). Tobin-Q and the importance of focus in firm performance. *American Economic Review, 78*(1), 246–250.

Whiteman, G., & Cooper, W. H. (2000). Ecological embeddedness. *Academy of Management Journal, 43*(6), 1265–1282.

Wood, M. (2009). Does One size Fit all? The multiple organizational forms leading to successful academic entrepreneurship. *Entrepreneurship Theory and Practice, 33*(4), 929–947.

World Commission on Environment and Development (WCED) (1987). *Our common future*. Oxford: World Commission on Environment and Development, Oxford University Press.

Yiu, D., & Chung-Ming, L. (2008). Corporate entrepreneurship as resource capital configuration in emerging market firms. *Entrepreneurship: Theory & Practice, 32*(1), 37–57.

Zahra, S. A., & Covin, J. G. (1995). Contextual influences on the corporate entrepreneurship performance relationship—a Longitudinal analysis. *Journal of Business Venturing, 10*(1), 43–58.

CHAPTER SIX

What Kinds of Photovoltaic Projects Do Lenders Prefer to Finance?

FLORIAN LÜDEKE-FREUND AND MORITZ LOOCK*

Discussions of innovation in renewable energy have covered different types of energy production (e.g., Nguyen, Gheewala, & Sagisaka, 2010; Sookkumnerd, Ito, & Kito, 2007; Yusoff, 2006), different industries (e.g., Narodoslawsky, Niederl-Schmidinger, & Halasz, 2008; Smyth, ó Gallachóir, Korres, & Murphy, 2010), or have focused on different geographic regions (e.g., Kaldellis, Simotas, Zafirakis, & Kondili, 2009; Smyth et al., 2010; Sookkumnerd et al., 2007; Zuluaga & Dyner, 2007). However, a crucial aspect has been rarely discussed: financing. Project financing is a central challenge for the diffusion of renewable energies, as it claims a major percentage of overall renewable energy investments (United Nations Environment Programme [UNEP], 2010) and requires huge amounts of debt capital: Debt ratios of 80 percent or even 90 percent are common (Johnson, 2009; Böttcher, 2009; Wuppertal Institute [WI], 2010).

Following the credit crunch in 2008, financial markets experienced a shortage of debt capital, giving rise to concerns about growth expectations (Jäger-Waldau, 2009; Schwabe, Karlynn, & Newcomb, 2009; New Energy Finance [NEF], 2009a, 2009c). Thus, in early 2009 the outlook was cautious (NEF, 2009a). The banking sector still provided large debt capital volumes, but lending had become more restrictive, i.e., more thorough, more risk-averse and more selective (Schwabe et al., 2009; WI, 2010). Consequently, third-party investments were channeled into absolutely effective, rigorously evaluated projects only, which created a new bottleneck for the diffusion of renewable energies: the availability of debt capital. In several energy studies, solar photovoltaic energy (PV, electricity from solar radiation) had been seen as one of the most important renewable energy technologies for future electricity production (e.g., Hoffmann, 2006; International Energy Agency

---

*This chapter is based on a study that is part of the authors' cumulative PhD dissertations. A different version has been accepted for publication in the *Journal of Cleaner Production*.

[IEA], 2011a; Poullikkas, 2010), but PV investments went down above average during the financial crisis (UNEP, 2010; PricewaterhouseCoopers [PwC], 2011). Our research focuses on the seemingly extra-tight bottleneck of PV project financing.

In 2009, US $162 billion were invested in renewable energies worldwide, from which project-based asset finance for new energy generation capacities totaled $101 billion or 62 percent (UNEP, 2010). Wind and solar were by far the largest asset classes with wind at US $67 billion dollars and solar at US $24 billion dollars. But although wind energy had become an established and mature industry that grew even during the financial crisis and the following economic downturn, solar energy was in a very challenging situation. Investments in 2009 were significant, but were in fact 27 percent below 2008 levels.

Taking a look at the two leading PV markets, Spain and Germany (IEA, 2011b), reveals great challenges for further market growth. Contingencies such as the global financial crisis and policy changes (such as reduced photovoltaic feed-in tariffs) lead to market consolidation. In this regard, photovoltaic was in a different situation than the wind energy industry, which already was consolidated. Solar thus faces unique challenges, as financiers are less experienced in setting up PV projects. Since the Spanish PV market saw a drastic slump in 2009 because of increasing deficits of public budgets (UNEP 2010), we focus on Germany, where from 2004 to 2009 nearly 50 percent of the world's new PV capacities were installed (IEA, 2011b; PwC, 2011). In Germany, PV technology production and application have become multibillion-dollar industries, with more than 60,000 employees (Bundesverband Solarwirtschaft, 2010).

Research Question. Two central aspects have to be considered when developing PV projects: First, PV, like any renewable energy, is politically determined. Second, the credit crunch changed the rules of financing. The market for PV projects is consolidating under these circumstances (NEF, 2009b, 2010b). Loan granting and dependence on bankability are important success factors for ongoing PV project development (Sarasin, 2009). However, current research provides very little detailed information about how debt capital providers evaluate loan applications—information that could help to handle uncertainties and risks from a PV project development perspective (Grell & Lang, 2008; WI, 2010). Photovoltaic projects are characterized by a multitude of parameters such as capacity, module and inverter technologies, maintenance concepts, economic indicators and stakeholder constellations (Grell & Lang, 2008). Loan commitments depend on how lenders evaluate project designs from a risk perspective. Therefore, we addressed the following research question: What kinds of photovoltaic projects do lenders prefer to finance? In search for answers to this question, we focus on medium- and large-scale ground-mounted installations subject to the German Renewable Energy Sources Act (EEG) as of 2009.

Research Approach. Our study follows an explorative market research approach. To answer our research question, we have developed an Adaptive

Choice-Based Conjoint experiment (ACBC), addressing German experts in PV project financing (see Chapter 3 of this book for another example of the use of this technique). Although conjoint experiments are widely used in marketing research (Louviere, Hensher, Swait, & Adamowicz, 2003; Train, 2003) and for exploring investment behavior (Clark-Murphy & Soutar, 2004), scholars in renewable energy investment have only just started utilizing this method (e.g., Oschlies, 2007). PV project developers may be able to use insights from our research to design projects according to lenders' preferences and thus increase the likelihood of fundraising success. In this way, our results may help in mainstreaming investments in green energy technologies.

Before applying conjoint analysis, it was necessary to examine PV project development practice. Explorative expert interviews and in-depth literature studies were combined to develop a conjoint experiment for financing professionals. The problem was to identify and conceptualize a set of attributes that help in understanding lenders' preferences and to reduce real complexities of project development at the same time. We studied different credit application procedures of relevant institutions, e.g., UmweltBank AG and GLS Gemeinschaftsbank eG, and analyzed industry-specific publications, e.g., guidelines, textbooks and journals, dealing with renewable energy project financing (e.g., Böttcher, 2004, 2009; Deutsche Energie-Agentur [dena], 2004; Grell & Lang, 2008; Oschlies, 2007; Sarasin, 2009; Schwabe et al., 2009; WI, 2010 and regular releases from New Energy Finance). Additionally, we conducted telephone interviews with financing consultants from different institutions (UmweltBank AG, GLS Gemeinschaftsbank eG, Windwärts Energie GmbH, and SunEnergy Europe GmbH).[1] The first steps of this iterative process were the expert interviews and parallel literature studies to develop an initial set of attributes. The second step, which started with 10 PV project attributes, was a further round of consultations, resulting in a reduced list of 6 attributes with three, four or five levels each. Our basic hypothesis was that these 6 attributes are essential for credit granting. With regard to individual attribute importance, our research approach is explorative; i.e., no hypotheses were developed referring to the attributes' individual weights and significance for loan commitments. Hence, the primary objective of the conjoint experiment was to make lenders' decisions more transparent.

Aspects of PV Project Development

Project Financing. Project financing is crucial to renewable energies (dena, 2004; Grell and Lang, 2008; Böttcher, 2009; NEF, 2009d; UNEP, 2010). This financing method has been established for decades for one-time ventures such as infrastructure projects (Backhaus, Sandrock, Schill, & Uekermann, 1990; Reuter & Wecker, 1999). Three significant characteristics of project financing are often discussed in the literature. The first is off-balance-sheet financing, i.e., a financing method that is separated from the individual or corporate books of project shareholders.The second characteristic is orientation toward future project cash flows, which are the only source of economic performance and security. The third characteristic is a complex network of stakeholders and a network of contracts to provide for broad risk-sharing and risk-reduction (Grell & Lang, 2008; Reuter & Wecker, 1999; Nevitt & Fabozzi, 2000; Böttcher, 2009).

The financial strengths and interests of potential shareholders are decisive for project development. According to authors like Reuter and Wecker (1999), Nevitt and Fabozzi (2000) and Böttcher (2009), project financing is a very flexible method. Nevertheless, meeting different shareholders' interests (equity versus debt capital) simultaneously is more challenging. As there are no universally applicable debt/equity ratios, we analyzed diverse practitioner literature such as product development guidelines and conducted expert interviews to define adequate ratios.[2] For the conjoint experiment, equity shares of 10 percent, 20 percent, and 30 percent were considered to be suitable.

Technical Components: Modules and Inverters. The generator is the heart of each PV installation. It consists of a variable number of modules, which are made from solar cells based on, e.g., crystalline silicon or different kinds of thin-film materials. The modules produce direct current (DC), which has to be transformed into alternating current (AC) by the DC-to-AC inverter, which feeds the electricity into the grid. Another basic component is the mounting system, which has to guarantee stability in cases of stress, caused by wind or snow, for example. It is sometimes also used as a tracker system to follow the sun. Technical quality is decisive for an installation's performance in terms of effectiveness, efficiency, and long-term reliability. Brands, certificates, producers' track records and long-term experience are quality indicators (Grell & Lang, 2008; Böttcher, 2009). For technical quality, two generic choices are possible. First, one can decide in favor of technology of superior quality, for which a price premium has to be paid. This option may be referred to as "premium brand." The second option is to save the price premium and use "low-cost" technology, accepting the risk of additional costs due to inferior quality. To enable quality-related choices, we included a premium brand/low-cost attribute.

Capacity of PV Power Plant. Capacity is another crucial physical characteristic determining not only financing needs but also efficiencies of scale and thus cost effectiveness. We refer to Lenardič's classification of PV power plant sizes (Lenardič, 2009).[3] In his annual review, he defines seven classes from 200 kWp to 20 MWp and above.

Another clue for attribute construction might be the German Renewable Energy Sources Act (EEG). The EEG distinguishes installations which are ground-mounted (lower tariff)[4] from those installed on roofs (higher tariff)[5]. For ground-mounted PV plants, a general tariff is applied, that is, the funding scheme does not trigger decisions for specific capacities. It can be assumed that efficiencies of scale generally lead to increasing system sizes. We follow Lenardič's classification in a slightly modified way. Our attribute classifies medium- and large-scale ground-mounted PV systems into four categories: 200 kWp to 1 MWp, 1 MWp to 5 MWp, 5 MWp to 10 MWp, and greater than 10 MWp.

Maintenance Concept. Following Grell and Lang (2008), an extensive quality-assurance concept is central to applications for credit, since constant cash flows have to be secured. The task from a financial point of view is to guarantee rates of return (for sponsors and further equity investors) and debt coverage ratios (for lenders). Instruments to guarantee quality include revenue forecasts, performance assessments, inspections, and monitoring and operations control. Inspections and assessments of activated systems are necessary as PV installations face circumstances different from standard test conditions. Such inspections can be enhanced, e.g., by thermal imaging to identify damaged modules, incorrect wiring, or insufficiently calibrated inverters. Quality-assurance also requires permanent monitoring and automated operations control to monitor actual performance ratios and to recognize malfunctions immediately. Thus, system inspection and system monitoring stand for quality assurance within our survey.

Economic Requirements. According to the concept of project financing (off-balance-sheet financing, cash-flow-related lending, risk sharing), a project's bankability depends on the project itself and its cash flows; that is, with regard to negotiated recourse (full-recourse, limited recourse, nonrecourse), project cash flows can be the only security for debt capital providers. Therefore, to evaluate a project from a lender's perspective, a special indicator is used (Grosse, 1990; Grell & Lang, 2008; Böttcher, 2009).[6]

Debt Service Cover Ratio (DSCR) refers to the ratio of gross cash flow to the debt service on a yearly basis, and thus varies with different project phases (Böttcher, 2009). This indicator has to be applied to prevent annual shortages; it is even acceptable to use DSCR alone to evaluate a project's economic viability (ibid). Basically, a ratio of 1.0 indicates exact coverage of debt service. If cash flows suffice, the ratio exceeds 1.0; if not, it falls below. For renewable energy projects, Böttcher (2004) as well as Grell and Lang (2008) refer to a minimum average DSCR of 1.3; lenders always charge a minimum contingency reserve. Practical examples of PV project calculations indicate the possible range of DSCR as being from roughly 1.0 to 3.0 and above. To create a DSCR attribute, we use three average DSCRs to offer different degrees of bankability (1.2, 1.5, and 1.8).

Sponsor Types. The project initiator generates the project idea, identifies further project parties, negotiates, concludes contracts, and thus actively designs the PV value network. He can contribute equity capital (sponsor), often acting in concert with a closed-end fund for private and institutional investors (Grell & Lang, 2008). An initiator can play different roles and may be differently motivated. He may be some kind of investor who is interested in maximizing return on equity. He can also be a service provider; in this case, his interest is to offer services such as consulting and project development (Schoettl & Lehmann-Ortega, 2011). If utilities set PV projects in motion, their strategic interests may include a blend of political, technological, and financial aspects. Lenders probably consider the initiator's background to be noteworthy, since, from a financial point of view, both lenders and initiators can have diametrically opposed motivations that have to be matched (Reuter & Wecker, 1999). Two categories of project initiators can be defined: those who will own the PV facility and those who will not. Current studies on PV-related value networks and business models consider ownership status to be a central actor characteristic (Frantzis, Graham, Katofsky, & Sawyer, 2008; Schoettl & Lehmann-Ortega, 2011). The nonowner group of project initiators is represented by service providers, since their core business is providing construction, installation and other value-added project services.[7] Finally, due to current discussions throughout the PV industry, four different prototypical initiator types can be identified as potential facility owners: regional utilities, multinational utilities, financial investors, and vertically integrated PV manufacturers.

Photovoltaic Projects and Business Models

From PV Project to PV Business Model. The contractual, financial, and operational structures among stakeholders finally develop into a project company ("special-purpose vehicle" or SPV; Grell & Lang, 2008), which is the predecessor of the operating company. Referring to the PV facility life cycle, it follows that a project, or its SPV, is a bridge to the setting up of an operating company.[8] At the end of the life cycle, deconstruction can also be managed as a separate project.

In contrast to a regular company, a project is based on a singular, noncyclical undertaking—in our case the construction of a PV facility. It is limited in lifetime and funding, serves unique project targets, and has individual resources brought in by diverse stakeholders (Backhaus et al., 1990; Reuter & Wecker, 1999; Nevitt & Fabozzi, 2000; Kerzner, 2001). From an organizational point of view, the project, as "temporary company" (Nausner, 2006), must be distinguished from its successor—the operating company. The operating company is an independent, legally responsible and creditable entity, which conducts regular tasks like technical and financial operations and thus secures long-term cash flows. It follows that the initial project creates the basis for cash flows, whereas the operating company handles their long-term realization. In the following, we refer to this approach of value creation and value capture as the essence of every PV business model. For our research, we broadly define a PV business model as the logic of how economic value is created and captured with a PV facility.

Photovoltaic projects and business models interrelate: Since the initial project defines essential parameters such as facility characteristics and the surrounding value network layout, it also determines the resulting PV business model (Frantzis et al., 2008). Thus, in the project phase the overall business model is explicitly or implicitly shaped. For example, variations of a parameter such as project initiator lead to different approaches of value creation and capture. A financial investor might develop or even complete a PV project in order to sell it immediately, charging a profit margin. In contrast, a regional utility could instead be interested in the technical aspects of integrating PV facilities into its grid. Different motivations lead to different PV projects and different PV business models (Frantzis et al., 2008). Finally, we can add a crucial task of project development that has been neglected to date: business model design (e.g., Chesbrough, 2007; Zott & Amit, 2007, 2008).

PV Business Model Design. Based on the identified attributes, different PV projects and business models can be designed. In the disruptive and competitive project financing market, this approach might turn out to be of strategic value. We add this aspect to our research question: In the face of changing political and dynamic market conditions, what are, from a lender's perspective, promising PV projects and business models? Therefore, our research includes a second investigation: After defining attributes for the conjoint experiment (Table 6.1), we use these attributes and the empirical findings from the experiment to evaluate different PV business models in a market simulation to discern lenders' preferences for different designs (see "Simulation Results" section).

Table 6.1 Photovoltaic business model attributes and levels

Attribute	Levels
Debt Service Cover Ratio, DSCR (Average)	1.2 1.5 1.8
Capacity	200 kWp to 1 MWp 1 MWp to 5MWp 5 MWp to 10 MWp >10 MWp
Brand	Low-cost modules and low-cost inverters Low-cost modules and premium-brand inverters Premium-brand modules and low-cost inverters Premium-brand modules and premium-brand inverters
Initiator	Vertically integrated manufacturer Regional utility Multinational utility Financial investor Service provider
Maintenance concept	System inspection Constant system monitoring System inspection and system monitoring
Equity	10% 20% 30%

Table 6.1 summarizes the above-derived photovoltaic attributes that were used in the conjoint experiment and market simulations on PV business models.

Empirical Evidence on Project Financing

Method. The data for this study have been collected online within an Adaptive Choice-Based Conjoint (ACBC) survey of German bank managers who are responsible for loan decisions for PV projects. In this experiment, we asked the participants to choose from different fictitious medium- and large-scale project proposals, based on the set of photovoltaic attributes we had developed (Table 6.1). The geographical scope was limited to Germany; participants were asked to consider the German renewable energy legislation of 2009. The experiment was conducted from January to March 2010.

Conjoint experiments have been frequently used for exploring investment behavior (Clark-Murphy & Soutar, 2004; Oschlies, 2007; Riquelme & Rickards, 1992; Shepherd & Zacharakis, 1999). Recently, scholars in renewable energy investment have started to apply conjoint experiments in order to investigate investors' preferences (e.g., Oschlies, 2007). This chapter investigates lenders' preferences and uses the ACBC (Adaptive Choice-Based Conjoint) tool from Sawtooth Software (www.softtoothsoftware.com) to perform and analyze choice tasks. ACBC combines Adaptive Conjoint Analysis (ACA) and Choice-Based Conjoint (CBC). Compared to the latter two methods, one important advantage of ACBC is that it provides greater information from a given number of choice tasks (Johnson, Huber, & Bacon, 2003; Johnson & Orme, 2007). This method is especially helpful in cases of small sample size and is therefore ideal for this research approach.

In more general terms, conjoint methods are usually used to analyze tradeoff decisions among the different features of a product (represented by attributes and their levels). The objective is to measure the perceived values of those features and their relations to prices. Therefore, so called *part-worth utilities* are estimated, which participants allocate to the attributes and their levels through their tradeoff decisions. These part-worth utilities are used to calculate the utility and thus the degree of acceptance of an alternative. Sawtooth Software developed its methods above all for product acceptance analyses and sees several advantages of this indirect measurement method: "Rather than directly ask survey respondents what they prefer in a product, or what attributes they find most important, conjoint analysis employs the more realistic context of respondents evaluating potential product profiles. Each profile includes multiple conjoined product features (hence, conjoint analysis)."[9] We transfer the approach of measuring tradeoffs among different features to the situation of project evaluation. In our experimental setup, we ask bank managers to decide between different PV projects (instead of products), which are generally comparable but differ in some aspects—these are our attributes and levels (Table 6.1).

Before confronting the participants with choice tasks, we explained some technical aspects concerning the survey and the underlying assumptions to them in order to clearly frame the decision situation. Participants were asked to apply their evaluation criteria of the last one or two years for ground-mounted PV power plants in Germany. Then we stated that the projects within our survey were approved and all legal and project planning tasks were fulfilled. Regarding the brand attribute, we explained the following: "Within the questionnaire, we distinguish between premium brand solar cells (e.g., Sharp, First Solar, etc.) and inverters (e.g., SMA) and low-cost solar cells and inverters (e.g., from young Chinese manufacturers)." That is, we did not define the type of PV technology, for instance, crystalline silicon or thin-film, as this was identified as a potential systematic bias in our pretests. The different project initiators' business models were defined. This information was given to create a transparent and unbiased framing—knowing that decisions are always more or less biased and that situational framing can have significant influences on people's decisions (see e.g., Kahneman & Tversky, 1979; Tversky & Kahneman, 1981).

After the introduction, the experiment had three stages. In the *Build Your Own* section, we asked the interviewee to design the PV project he or she would be most likely to finance. In the *Screening* section, four different projects had to be evaluated as being "a possibility" or "won't work for me." In the *Choice Task* section, three different projects were presented, of which only one could be chosen. Finally, based on participants' choices throughout the three stages, we were able to estimate the part-worth utilities, i.e., the value they allocate to certain attributes and levels, which allowed for analyses of their preferences.

Sample. More than 40 experts took part in our conjoint experiment. The sample size is small because of the participants' professional expertise. Nevertheless, this circumstance contributes to consistency and is beneficial for our findings. Our sample was exclusively compiled for this study and consisted of 141 companies. Although most of the companies were from the finance industry, the fields of sustainable finance, independent financial consulting, and renewable energy project development were also represented. We contacted the companies by phone and e-mail to identify individual experts in PV project evaluation. In 55 cases, experts could be identified, and in 31 cases they agreed to participate right away.[10] The Internet link to the survey and additional information were sent. In February, the sample of experts was contacted again via e-mail to motivate the remaining 24 respondents. When the website was closed on March 31, 2010, 43 experts had participated in the conjoint experiment.

The following are some socioeconomic data that describe the participants of our study:[11] From 2008 to 2010, 28.2 percent of the respondent companies financed PV projects exceeding €500 million total volume. A volume of €100 to €500 million was financed by another 28.2 percent. Of the respondent companies, 43.6 percent financed PV projects with a total volume of up to €100 million. Among the companies, 38.5 percent operate in Europe; 38.5 percent operate in Germany, Austria, and Switzerland only; 23.1 percent operate within a global context. Nearly all of the companies have their headquarters in Europe (97.4 percent). The interviewees work in various positions in renewable energy project financing (e.g., executive director of renewable energies, head of project financing, project manager, structured finance specialist). Although 43.6 percent of the respondents have more than five years of personal experience in renewable energy financing, 33.3 percent have two to four years, and 23.1 percent have less than two years of experience.

Conjoint Experiment Results. Our report is based on 1,698 choice tasks, conducted by 43 survey participants (39.5 tasks per respondent on average). Table 6.2 displays the interval data of the conjoint results as average utilities based on Hierarchical Bayes (HB) Estimation, which is a statistical method that improves conjoint analyses when only a limited amount

Table 6.2 ACBC (Adaptive Choice-Based Conjoint) analysis—Heirarchical Bayes (HB) summary of results

Attribute	Average Utilities (Zero-Centered Diffs)	Average Utilities	Standard Deviation	t-value
Debt Service Cover Ratio, DSCR (Average)	1.2	−5.99	20.90	−0.29
	1.5	−1.87	14.70	−0.13
	1.8	7.86	17.95	0.44
Capacity	200 kWp to 1 MWp	−33.84	67.15	−0.50
	1 MWp to 5 MWp	30.77	28.47	1.08
	5 MWp to 10 MWp	8.40	33.99	0.25
	> 10 MWp	−5.33	53.72	−0.10
Brand	Low-cost modules and low-cost inverters	−93.56	39.53	−2.37
	Low-cost modules and premium-brand inverters	−18.60	34.59	−0.54
	Premium-brand modules and low-cost inverters	9.52	28.35	0.34
	Premium-brand modules and premium-brand inverters	102.65	46.03	2.23
Initiator	Vertically integrated manufacturer	7.21	23.64	0.31
	Regional utility	17.74	17.71	1.00
	Multinational utility	3.06	20.93	0.15
	Financial investor	−20.61	22.71	−0.91
	Service provider	−7.41	21.59	−0.34
Maintenance concept	System inspection	−19.89	20.25	−0.98
	System monitoring	−23.43	21.61	−1.08
	System inspection and system monitoring	43,32	25.35	1.71
Equity	10%	−54.75	46.46	−1.18
	20%	31.75	22.28	1.43
	30%	23.00	31.57	0.73

Source: Our own calculations.

of data are available. The relatively high standard deviation reflects the sample size.

By focusing on the values of different attribute levels, we gain detailed insight into lenders' preferences (Table 6.2). Positive values in Table 6.2 indicate positive utilities and thus a positive impact on choices, whereas negative values point to aversion to attribute levels. Overall, lenders favored premium brands. Additionally, they appreciated an all-inclusive maintenance concept with system inspection and system monitoring. Moreover, they opted for project initiators who possibly would provide for disposal of generated electricity. Hence, they prefered regional and multinational utilities to be involved in projects. Project initiators such as service providers, vertically integrated manufacturers, and financial investors even deter lenders. Regarding capacity, we learn that project sizes of 1 MWp to 5 MWp were the most attractive, followed by projects with greater than 5 MWp to 10 MWp capacity. Small projects of 200 kWp to 1 MWp and projects above 10 MWp have a negative impact on choices. Finally, we see an inverted U-curve relationship for the optimal equity ratio, peaking at 20 percent.

Displaying the results for attributes only (without utilities of the individual levels), we see that DSCR (Debt Service Cover Ratio), initially assumed to be a decisive hard fact, is of lowest importance for lenders' choices. Of superior importance is the premium brand/low cost attribute (Figure 6.1).

Simulation Results. The empirically derived utility-values allow for the composition of different PV projects and business models. To measure how investors prefer these, the package from Sawtooth Software offers a market simulator. Each of the following three simulations is based on two different PV projects, which stand for specific business model "themes" (Table 6.3). The simulation results reveal investors' preferences with regard to different designs. Overall, we find that lenders prefer PV busi-

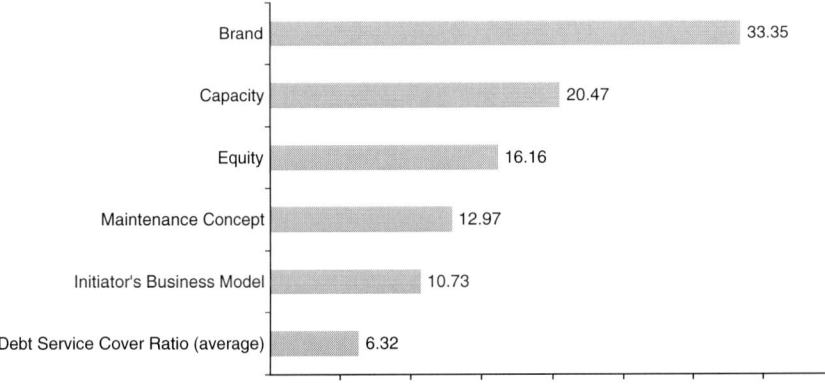

Figure 6.1 Graph showing relative importance of PV business model attributes to project lenders.

Table 6.3 Overview of simulation results

Simulation	Result
Simulation 1: Project with low-cost business model vs. project with premium brand business model	96.2% of lenders would choose the premium brand business model
Simulation 2: Project with low-cost business model and high DSCR (debt service cover ratio) vs. project with premium brand business model and low DSCR	83.2% of lenders would choose the premium brand, low DSCR business model
Simulation 3: Project with low-cost business model, high DSCR and "low risk" vs. project with premium brand business model, low DSCR and "high risk"	53.3% of lenders would choose the premium brand, low DSCR, high-risk project

ness models with premium brands, even if other attributes like DSCR would lead to different expectations about lenders' choices.

In Simulation 1, we created a project with a low-cost business model and a second project with a premium brand business model. Both projects were equal in all attributes (e.g., DSCR, capacity) but differed in terms of brands of equipment. Whereas the project with a low-cost business model had low-cost modules and low-cost inverters, the project with a premium brand business model applied premium brand models and premium brand inverters. The results of this simulation are unambiguous: Investors by far prefer the project with premium brand business models; 96.2 percent would choose this project.

For Simulation 2, we also varied the project attribute DSCR in addition to brand. Our initial assumption was that lenders would prefer projects with higher DSCR in comparison to projects with a lower value, since higher DSCR indicates a greater contingency reserve for debt service. In our simulation, the project with a low-cost business model has the highest DSCR and the project with a premium brand business model has the lowest DSCR. The result is counterintuitive: As soon as projects incorporate a premium brand (premium brand models), 83.2 percent of lenders would choose the low DSCR project. The supposedly rational choice of the project with the highest DSCR is biased; thus, lenders would prefer premium brand business models to those with high DSCR. We call this bias "debt for brands."

In Simulation 3, we modeled both projects in order to additionally account for risk. For the first project in Simulation 3, which had a low-cost business model, we posited not only the highest DSCR, but also attributes indicating low risk. For that purpose, we defined a multinational utility as initiator, which we could assume accounted not only for security against loan defaults but also promised energy feed-in and distribution. Additionally, this project was characterized by an all-in maintenance concept, which reduces the risk of operating failures. Finally, a high equity share served as additional security. The second project in Simulation 3, with the premium brand business model, was configured with attributes indicating a comparatively higher risk. Our basic assumption for this simulation was that debt investors would prefer the first project as it was of lower risk and promised stronger debt service. However, the third simulation also supported the debt-for-brands bias, as 53.3 percent of lenders would choose the project that assumed a high risk.

Discussion of Results

We found a bias in financing PV projects that we call "debt for brands." Simulations based on our empirically derived results revealed that lenders prefer PV projects leading to business models with premium brand technology rather than low-cost technology. Although we assumed that lenders would always favor project proposals with the highest Debt Service Cover Ratios (DSCR), our study revealed that they also chose inferior proposals with comparably lower DSCR as long as these projects included premium brand solar modules and/or premium brand inverters. Finally, we found that seemingly risk-averse lenders would also choose comparably inferior projects, even with comparably higher risk, as long as such projects were developed with premium-brand modules and/or premium-brand inverters. How can this brand effect be explained from a psychological perspective? Theories from the field of behavioral finance might be helpful.

Kahneman and Tversky, who empirically analyzed decisions and judgments under uncertainty and developed the influential Prospect Theory (Kahneman & Tversky, 1979; Tversky & Kahneman, 1974, 1981), point to different psychological biases in decision behavior. The brand effect as observed in our experiment can be interpreted within their framework as an expression of overconfidence: Events with high probabilities (but not guaranteed) are taken for granted, while events with low probabilities (but not impossible) are seen as unlikely. That is, when people allocate decision weights (representing their subjectively perceived probabilities) they tend to exaggerate true probabilities (Montier, 2007). This psychological bias was observed empirically by Kahneman and Tversky and became a central element of their theory. The effect that moderate probabilities, the area somewhere between "guaranteed" and "impossible," are significantly underestimated, which leads to rather binary decisions for or against something, is crucial in our decision-making context. We know from our experiment that lenders preferred premium brand modules and inverters to high equity ratios, for example. In terms of risk management, they decided to accept a higher risk in terms of smaller equity ratios in order to get premium brand technologies. This does not only mean that from a risk perspective the brand vs. low-cost decision was used as dominant heuristic for risk minimization, but also that lenders seemed to act overconfident with regard to expected premium brand technology performance. By nearly excluding low-cost components from their decisions (see Simulation 1 in Table 6.3), they also excluded the possibility that low-cost components might perform sufficiently and thus eventually overcompensate for small equity ratios or different maintenance concepts. In turn, it seems like the survey participants overly relied on their trust in premium brands. The participants of our survey thus behaved according to the overconfidence effect—that is, overconfident behavior is an element of the observed brand bias.

This interpretation has implications for diverse groups of practitioners. Based on our results and knowledge about natural decision biases such as overconfidence, we first encourage project managers to design PV projects and business models either with established premium brands or brands that are known for different (positive) reasons. Chinese manufacturer Yingli, for example, used the last World Soccer Cup to introduce its brand to a global audience, which might be a strategy to gain trust by diverse project stakeholders. Even if initial costs for branded projects are higher, the price premium for brands serves as an investment that will positively influence lenders' willingness to allow credits. Second, we find that debt investors should reevaluate their decision-making process: They should investigate whether it is biased, and whether inferior projects are possibly accepted just because of proposals that integrate brands. Third, we encourage technology managers, especially those of premium brand technology companies, to promote our findings as important selling arguments for premium brand PV components.

Regarding the bottleneck for the diffusion of photovoltaic technologies, researchers and practitioners might take our exploratory study as a starting point for further investigations. However, such research always has to cope with some limitations.

We designed our study according to the results of expert consultations and pretests. Moreover, participants had the opportunity to comment on the experiment. The feedback supported the appropriateness of our experimental setting. Nevertheless, we find the main limitation of our work in the experimental setup of the conjoint method. Experiments reduce real-world complexities to guarantee consistent results, which automatically leads to the exclusion of interesting aspects such as the question of whether the debt for brands effect is unique for photovoltaic. We are not able to conclude from our study whether it is unique for PV or not, but this or similar effects might be found in any technology-focused industry. Brand-oriented marketing, as well as behavioral finance, deal with these issues. Further research could analyze the brand effect for other renewable energy technologies such as wind or solar thermal energy.

Moreover, regarding the strength of the brand effect, further questions arise. One might interpret our findings in a way that lenders are sophisticated enough to not only rely on quantitative measures like DSCR and thus refer to qualitative signals. We assume that the degree to which they rely on brands depends on the perceived overall risk of a project. In our experiment, this overall risk might have been small because of the exclusion of political risk and the application of potentially noncritical DSCR scores, which might have lead to an increased reliance on brand. That is, we need to pay attention to the other attributes of the experiment, too, to interpret the debt-for-brands effect in concert with the other features and to avoid an underestimation of factors such as DSCR and equity ratio. Follow-up research should test the reported bias to ensure that no other effects artificially reinforce the debt-for-brands effect; for instance, we encourage further investigations to test different DSCR scales and to include real brands.

Aware of these limitations, we would like to offer recommendations for future research. First, our experiment can be a first step toward understanding lenders' preferences for renewable energy projects and business models. Future research may build on that and may consider further determinants of decision-making and thus extend our understanding of how banks involve themselves in project financing. Second, drawing comparisons between debt capital providers' preferences from different cultural and policy backgrounds is of interest, as understanding such determinants could be decisive in contexts of global project financing. Third, we encourage research to conduct an ex-post analysis to investigate how premium brand business models perform, compared to those that apply low-cost technology. Finally, comparisons of whether and how preferences of project financiers and project developers differ would be of interest (e.g., in a gap analysis). Identifying ways of bridging differences in preferences and therefore facilitating renewable energy project financing could possibly be based on such insights. Consequently, further research on project financing and debt capital provision could significantly contribute to the diffusion of renewable energy.

Notes

1. http://www.umweltbank.de; http://www.gls.de; http://www.windwaerts.de; www.sunenergy-gmbh.de
2. Interviews were done with: Mr. Oliver Thominsky, Director of Finance and Administration, and Mr. Günther Störmer, Head of Corporate Strategy, both SunEnergy Europe GmbH, Hamburg, Germany; also with Ms. Tanja Finke, Head of Project Financing, Windwärts Energie GmbH, Hanover, Germany. We conducted interviews in December 2009 and January 2010.
3. pvresources.com lists the 1,000 largest installations, ranging from 2 to 97 MWp capacity (as of August 2011).
4. EEG 2009 Section 32 (1) defines the tariff as follows: (1) the tariff paid for electricity from installations generating electricity from solar radiation shall amount to 31.94 Euro cents per kilowatt-hour. (Note: All tariffs are subject to the digression rules of section 20. The tariffs mentioned are only valid for installations put into operation in 2009.)
5. EEG 2009 Section 33 (1) structures the tariff as follows: (1) 43.01 Euro cents per kilowatt-hour for a total output of 30 kilowatts; (2) 40.91 cents per kilowatt-hour for a total output of 100 kilowatts; (3) 39.58 cents per kilowatt-hour for a total output of 1 megawatt; and (4) 33.0 cents per kilowatt-hour for output over 1 megawatt.
6. Further coverage ratios are: Loan Life Cover Ratio (LLCR), in which the focus is on debt service during the life of the loan, and Project Life Cover Ratio (PLCR), which asks for cash flows during the project's whole lifetime (Grosse, 1990; Böttcher, 2009).
7. Nevertheless, service providers such as project developers sometimes also invest in projects. Their revenue primarily comes from consulting and local project management activities as well as their exclusive access to specific resource markets (Schoettl & Lehmann-Ortega, 2011).

8. In Germany, medium- and large-scale installations (e.g., solar parks) are often managed by such operating companies, for example, Solarpark Straßkirchen GmbH & Co KG, which is the operating company behind Germany's biggest ground-mounted PV facility.
9. See http://www.sawtoothsoftware.com/conjoint-analysis-software
10. Two aspects were critical to the sample size: first, many of the identified institutions are connected in some manner with each other (e.g., different branches of Sparkasse, Sparda Bank, Landesbausparkasse, Hypo- und Vereinsbank). If one of their branches agreed to participate, others generally refused to. Second, due to the many requests institutions received from different fields, the invitation to our survey was immediately declined, either for reasons of data security or just to avoid additional work.
11. Socioeconomic data were reported by 39 interviewees; 4 completed the choice tasks without answering our additional questions.

References

Backhaus, K., Sandrock, O., Schill, J., & Uekermann, H. (Eds.) (1990). *Projektfinanzierung—Wirtschaftliche und rechtliche Aspekte einer Finanzierungsmethode für Großprojekte* [Project finance—Economic and legal aspects of financing large-scale projects]. Stuttgart, Germany: Poeschel.

Barradale, M. (2010). Impact of public policy uncertainty on renewable energy investment: Wind power and the production tax credit. *Energy Policy, 38*, 7698–7709.

Bundesverband Solarwirtschaft (BSW Solar) (German Solar Industry Association) (2010). *Statistic data on the German photovoltaic industry*. Berlin: BSW-Solar (http://en.solarwirtschaft.de).

Böttcher, J. (2004). Projektfinanzierung [Project finance]. In Deutsche Energie-Agentur (German Energy Agency) (Ed.), *Finanzierungs-Know-how—Handbuch für Erneuerbare Energien im Ausland* [Financing know-how for renewable energies in foreign countries] (pp. 134–181). Berlin: dena.

Böttcher, J. (2009). *Finanzierung von Erneuerbare-Energien-Vorhaben* [Financing of renewable energy projects]. Munich: Oldenbourg.

Chesbrough, H. (2007). Business model innovation: It's not just about technology anymore. *Strategy & Leadership, 35*, 12–17.

Clark-Murphy, M., & Soutar, G. (2004). What individual investors value: Some Australian evidence. *Journal of Economic Psychology, 25*, 539–555.

Deutsche Energie-Agentur (dena) (German Energy Agency) (Ed.) (2004). *Finanzierungs-Know-how—Handbuch für Erneuerbare Energien im Ausland* [Financing know-how for renewable energies in foreign countries]. Berlin: dena.

Frantzis, L., Graham, S., Katofsky, R., & Sawyer, H. (2008). *Photovoltaic business models*. Golden, CO: National Renewable Energy Laboratory.

Grell, A., & Lang, T. (2008). *Photovoltaik Leitfaden für Kreditinstitute—Handbuch zur Prüfung und Finanzierung von Photovoltaikprojekten* [A bank guide for assessing and financing photovoltaic projects]. Freiburg im Breisgau, Germany: forseo. (www.forseo.de)

Grosse, P. B. (1990). Projektfinanzierung aus Bankensicht [Project finance from a bank perspective]. In K. Backhaus, O. Sandrock, J. Schill, & H. Uekermann (Eds.), *Projektfinanzierung—Wirtschaftliche und rechtliche Aspekte einer Finanzierungsmethode für Großprojekte* [Project finance—Economic and legal aspects of financing large-scale projects] (pp. 41–62). Stuttgart, Germany: Poeschel.

Hoffmann, W. (2006). PV solar electricity industry: Market growth and perspective. *Solar Energy Materials & Solar Cells, 90*, 3285–3311.

International Energy Agency (IEA) (2011a). *Interactions of policies for renewable energy and climate (IEA Working Chapter)*. Paris: IEA.

International Energy Agency (IEA) (2011b). *Technology roadmap—Solar photovoltaic energy*. Paris: IEA.

Jäger-Waldau, A. (2009). *PV Status report 2009—Research, solar cell production and market implementation of photovoltaics*. Luxembourg: European Commission.

Johnson, G. L. (2009). *PV sector from an investor's perspective*. Presentation at the 3rd EPIA International Conference on Solar Investments, 7 April 2009, Frankfurt, Germany.

Johnson, R., & Orme, B. (2007). *A new approach to Adaptive CBC*. Sequim, WA: Sawtooth Software.

Johnson, R., Huber, J., & Bacon, L. (2003). *Adaptive Choice Based Conjoint. Sawtooth Software Research Papers*. Sequim, WA: Sawtooth Software.

Kahneman, D., & Tversky, A. (1979). Prospect Theory: An analysis of decision under risk. *Econometrica, 47*, 263–292.

Kaldellis, J., Simotas, M., Zafirakis, D., & Kondili, E. (2009). Optimum autonomous photovoltaic solution for the Greek Islands on the basis of energy pay-back analysis. *Journal of Cleaner Production, 17*, 1311–1323.

Kann, S. (2009). Overcoming barriers to wind project finance in Australia. *Energy Policy, 37*, 3139–3148.

Kerzner, H. (2001). *Project management—A systems approach to planning, scheduling, and controlling*. Hoboken, NJ: Wiley.

Lenardič, D. (2009). *Annual review 2008—Large-scale photovoltaic power plants: Annual and cumulative installed power capacity—Key statistical indicators (Free Edition)*. Jesenice, Slovenia: pvresources.com.

Louviere, J. J., Hensher, D. A., Swait, J. D., & Adamowicz, W. (2003). *Stated choice methods analysis and applications*. Cambridge: Cambridge University Press.

Masini, A., & Menichetti, E. (in press). The impact of behavioural factors in the renewable energy investment decision making process: Conceptual framework and empirical findings. *Energy Policy*, from 10.1016/j.enpol.2010.06.062.

Montes, G., & Martín, E. (2007). Profitability of wind energy: Short-term risk factors and possible improvements. *Renewable and Sustainable Energy Reviews, 11*, 2191–2200.

Montier, J. (2007). *Behavioural finance. Insights into irrational minds and markets*. Chichester: Wiley.

Narodoslawsky, M., Niederl-Schmidinger, A., & Halasz, L. (2008). Utilising renewable resources economically: New challenges and chances for process development. *Journal of Cleaner Production, 16*, 164–170.

Nausner, P. (2006). *Projektmanagement—Die Entwicklung und Produktion des Neuen in Form von Projekten* [Project management—Developing and creating the new through projects]. Vienna, Austria: WUV Universitätsverlag.

Nevitt, P. K., & Fabozzi, F. J. (2000). *Project financing*. London: Euromoney Books.

New Energy Finance (NEF) (2009a). *2008—A year of two halves for clean energy investment: NEF Press Release 14 January 2009*. Retrieved April 23, 2010, from http://www.newenergyfinance.com/Download/pressreleases/30/pdffile/.

New Energy Finance (NEF) (2009b). *New Energy Finance sees year of consolidation for clean energy in 2009: NEF Press Release 23 January 2009*. Retrieved April 23, 2010, from http://www.newenergyfinance.com/Download/pressreleases/31/pdffile/.

New Energy Finance (NEF) (2009c). *44% plunge in investment as crisis catches up with clean energy: NEF Press Release 2 April 2010*. Retrieved April 23, 2010, from http://www.newenergyfinance.com/Download/pressreleases/48/pdffile/.

New Energy Finance (NEF) (2009d). *Global trends in clean energy investment (Journalist fact-pack—WEF Meeting: Davos, January 2009): NEF Press Release 2 April 2010*. Retrieved April 23, 2010, from http://www.newenergyfinance.com/Download/pressreleases/48/pdffile2/.

New Energy Finance (NEF) (2010a). *Clean energy investment down just 6.5% in 2009; Asia outstrips the Americas: NEF Press Release 7 January 2010*. Retrieved April 23, 2010, from http://www.newenergyfinance.com/Download/pressreleases/105/pdffile/.

New Energy Finance (NEF) (2010b). *Bounce-back in global clean energy investment continues, with first quarter total up 31% year-on-year: NEF Press Release 12 April 2010*. Retrieved April 23, 2010, from http://www.newenergyfinance.com/Download/pressreleases/115/pdffile/.

Nguyen, T., Gheewala, S., & Sagisaka, M. (2010). Greenhouse gas savings potential of sugar cane bio-energy systems. *Journal of Cleaner Production, 18*, 412–418.

Oschlies, M. K. (2007). *A behavioral finance perspective on sustainable energy investment decisions*. St. Gallen: University of St. Gallen, Switzerland.

Poullikkas, A. (2010). Technology and market future prospects of photovoltaic systems. *International Journal of Energy And Environment, 1*, 617–634.

PricewaterhouseCoopers (PwC) (2011). *Germany's photovoltaic industry at the crossroads*. Hamburg: PwC.

Reuter, A., & Wecker, Claus (1999). *Projektfinanzierung—Anwendungsmöglichkeiten, Risikomanagement, Vertragsgestaltung, bilanzielle Behandlung* [Project finance—Application, risk management, contract design, financial treatment]. *Schriftenreihe Der Betrieb* [Series on The Firm]. Stuttgart, Germany: Schäffer-Poeschel.

Riquelme, H., & Rickards, T. (1992). Hybrid conjoint analysis: An estimation probe in new venture decisions. *Journal of Business Venturing, 7*, 505–518.

Sarasin (2009). *Solarwirtschaft—Grüne Erholung in Sicht* [Solar economy—Green recovery]. Basel: Bank Sarasin.

Schoettl, J. M., & Lehmann-Ortega, L. (2011). Photovoltaic business models: Threat or opportunity for utilities? In R. Wüstenhagen & R. Wuebker (Eds.), *Handbook of research on energy entrepreneurship* (pp. 145–171). Cheltenham: Edward Elgar.

Schwabe, P., Karlynn, C., & Newcomb, J. (2009). *Renewable energy project financing: Impacts of the financial crisis and federal legislation*. Golden, CO: National Renewable Energy Laboratory.

Shepherd, D., & Zacharakis, A. (1999). Conjoint analysis: A new methodological approach for researching the decision policies of venture capitalists. *Venture Capital, 1*, 197–217.

Smyth B.M., ó Gallachóir B.P., Korres N.E., & Murphy J.D. (2010). Can we meet targets for biofuels and renewable energy in transport given the constraints imposed by policy in agriculture and energy? *Journal of Cleaner Production, 18*, 1671–1685.

Sookkumnerd, C., Ito, N., & Kito K. (2007). Feasibility of husk-fuelled steam engines as prime mover of grid-connected generators under the Thai very small renewable energy power producer (VSPP) program. *Journal of Cleaner Production, 15*, 266–274.

Train, K. E. (2003). *Discrete choice methods with simulation*. Cambridge: Cambridge University Press.

Tversky, A., & Kahneman, D. (1974). Judgment under uncertainty: Heuristics and biases. *Science, 185*, 1124–1131.

Tversky, A., & Kahneman, D. (1981). The framing of decisions and the psychology of choice. *Science, 211*, 453–458.

United Nations Environment Programme (UNEP) (2010). *Global trends in sustainable energy investment 2010: Analysis of trends and issues in the financing of renewable energy and energy efficiency*. Nairobi: UNEP.

Wuppertal Institute (2010). *Potenziell treibende Kräfte und potenzielle Barrieren für den Ausbau erneuerbarer Energien aus Integrativer Sichtweise* [Drivers and barriers of renewable energies from an integrative perspective]. Wuppertal: Wuppertal Institute.

Wüstenhagen, R., & Wuebker, R. (Eds.) (2011). *Handbook of research on energy entrepreneurship*. Cheltenham: Edward Elgar.

Yusoff, S. (2006). Renewable energy from palm oil: Innovation on effective utilization of waste. *Journal of Cleaner Production, 14*, 87–93.

Zott, C., & Amit, R. (2007). Business model design and the performance of entrepreneurial firms. *Organization Science, 18*, 181–199.

Zott, C., & Amit, R. (2008). The fit between product market strategy and business model: Implications for firm performance. *Strategic Management Journal, 29*, 1–26.

Zuluaga, M., & Dyner, I. (2007). Incentives for renewable energy in reformed Latin-American electricity markets: The Colombian case. *Journal of Cleaner Production, 15*, 153–162.

CHAPTER SEVEN

Path Dependence and Creation in Venture Capital Investment

ALFRED A. MARCUS, SHMUEL ELLIS, JOEL MALEN,
ISRAEL DRORI, AND ITAI SENED

The purpose of this chapter is to describe cleantech venture capital investment decisions in innovative renewable and energy efficiency start-up companies as a process of path dependence and creation. Path dependence implies steadiness of investment choice and lack of change, though not necessarily outcomes that are uninformed or suboptimal (Sydow, Schreyögg, and Koch, 2009; Wüstenhagen and Teppo, 2006). Path creation implies charting a new investment course, based on feedback from external events and knowledge of prior venture outcomes (Garud, Kumaraswamy, & Karnoe, 2010). The focus of the chapter is the renewable energy (RE) and energy efficiency (EE) segments of the nascent cleantech industry. This industry is composed of companies that produce products and services that reduce energy consumption, waste, or pollution while they also try to improve operational performance, productivity, and efficiency. Different forms of renewable energy, such as solar power, wind power, and biofuels, as well as energy efficiency firms, are considered to be important part of this nascent industry. In the first decade of twenty-first century, this industry experienced a mini-investment boom (O'Rourke, 2009). The aim of this chapter is to determine the extent to which 2003 to 2009 venture capital investments in solar power, wind power, biofuels, and energy efficiency stuck to a path based on the initial conditions that prevailed at the start of this period or altered their direction in response to changing economic and political circumstances and the number of industry "exits," that is, the number of mergers and acquisitions (M&As) and initial public offerings (IPOs).

Rather than one path being common to the three forms of RE (solar, wind, and biofuel) and EE, we find that that each segment of the cleantech industry was on a slightly different path. Energy efficiency was the most path-dependent. It did not deviate much from its initial path. It was relatively immune to the influences associated with path deviation—the price of energy, the rate of world economic growth, changes in public policy, and exits. These factors affected solar, wind, and biofuels more than they affected energy efficiency. Solar power, wind power, and biofuels deviated from their initial paths, with the most extensive deviation evident in the case of solar power. We attribute the relative insensitivity of energy efficiency to such deviations to its comparative maturity. By comparative maturity we mean that the levelized costs of energy efficiency were low in comparison to conventional forms of energy generation (Lazard Ltd., 2009)—that is nuclear, coal, and natural gas. In contrast, the levelized costs of solar power, wind power, and biofuels were high in comparison to these conventional forms of energy generation. As the most mature alternative, energy efficiency was subject to less venture capital investment volatility than solar, wind, and biofuels. Of course, energy efficiency encompasses a variety of technologies, some of which were more mature and some of which were less mature than others, but overall it had a better profile than the alternatives during the period under consideration. For most applications, solar, wind, and biofuels were not yet competitive with nuclear energy, coal, and natural gas, while energy efficiency was competitive. Wind was very close to price parity, but it was still in the process of achieving it during the time period under consideration. It was moving in this direction, but it was not yet there. Though progress took place and forecasts suggested that solar and biofuels could achieve parity sometime in the future, during the period of our study, 2003 to 2009, their costs for most applications were higher than those of nuclear, coal, and natural gas. The cost disparity was especially true in the case of solar power, where venture capital (VC) investment advanced rapidly during this period, but in a highly uneven manner, one that was more volatile than the cases of biofuels and wind power.

These findings contribute to our understanding of the role of venture capital in alternative energy development and help clarify the meaning of path dependence. This concept still is not well understood (Vergne & Durand, 2010). With respect to path dependence, is it initial conditions or the events that follow that have more influence? We find that energy efficiency, being the least speculative of the alternative energy segments we analyzed, was the least subject to path deviation, while solar power, the most speculative of these segments, was the most subject to it. Though solar's costs were furthest from parity, as an abundant and ubiquitous energy form that might experience large technological leaps, its potential was great. Thus it attracted the most interest from venture capitalists, but this interest was variable. Because of the uncertain path ahead, investors regularly readjusted their assessment of solar power's potnetial. Among the alternative energy sources we examined, therefore, it was the most subject to path deviation based on external influence and feedback. Wind and biofuels, being more speculative than energy efficiency but less speculative than solar, were in between; that is, they were subject to more path deviation than energy efficiency but less path deviation than solar. This analysis provides evidence for the point of view that initial conditions *and* events that follow influence subsequent path development. Initial conditions were dominant in the case of energy efficiency, where price parity had been achieved, but in the case of solar power, wind, and biofuels, where price parity had not been achieved, subsequent events were more influential.

As Sydow, Scheryögg, and Koch (2009) describe it, path dependence is a three-stage process by which systems become more rigid, inflexible, and locked in over time.

- In Phase I, their scope of action is broad. They have yet to experience a critical juncture or a bifurcation where they start to automatically reproduce themelves. They are fluid and unfixed and wide open to external influences. No choice yet has been made that sets off self-reinforcing processes.
- In Phase II, on the other hand, the dynamics of self-reinforcing processes start to take hold. A more dominant and irreversible pattern gets set in place. Systems are not as likely to go back and change direction. Decision processes, however, have not closed entirely. They have not yet entirely converged on a fixed point. They are still partly open to external influences.
- Phase III, in contrast, connotes ever greater tightening, with less choice possible. Lock-in is greater and there is more complete commitment to a single path and less chance that outside influences will yield to deviation from the course in which the system is headed.

Based on our analysis in this chapter, we argue that solar power was in Phase I, wind power and biofuels were in Phase II, and energy efficiency was in Phase III. With more maturity, and by maturity we mean price parity, comes a higher level of lock-in, and with less maturity, a reduced level of lock-in. With less maturity, the system remains more fluid and open to outside influences; the past is less determinative of the future. History plays less of a role. Thus, we hold that whether initial conditions or subsequent events most influence [the direction?] that investment takes depends on the degree of an investment's market maturity—the degree to which the investment is competitive with incumbent technologies.

As Sydow et al. (2009) comment, the notion of lock-in is not necessarily negative and does not automatically imply that a system is suboptimal (David, 1985). We would say that given the relative certainty that prevails about energy efficiency, investors' choices to stick to a path despite external perturbations was logical, and given the open nature of the road that lies ahead for solar power, investors' flexibility also was logical. Their inflexibility with regard to the one and their flexibility with regard to the other were appropriate. These patterns of exploitation and exploration made sense, given that there were different degrees of maturity among these different types of energy savings and generation (Adner & Levinthal, 2004; Farjoun, 2010).

Our chapter develops these arguments about path dependence and creation in the cleantech venture capital arena. Starting with a description of the cleantech VC investment environment, it moves to a more complete model of how initial conditions and subsequent events lead to path dependence. The initial conditions we consider are ideology and culture, capital (physical, social, and intellectual/technical), economic conditions, public policy, and prior VC experience. The subsequent events we consider are prior deals, economic conditions, public policies, and number of exits (mergers and acquisitions plus initial public offerings). On this theoretical basis, we carry out an exploratory empirical analysis of the experience of 15 nations with VC cleantech investments from 2003 to 2009 that suggests that energy efficiency, the most mature segment in terms of its capacity to compete in the market with conventional energy, was the most affected by initial conditions. Thus, it was in Phase III of the Sydow et al. (2009) categories. Solar power, the least mature and the least capable of competing with conventional energy, was the least affected by initial conditions. As it was the most affected by subsequent events, it was in Phase I of the Sydow, et al. (2009) categories (also see Garud et al., 2010). Wind power and biofuels, we found, were in the middle; they were partially affected by initial conditions and partially affected by subsequent developments. The implications of the findings are discussed in our conclusion.

The Cleantech Investment Environment

Venture capital includes seed money for initial research and development, start-up money to begin a business, and growth money to sustain a business once established (Jeng & Wells, 2000). Across countries, those involved as venture capitalists differ, with venture capitalists in the United States taking larger stakes in companies than venture capitalists in other countries. Often U.S. venture capitalists also are more involved in managing their investments. They hold board positions, unlike venture capitalists in other nations, and play a bigger role in overseeing the companies in which they invest. Funding sources also tend to differ. U.S. venture capitalists obtain more money from pension funds, insurance companies, and endowments; venture capitalists abroad get more of their funding from banks (Jeng & Wells, 2000).

Some industries (software, biotechnology, and telecommunications) have received disproportionately larger shares of total venture capital investing (Brandera & De Bettignies, 2009). Cleantech has increasingly gained ground (O'Rourke, 2009). Starting in 2002 with a base below 5 percent, it had become the largest segment of U.S. funding by 2010, constituting close to 25 percent of U.S. venture capital investment and surpassing such sectors as biotech (20 percent), software (15 percent), and medical devices (12 percent); these data are from Cleantech Group, 2010. In 2010, overall global cleantech investment was nearly $8 billion (Cleantech Group, 2010). After the financial meltdown, cleantech made a strong comeback, reaching new heights in terms of the number of deals. A high percentage of this investment was made by U.S. venture capitalists (about $5 billion), but other countries had significant stakes, including the UK ($45 million), Canada ($31 million), and France ($3 million), according to Cleantech Group, 2010. The leading U.S. state was California, with about $3 billion invested in 2010, followed by Massachusetts, Texas, Oregon, and Colorado in terms of venture capital investments in cleantech.

Cleantech venture capital investments have many segments including agriculture, air quality, and the environment, recycling and waste, and water, but energy is the dominant segment. By amount invested, the largest investment in 2010 was in solar power. Biofuels, wind power, and energy efficiency followed. Venture capital investments in cleantech, however, constituted a relatively small proportion of all global clean energy transactions ($3 billion of nearly $700 billion in 2009) (Bloomberg New Energy Finance, 2010). Established companies and governments far outspent the venture capitalists. About a third of their money was spent on large-scale energy projects, equipment, and manufacturing.

We focus on the proportion of national venture capitalist investing (see Fulghieri & Sevilir, 2005; and Hochberg & Wester, 2010) devoted to solar power, biofuels, wind power, and energy efficiency. Taking the 15 most active countries, we computed their average annual solar power, biofuels, wind power, and energy efficiency portfolios in the years 2003 to 2009. The 15 most active countries during those years were: Australia, Canada, Denmark, Finland, France, Germany, Ireland, Israel, Netherlands, Norway, Spain, Sweden, Switzerland, the United Kingdom, and the United States. China and India came close, but were excluded from our analysis because they only started to seriously invest in about 2005, and we wanted to cover the entire period for each country. The analysis we have done (see Figure 7.1) shows the uneven portfolio paths of the different energy portfolios.

Solar had the most uneven of paths (top line figure with triangular boxes), with wind (the line with xes) and biofuels (the line with rectangles) being in the middle; the energy efficiency path (the bottom line with diamonds) was the steadiest. Solar power moved from about 20 percent of the total portfolio in clean energy investments in 2004 to over 30 percent in 2005, advancing again in 2007 to become more than 50 percent of the total clean energy investment portfolio in 2008, but losing ground in 2009, when it fell to about 38 percent of the total. Wind showed a sharp decline in 2005, from more than 40 percent of total clean energy investments in 2004 to 15 percent in 2005, but then it held fairly steady. Biofuels advanced in 2005 from about 17 percent of total clean energy investment to nearly 30 percent, but declined in 2006 to under 20 percent. Energy

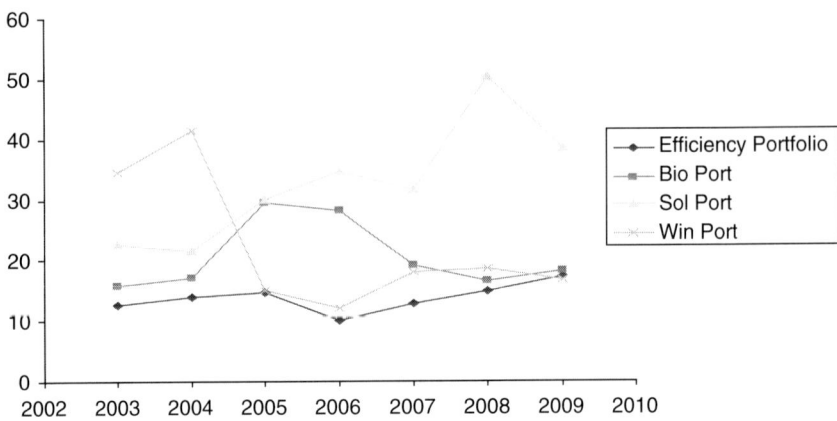

Figure 7.1 The uneven paths in the development of solar power, biofuels, wind power, and energy efficiency portfolios for 15 countries, 2003 to 2009.

Source: Clean Tech Group data.

efficiency fell below 10 percent of the total and never exceeded 17 percent. As is evident from Figure 7.1, its ups and downs were smaller than those of solar power, wind power, and biofuels.

Initial Conditions and Subsequent Events

We argue that the relatively steady path of the energy efficiency portfolio in comparison to the other portfolios was determined by initial conditions. The less steady paths of wind power and biofuels were determined by both initial conditions and subsequent events, while the least steady path, solar power, was determined mainly by subsequent events (see Figure 7.2). Our full model of the initial conditions and subsequent events that drove these results has a number of elements, which are discussed next (see Figure 7.3).

The initial conditions that affected subsequent path evolution were ideology and culture, capital (physical, social, and intellectual/technical), economic conditions, public policy, and prior venture capital experience.

Ideology and Culture. New fields like alternative energy and energy efficiency lack full legitimacy. To achieve it, they face normative and cognitive challenges (Jacobbsson & Bergek, 2004). Unless vigorous steps are taken to fill cognitive gaps, they are incompletely defined and are lacking in the necessary definitions to move forward (Santos & Eisenhardt, 2009). Large-scale social movements often are influential in creating the momentum they need to move forward (Sine & Lee, 2009). These movements propagate cognitive frameworks and norms, influence governments, consumers, and potential employees, and help solve collective action problems that new fields face (Hargrave & Van de Ven, 2006). The environmental movement has been a major shaper of wind energy

Segment	Price or Grid Parity	Initial Conditions	Subsequent Events
Energy efficiency	surpassed	x	
Wind power/biofuels	achieved	x	x
Solar power	behind		x

Figure 7.2 The impact of initial conditions and subsequent events. An x in the box means that we interpret that either initial conditions or subsequent events had a significant impact. An empty box means that we believe that they did not have a significant impact. Whether initial conditions or subsequent events dominate depends on the degree of price parity.

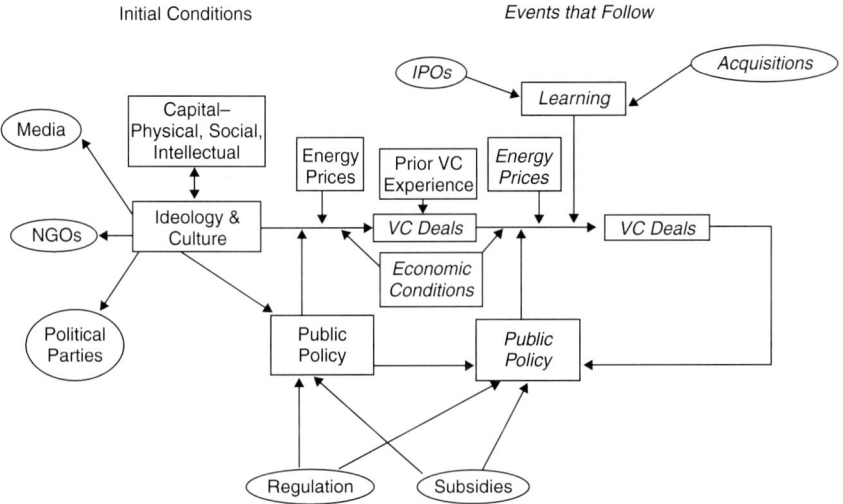

Figure 7.3 A full model of the factors that affect path dependence.

development, with fastest growth occurring in places with both a high density of environmental groups and sympathetic political parties (Vasi, 2009). Businesses of this type tend to cluster in regions where social and political values are supportive. The supportive values affect the availability of capital. Cleantech investors migrate to regions that show affinity with their cultural tastes (Russo & Earle, 2010; O'Rourke, 2009).

Physical and Social Capital. An abundance of physical assets (such as wind or sunlight) may affect the rate at which alternative energy projects are done (Russo, 2003). But physical assets by themselves are not sufficient—other assets must exist. Among the most important is social capital. It consists of supporting organizations. Their number and type can be large and the relationships among them can be complicated (Mitchell & Welch, 2009). Supporting organizations include public agencies, quasi-public and public-private ventures, private for-profit and nonprofit organizations, task forces, councils, trade offices, technical and business assistance organizations, business parks, and various types of incubators and university technology centers involved in activities that range from consumer education to networking, setting standards, certifying products, developing supply chains, improving the workforce, and attracting venture capital. Some method for organizing the relationships among these organizations and guiding their actions may be needed. Informal networks in which tacit and explicit knowledge is transferred about what is possible and how the future might look is not sufficient (Jacobsson & Bergek, 2004). Formal institutions may be required to establish norms and to guide, direct, and govern behavior. Theoretically, the problem these organizations face is a collective action problem (Olson, 1965). Because opposition from incumbent organizations and technologies must be overcome, spontaneous association without hierarchy is not necessarily sufficient (Axelrod, 1997). Given this opposition, institutional weakness can persist for long periods of time (Hockerts & Wüstenhagen, 2009; Ostrom, 2000.)

Intellectual and Technical Capital. Economic innovation depends not just on ideology, culture, and physical and social capital. It also requires that new technology be understood and that commercializers of this technology be accessible. Venture capitalists carefully assess the risk of this technology, but only if it is brought to their attention (Wüstenhagen & Teppo, 2006). University technology transfer offices may have to play a very important role (Siegel, Waldman, & Link, 2003). It has been found that disproportionately large shares of venture capital are likely to migrate to regions that that have strong universities and high rates of scientific progress (Brandera & De Bettignies, 2009).

Economic Conditions (Including Energy Prices). Macroeconomic fluctuations also have an effect. When the economy picks up, the number of funded start-ups should grow. With more start-ups, the demand for venture capital also will grow (Jeng & Wells, 2000). Higher energy prices also will stimulate venture capital growth in alternative energy and efficiency (O'Rourke, 2009).

Public Policies. Consistent government support in the form of regulatory devices, tax incentives, investment credits, public equity, renewable energy goals, and standards is needed to lower the perception of risk (Burer & Wüstenhagen, 2008). A study of venture capital in 21 countries found that government policies play an important role (Brandera & De Bettignies, 2009). The German Electric Feed-In Law of 1991 was very influential (Lüthi & Wüstenhagen, 2010). Tax policy also has been important. By themselves, government policies, however, may not be sufficient. For government policies to work well, a host of other quasi-public and public-private organizations, which intervene between government policies and their actual impacts, may be needed (York & Lenox, 2009).

Prior VC Experience. The experience that venture capitalists gain in helping client companies find funding, develop personnel, improve governance, and bring products to market is often transferred from one realm of activity—for example, information technology (IT)—to another, like cleantech (Brandera & De Bettignies, 2009). The past performance of the IT venture capitalists in backing companies with innovative technologies and growth potential like Apple, Intel, Microsoft, and Sun Microsystems helps to create a vibrant venture capital culture in which cleantech investment activity is hatched. California, in particular, has played this role globally (Jeng & Wells, 2000). It has been the hub of global cleantech investment. But it is not just Californian venture capital that migrated from IT to cleantech. Similar arguments have been made that the cleantech involvement of Israeli venture capital (Fiegenbaum, 2007) also derived from prior experience with software and other IT technologies. Thus cleantech benefits when venture capitalists with past achievements move some of their money from one area to another.

Subsequent events, then, build on these prior conditions. Those of import are the succession of prior deals that preceded current deals; changing economic conditions, including changing energy prices; changes in public policies; and successful exits—mergers and acquisitions (M&As) and initial public offerings (IPOs).

Past Deals. Past deal experience continues to be relevant. Venture capitalists carefully monitor their existing portfolios in light of the future investments they might make, evaluating the marginal costs and benefits of adding to or subtracting from their current portfolios (Cumming & Johan, 2010). They rely on the knowledge they obtain from managing and giving advice to existing clients and in this way their future choices are influenced by their past deal experience (Lerner, 2002).

Economic Conditions (Including Energy Prices). Venture capitalists also are keenly aware of opportunities and threats presented by changes in the overall economic climate and in energy prices. These changes affect the number of start-ups they choose and their sources of revenue (Cumming & Johan, 2010; Jeng & Wells, 2000).

Public Policies. Changes in public policies continue to provide venture capitalists with information that they can use to recalibrate their portfolios. The perception of consistent policy support for renewables under various European feed-in laws has been an investment driver (Wüstenhagen & Bilharz, 2006). Government incentives and renewable portfolio standards may have had an impact (Haji, 2011). During downturns in the economy, government policies tend bolster what otherwise might be lackluster investing. When policy advocates have pushed governments to put a price on carbon emissions, it has captured investors' interest. Their desire to invest in alternative energy and energy efficiency has grown. Until cost competitiveness with conventional power (grid parity) is reached, government policy is relevant (Kirkegaard, Hanemann, Weischer, & Miller, 2010).

Exits. The main risk venture that venture capitalists face is "not getting their money back." Thus, the existence of viable exit mechanisms is critical to sustaining their investments (Wüstenhagen & Teppo, 2006). Prior studies suggest that viable exits are among the strongest drivers of their investment choices (Jeng & Wells, 2000). Venture capitalists learn where to invest their money from past exits. Shares of venture capital money have been found to be distributed to sectors where the potential for exit is the highest (Brandera & De Bettignies, 2009).

An Exploratory Analysis

Here we do an exploratory analysis of the sensitivity of venture capitalists in different countries to subsequent events. We assume the initial conditions are in place and examine what takes place when subsequent conditions take hold. How do changes in world GDP, energy prices (oil/barrel), the number of clean energy-related public policies, and exits (mergers and acquisitions plus IPOs) affect the number of solar power, wind power, biofuels, and energy effiiciency investments? The period we examine is 2003 to 2009 (see figures 7.4 through 7.7). Figure 7.4 shows that the number of solar power deals grew in 2004–2005 and again in 2007–2008. This growth took place when oil prices and the number of solar power mergers and acquisitions and

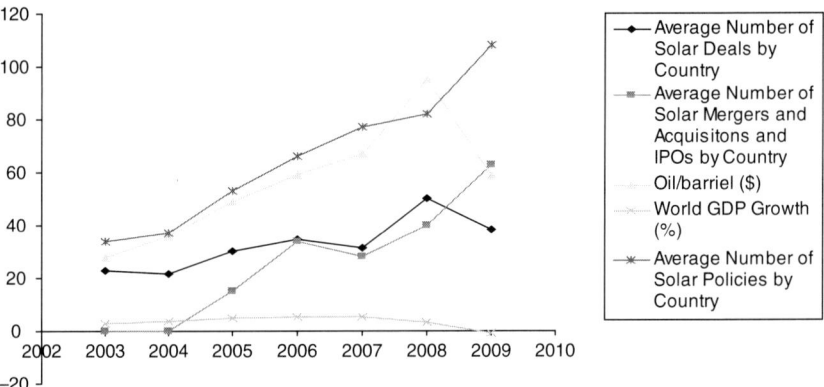

Figure 7.4 Growth of solar power and concurrent economic and policy data, 2003–2009.
Source: Clean Tech Group data.

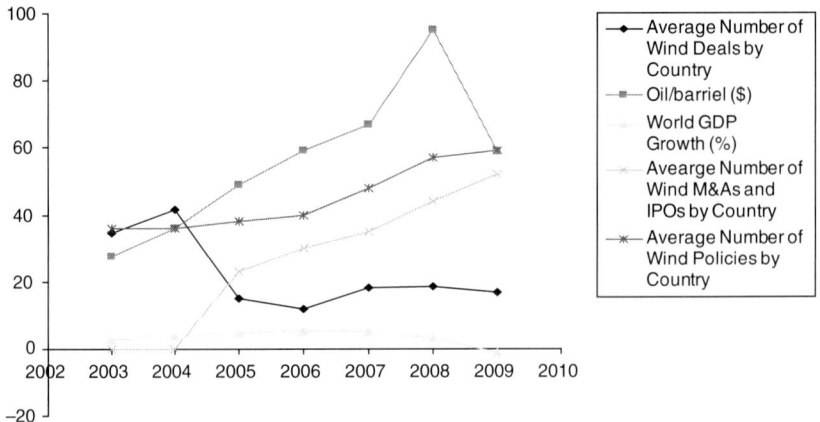

Figure 7.5 Growth of wind power and concurrent economic and policy data, 2003–2009.
Source: Clean Tech Group data.

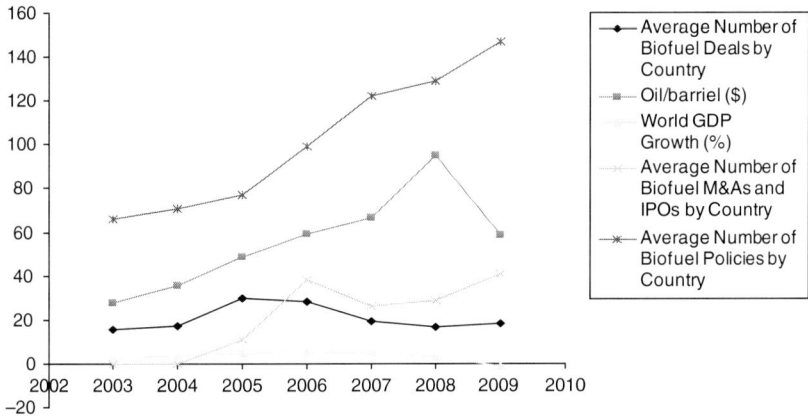

Figure 7.6 Growth of biofuels and concurrent economic and policy data, 2003–2009.
Source: Clean Tech Group data.

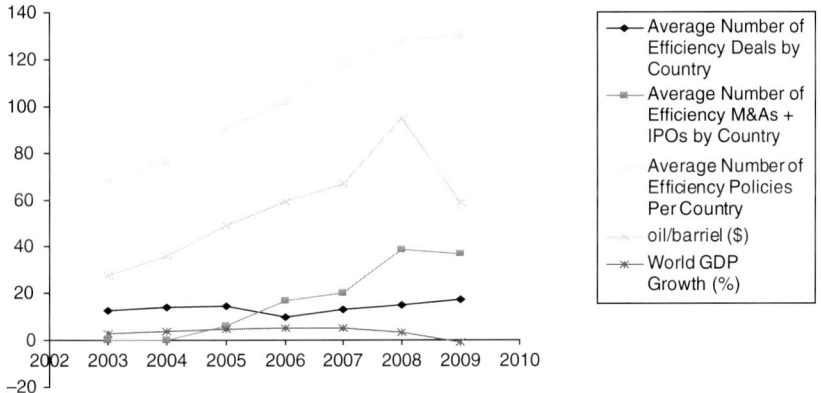

Figure 7.7 Growth of energy efficiency and concurrent economic and policy data, 2003 to 2009.
Source: Clean Tech Group Data.

initial public offerings (M&As + IPOs) increased. In 2008-2009, the average number of solar policies per country rose, nonetheless the number of deals declined when the world economy and oil prices fell.

Figure 7.5 shows that a steep 2004–2005 drop-off in the average number of wind power deals occurred in the same time period as a rise in the number of M&As + IPOs. The number of wind M&As+IPOs continued to increase from 2007–2009, but the number of wind power deals stayed steady. Changes in the number of wind power deals did not closely correspond with changes in the economy (world GDP growth), oil prices (oil/barrel), or the number of public policies.

Figure 7.6 shows that the 2004–2005 rise in the average number of biofuel deals per country was concurrent with an increase in oil prices (oil/barrel) and an increase in the avearge number of biofuel M&As + IPOs per country. The 2006–2007 decline in the average number of deals per country coincided with a reduction in the average number of biofuels M&As + IPOs per country. Changes in the number of biofuel power deals did not closely correspond to changes in the economy (world GDP growth), oil prices (oil/barrel), or the number of public policies.

Figure 7.7 shows that the average number of energy efficiency deals per country remained relatively steady throughout this period, seemingly impervious to concurrent economic events (world GDP growth) and energy prices. The only exception might be a small growth in the portfolio that takes place in 2007–2008, which coincides with an increase in the average number of energy efficiency M&As + IPOs (efficiency M&As + IPOs). Again, as in the case of solar, wind, and biofuels, energy efficiency saw a steady growth in the average number of public policies. The growth of public polices seemed to have a steadying effect in all instances, adding to certainty rather than contributing to uncertainty as is often claimed (Marcus, 1984).

This exploratory analysis suggests that there was a pattern of lock-in in the case of energy efficiency. In the case of wind power and biofuels, there was some reaction to external feedback, but solar power was most affected by feedback from subsequent events. Energy efficiency was the most path-dependent segment and solar power was the least dependent. These results are important, as investors are responsible for the level and nature of financial resources available to these ventures. What takes place in this nascent field involves a sorting process among investors, in which learning from changing circumstances can influence subsequent choices. Initial conditions establish basic patterns that are followed by events that shape each segment differently. When economic returns are far off and speculative, as in the case of solar, investors are in Phase I of the path-dependent process (Sydow et al., 2009). This stage is characterized by a broad scope of action. Subsequent events may have a substantial influence on path evolution. Paths are created and re-created in response to external stimuli (Garud, et al., 2010). Depending on how investors believe events will unfold, they are ready to adjust what they plan to do. Sensitive to external cues, they do not exclude the possibility of change, based on what they subsequently learn.

On the other hand, initial conditions led to Phase III path dependence in energy efficiency. Investors see a steady stream of predictable returns ahead. Energy efficiency has achieved parity with conventional ways of generating energy. It is stable, locked-in, such that its further evolution is not determined by current and future contingencies (Sydow et al., 2009). Investors have no need to diverge from the path they have chosen based on feedback.When price parity is achieved, as is the case with energy efficiency, lock-in is more complete. When it is only moving in this direction, as in the instances of wind power and biofuels, then both initial conditions and subsequent events are influential. The system is not entirely determined by initial conditions. Investors still have reason to react to subsequent events (Sydow et al., 2009).

Wind power and biofuels are in Phase II of the path dependence process, but their behavior has been different. In the case of wind power, investors who obtained feedback from knowledge of successes (the number of exits) searched for alternatives to their current investments (March, 1991). They tried to find more promising returns in new domains, different from those occupied by ventures whose outcomes they now knew to be successful. When a specific area receives a large amount of venture capital, promising opportunities already are taken. The high levels of successful venture capital investment in a particular area can deplete the set of unexploited opportunities within that domain (Brandera & De Bettignies, 2009). In the case of wind, investors searched for new, unexploited opportunities outside this domain. The set of investments they considered capable of delivering adequate returns broadened. They were looking to expand their investment portfolios outside of wind.

In the case of biofuels, investors who obtained feedback from knowledge of successes (the number of exits) imitated each other. They did not search for new, unexploited opportunities outside these domains. Instead, they clustered around the successes that had been achieved in an effort to limit future risk (Offerman & Sonnemans, 1998). Their clustering may have something to do with the nature of venture capitalists, especially those in the U.S., who play intensive roles in helping to manage the ventures in which they invest, roles that are way beyond those of conventional financial intermediaries (Hellmann & Puri, 2002). These intensive management roles may bring about greater domain specific understanding and attachment to the domain than otherwise would be the case.

Summary and Conclusions

Path dependence implies that a dominant action pattern is likely to emerge in a system, which renders the whole process more and more irreversible (David, 1985). The range of options starts to favor a particular type of decision or action pattern, which begins to replicate. In this chapter, we suggest that the critical factor that initiates this type of congealing in venture capital investments in cleantech is price parity. The greater the price parity, the more stable a cleantech path appears to be. The less the parity, the less stable the path. Price parity brings on a positive feedback loop of increasing returns, which reinforces and amplifies stability. Absent price parity, exogenous events are likely to influence these systems. Stickiness of a path is less likely.

Path dependence implies steadiness of investment choice and lack of change in response to external feedback, but not necessarily inefficient or less-than-optimal outcomes. The lock-in of energy efficiency, as a result of its price superiority, is not suboptimal. Nor do all paths quickly congeal.

Path dependence predicts, everything else being equal, that initial patterns persist into the future and that investors will find it hard to alter their strategies. The past character of investments continues into the future without much alteration. In accord with Sydow et al. (2009), our model suggests that while paths are set in motion by initial conditions, these patterns are still broken and then under some circumstances re-created, based on feedback from external circumstances and news about past investment outcomes. The continued charting of new courses based on feedback implies that there is continuous creation of a path, as well as dependence. There are adjustments in strategic choices based on investors' awareness of changes in economic conditions, public policies, and successful rounds of prior VC funding.

References

Adner, R., & Levinthal, D. (2004). What is not a real option: Considering boundaries for the application of real options to business strategy. *Academy of Management Review*, 29(1), 74–85.

Axelrod, R. (1997). *The complexity of cooperation*. Princeton: Princeton University Press.

Bloomberg. (2010). New Energy Finance, UNEP, SEFI.

Brandera, J., & De Bettignies, J. (2009). Venture capital investment: The role of predator–prey dynamics with learning by doing. *Economics of Innovation and New Technology*, 18(1), 1–19.

Burer, M. J., & Wüstenhagen, R. (2008). Cleantech venture investors and energy policy risk: An exploratory analysis of regulatory risk management strategies. In R. Wüstenhagen, J. Hammischmidt, S. Sharma, and M. Starik (Eds.), *Sustainable innovation and entrepreneurship* (pp. 290–309). Northampton, MA: Edgar Elgar.

Cleantech Group. (2010). www.cleantech.com http://cleantech.com/cleantech3lite/main.html http://cleantech.com/about/pressreleases/Q2-2010-release.cfm

Cumming, D., & Johan, S. (2010). Venture capital investment duration. *Journal of Small Business Management*, 48(2), 228–257.

David, P. A. (1985). Clio and the economics of QWERTY. *American Economic Review* 75, 332–337.

Farjoun, M. (2010). Beyond dualism: Stability and change as a duality. *Academy of Management Review*, 35(2), 202–225.

Fiegenbaum, A. (2007). *The take-off of Israeli high-tech entrepreneurship during the 1990's*. Elsevier: Amsterdam.

Fulghieri, S., & Sevilir, M. (2005). Size and focus of a venture capitalist's portfolio. University of North Carolina, University of North Carolina, unpublished working paper.

Garud, R., Kumaraswamy, A., & Karnoe, P. (2010). Path dependence or path creation. *Journal of Management Studies*, 47(4),760–774.

Haji, S. (2011). Webinair Cleantech Group January 7, CleanTech Group [should this be Webinar? is it available online?]

Hargrave T., & Van de Ven, A. (2006). A collective action model of institutional innovation. *Academy of Management Review*, 31(4), 864–888.

Hellmann, T., & Puri, M. (2002). Venture capital and the professionalization of start-up firms: Empirical evidence. *The Journal of Finance*, 57(1), 169–197.

Hochberg, Y., & Wester, M. (2010). The size and specialization of direct investment portfolios, Northwestern University. unpublished working paper.

Hockerts, K., & Wüstenhagen, R. (2009). Greening Goliaths vs. emerging Davids—Incumbents and new entrants in sustainable entrepreneurship. Paper presented at the Academy of Management Annual Meeting. Chicago, IL.

Jacobsson, S., & Bergek, A. (2004). Transforming the energy sector: The evolution of technological systems in renewable energy technology. *Industrial and Corporate Change*, 13(5), 815–849.

Jeng, L., & Wells, P. (2000). The determinants of venture capital funding: Evidence across countries. *Journal of Corporate Finance*, 6, 241–289.

Kirkegaard, J. F., Hanemann, T., Weischer, L., & Miller, M. (2010). Toward a sunny future? Global integration in the solar PV industry. WP10-6, May 2010. Working Series. Washington DC: World Resources Institute, Peterson Institute for International Economics. http://pdf.wri.org/working_papers/toward_a_sunny_future.pdf

Lazard Ltd. (2009). *Levelized Cost of Energy*. New York: Lazard, Ltd. http://blog.cleanenergy.org/files/2009/04/lazard2009_levelizedcostofenergy.pdf.

Lerner, J. (2002). Boom and bust in the venture capital industry and the impact on innovation 2002. Harvard NOM Unpublished Working Paper No. 03-13. Available at SSRN: http://ssrn.com/abstract=366041 or doi:10.2139/ssrn.366041.

Lüthi, S. & Wüstenhagen, R. (2010). The price of policy risk—Empirical insights from choice experiments with European photovoltaic project developers. University of Saint Gallen, Switzerland. Unpublished working paper.

March, J. (1991). Exploration and exploitation in organizational learning, *Organization Science*, 2, 1.

Marcus, A. (1984). *The adversary economy: Business responses to changing government requirements*. Westport, CT: Quorum Books.

Mitchell, K., & Welch, A. (2009). Current structures, strategies, and examples for green economic development. Minneapolis, MN: Blue Green Alliance.

Offerman, T., & Sonnemans, J. (1998). Learning by experience and learning by imitating successful others. *Journal of Economic Behavior & Organization*, 34(4), 559–575.

Olson, M. (1965). *The logic of collective action*. Cambridge, MA: Harvard University Press.

O'Rourke, A. (2009). The emergence of cleantech. Unpublished dissertation. Yale University.

Ostrom, E. (2000). Collective action and the evolution of social norms. *Journal of Economic Perspectives*, 14(3), 137–158.

Russo, M. (2003). The emergence of sustainable industries: Building on natural capital. *Strategic Management Journal*, 24, 317–31.

Russo, M., & Earle, A. (2010). The geography of sustainable enterprise and the concentration of mission-driven companies. Chapter presented at Academy of Management Annual meeting, Montreal.

Santos, F., & Eisenhardt, K. (2009). Constructing markets and shaping boundaries: Entrepreneurial power in nascent fields. *Academy of Management Journal* 52(4), 643–671.

Siegel, D., Waldman, D., & Link, A. (2003). Assessing the impact of organizational practices on the relative productivity of university technology transfer offices: an exploratory study. *Research Policy*, 32, 27–48.

Sine, W., & Lee, B. (2009). 'Tilting at windmills?' The environmental movement and the emergence of the U.S. wind energy sector. *Administrative Science Quarterly*, 54, 123–155.

Sydow, J., Scheryögg, G., & Koch, J. (2009). Organizational path dependence: Opening the 'black box.' *Academy of Management Review*, 34, 689–709.

Vasi, I. (2009). Social movements and industry development. *Mobilization: An International Journal*, 14(3), 315–336.

Vergne, J., & Durand, R. (2010). The missing link between the theory and empirics of path dependence: Conceptual clarification, testability issue, and methodological implications. *Journal of Management Studies* 47(4), 736–756.

Wüstenhagen, R., & Bilharz, M. (2006). Green energy market development in Germany: E public policy and emerging customer demand. *Energy Policy* 34, 681–696.

Wüstenhagen, R., & Teppo, T. (2006). Do venture capitalists really invest in good industries? Risk-return perceptions and path dependence in the emerging European energy VC market. *International Journal of Technology Management* 34, 1–2; 63–87.

York, J., & Lenox, M. (2009). Its not easy building green: The interaction of private and public institutions in the adoption of voluntary standards.unpublished working paper, University of Virginia Darden School.

CHAPTER EIGHT

High-Tech Cluster Revolution from an Organizational Ecology Perspective

DEBORAH E. DE LANGE

The green revolution offers a plethora of potential innovation, much of which will take place in high-tech clusters. These clusters thus become interesting settings for testing and extending theories of organizational ecology, a set of theories that explain how selection processes shape population level organizational adaptation to environmental variation. This theory can be used to explain the adaptation of a new industry to an incumbent one. What happens when a new industry encroaches on the incumbent industry's space? To what extent are the new and old industries able to survive and coexist? What is the effect of the second-generation industry's movement into a niche incumbents already occupy?

Familiar groups of first-generation high-tech geographical clusters are located in Silicon Valley in California and Route 128 in Boston (Bresnahan & Gambardella, 2007; Porter, 1998). They have changed in an evolutionary manner in recent decades, but the appearance of revolutionary change on the horizon is evident (Gersick, 1991; Tam, 2010; Tushman & Anderson, 1986). Green technologies have started to encroach upon their spaces (Hockerts & Wüstenhagen, 2010, p. 4). New firms with competence-enhancing and -destroying capabilities are entering (Tam, 2010; Tushman & Anderson, 1986). Can these clusters continue on a path of incremental innovation? To what extent must they change direction? Can they both exploit their existing advantages and explore for new ones? (O'Reilly & Tushman, 2004).

The emphasis in prominent clusters like Silicon Valley and Route 128 (Saxenian, 1994) is on computer and telecommunications technologies (ICT); other clusters have had a biotech or manufacturing emphasis (Bresnahan & Gambardella, 2007; Nair, Ahlstsrom & Filer, 2007). But what happens when these clusters evolve outside of their main domain? Can a second-generation industry be successful within an existing cluster dedicated to a different technology? Hockerts and Wüstenhagen (2010) view green firms as new entrants into existing incumbent spaces within the same industry; they have not examined what happens when new firms outside the industry of the incumbents move into the incumbent's space (Hockerts & Wüstenhagen, 2010).

This chapter examines what happens when a new high-tech population, unrelated to the first, impinges on an incumbent population. What are the interactions between new and existing firms? Having the choice to settle elsewhere, what does the new industry gain from coinhabiting the space of an existing industry? Indeed, there may be serious disadvantages from colocation. The second-generation industry may threaten the first-generation industry if resources are limited and both are struggling to obtain them. Yet we observe that new green technology clusters are often arising in the same places as first-generation clusters, either intruding on the space of an original group of companies (as in Silicon Valley) or acting in a way that might revive a region with a decaying cluster (Boston). Why shouldn't the second-generation industry locate itself elsewhere, in a different region (Arizona, for instance[1]), rather than settling in the places where technologies flourished in the past (LaMonica, 2009)?

Governments choose regions where industries will be located.[2] Japan's Ministry of Economy, Trade, and Industry (METI), for instance, has planned and organized IT, biotech, environmental, and manufacturing clusters. It is highly involved in facilitating networks among firms and universities (Ministry of Economy, Trade, and Industry, 2004). In some cases, it has considered co-location, putting a second-generation industry where an existing one is (ibid., 2004). In the U.S., cluster choice tends to be less dependent on government initiatives, though this does not mean that specific locations have not tried to attract clusters through packages of favorable subsidies and incentives. What are the advantages and disadvantages of co-location? When green technology moves into an existing cluster, it does not necessarily replace the existing firms. The existing firms may be mature but they may not have entirely lost momentum. On the one hand, it may not be in their interest to make room for a new set of firms. They will struggle against them. On the other hand, there may be complementary benefits. Green companies will not replace computer firms in Silicon Valley; they will coexist: iPods using solar cells, for instance. New green devices can capitalize on computer hardware and software knowledge in locales like Silicon Valley; for example, sensors to monitor and control home consumption of electricity may be the bases for the new companies. In this instance, the old and new firms coexist. They benefit from colocation.

Are resources sufficient to support first and second-generation companies? Venture capital investment and specialized employees may migrate from first-generation companies to second because they offer better growth opportunities (Tam, 2010). Incumbents will fight to retain their share of these resources. Through their control of resources and relationships, they can stifle and eliminate the fledgling competition. However, another outcome is possible—the incumbents may develop mutually beneficial relationships. Their interactions will lead to a more vibrant cluster. When there is competition between first and second-generation firms, the cluster may attract additional resources. Specialized employees may migrate to the cluster and bring with them new investment dollars, thus expanding the cluster's resources. The cluster is revitalized. Its survival is enhanced. The benefits accrued by new and old firms will dampen the need for them to compete.

This chapter explores the phenomenon of a second-generation population entering an existing industrial space from the perspective of organizational ecology (Carroll, 1984; Carroll & Delacroix, 1982; Freeman, Carroll, & Hannan, 1983; Hannan & Freeman, 1977, 1984). Green firms that enter a region like Silicon Valley are not in different niches of the same industry as the existing firms. Thus, it is not easy to partition resources. Rather than portioning resources, there is the potential for competition. Thus, the generalist-specialist interaction of organizational ecology's resource partitioning theory does not apply here (Carroll & Swaminathan, 2000). Most studies focus on a single industry (Bresnahan & Gambardella, 2007; Myint, Vyakarnam & New, 2005; Nair et al., 2007). This chapter asks whether firms representing a new industry are more likely to survive and experience rapid growth in a new cluster or an established one.

Organizational Ecology

A brief review of organizational ecology theory is in order. This theory describes how selection processes shape adaptation to environmental variations. Why is there a variety of organizational forms? How have they evolved? (Hannan & Freeman, 1989; Carroll, 1984; Amburgey. Kelly, & Barnett, 1993; Hannan & Freeman, 1977; Roughgarden, 1979). Organization ecology predicts organizational births and deaths and changes across time of organizational populations and communities. It explains the birth (or founding) and death (or failure) rates of organizations. The goal is to examine the forces that shape the population structure over time, to predict the net mortality of organizational forms and the change in populations of organizations based on environmental selection processes. The focus is on natural selection and competition among organizational forms and their replacement.

According to organization ecology, each population has a niche. These are spaces where a population may out-compete all other populations (Hannan & Freeman, 1989). Survival is achieved on the basis of fit. Similar to contingency theory (Burns & Stalker, 1961; Chandler, 1962), organization ecology proposes that variables such as age and size affect survival. Another important concept is density dependence. In a young population when the density of organizations is low, legitimation is the main evolutionary process; later, when density increases, competition becomes the main force of selection (Carroll & Swaminathan, 1991; Hannan, 1986).

According to organization ecology, organizations are black boxes that are limited by structural inertia. Structural inertia reflects the decreasing responsiveness of an organization to environmental forces with increasing age (Hannan & Freeman, 1984). More inertia means that it is harder to change, but also that structure reproduces itself with high fidelity (Hannan & Freeman, 1984). Inertia is generated by internal politics, forces of history, information constraints, fixed assets, entry and exit barriers, legitimacy constraints, and collective rationality (Hannan & Freeman, 1984).

The theory predicts that inertial organizations, those that are reliable and accountable, are favored by selection processes (Hannan & Freeman, 1984). A reliable organization generates collective actions with small variance in quality (Hannan & Freeman, 1984). It makes internally consistent moves based on rules and procedures that reproduce rational resource allocations and appropriate actions (Hannan & Freeman, 1984). When there is environmental turbulence, the survival of the inertial organizations is not certain, because traits needed survive in turbulence are not the same as those needed to survive in tranquility (Hannan & Freeman, 1989, p. 90).

New firms, on the other hand, face a liability of newness; their failure rates decline with increasing age and size (Hannan & Freeman, 1984). Stinchcombe (1965) distinguishes four reasons for the liability of newness. First, the firm and its constituents have new roles to play and they make mistakes in the beginning. Second, new firms need time to learn their roles and must rely on their wits and initiative to do this. Third, those involved in new firms are strangers at first and must build trust to work together. Lastly, building new external relationships is a challenge since existing firms have relationships with customers whose loyalties do not switch easily. New firms, on the other hand, must build a customer base from scratch (Carroll & Delacroix, 1982; Stinchcombe, 1965).

According to organization ecology, organizations are affected by random variation rather than by deliberate actions (Hannan & Freeman, 1977). Managers should avoid attempting frequent major reorganizations because core change takes the organization's clock back to zero so that it experiences a renewed liability of newness (Hannan & Freeman, 1984). However, interorganizational linkages may buffer firms from failure (Miner, Amburgey, & Stearns, 1990). Resource buffering may occur when an organization has a linkage with other organizations that can provide it with access to resources such as funding, information, and material goods (Miner et al., 1990). Institutional buffering is a benefit that is gained through associations with other respected or powerful organizations (Miner et al., 1990).

As members of a population, organizations have tendencies to generalize or specialize. A specialist population of organizations "flourishes because it maximizes its exploitation of the environment and accepts the risk of having that environment change" (Hannan & Freeman 1977, p. 948), whereas a generalist organization, "accepts a lower level of exploitation in return for greater security" (ibid.). Generalists use a wide variety of resources and maintain excess capacity that allows them to change to take advantage of more readily available resources (Hannan & Freeman, 1977). Specialists do not have this slack so they are more efficient and perform better in times of stability, but in turbulent times, generalists have higher survival rates because they can draw upon excess capacity to help them adjust (Cyert & March, 1963; Hannan & Freeman, 1977).

These concepts of generalists and specialists contribute to explaining "resource partitioning" that occurs when two trends are happening at the same time; while large generalists are consolidating, small specialist firms enter the mature industry (Carroll & Swaminathan, 2000). The specialists target resource spaces outside of the generalists' interests (Carroll & Swaminathan, 2000). These thin resource spaces allow only the existence of small specialists, but this partitions the resource space, because the two types of firms do not compete and instead coexist (Carroll & Swaminathan, 2000).

Two Scenarios for Green Clusters

New industries have choices about where to locate, in vibrant existing clusters, or in places where they are a first generation, carving out new clusters in areas that desire economic development. Let us consider new green technology firms—should they start up in an existing industrial cluster? Organizational ecology is a lens through which to examine location options for new green firms and to analyze potential interactions when a new industry locates where incumbents reside. Many factors influence the location decision; these factors have been studied extensively by scholars who ask what factors support growth of an industrial cluster or its agglomeration in a particular area.

Marshall's (1920) three explanations for the existence of positive agglomeration externalities are local information and knowledge spillovers, local supply of nontraded inputs, and a skilled local labor pool (Iammarino & McCann, 2006, p. 1021). Added to his list are the "geography of cooperation" argument and the social network view (Iammarino & McCann, 2006; Granovetter, 1973). Organizational ecology explains life, survival and death in agglomerations that are the spatial versions of a niche. Firms sharing a location share resources and similar environmental conditions. According to partitioning theory, specialists seek a different set of customers in the same geographic area as the generalists; this explains the success of specialists in what otherwise appears as generalist territory.

Founders of firms in a new cluster are also influenced in their choice of location by proximity to their homes; for help and support, they want to stay close to family and friends. Also, they are attracted to areas with universities that have special expertise in supporting research (Bramwell & Wolfe, 2008). Small firms need specialized employees; they must be in a convenient location to gain the attention of venture capitalists and angel investors. Moving to an established cluster, though, is expensive because rents may be prohibitively high. Living and working conditions may be expensive for employees, especially when new firms cannot pay high salaries as they are only launching themselves. Consequently, the benefits of moving to an existing cluster are unclear; a new firm in a different industry could be an unnoticed outsider, not able to make useful connections, so that the benefits of being in the existing cluster might not be initially available to it, yet it would pay the price of the location.

Scenario 1 depicts the start of a cluster (see Figure 8.1). Governments around the world often encourage clusters (Eisingerich, Bell, & Tracey, 2010; Fromhold-Eisebith & Eisebith, 2005; Iammarino & McCann, 2006). They expect them to be engines of economic growth (Porter, 1998). How can the government attract founders and firms so as to foster new cluster growth and their vitality? A location must display some uniqueness (Knutson, 2009). It must overcome the problem of inadequate funding; most regions do not have sufficient numbers of wealthy, entrepreneurially minded individuals. A risk-taking mentality is also important (Casper, 2007). The ingredients to start a cluster are special (Bresnahan & Gambardella, 2007).

Green firms face a liability of newness (Hannan & Freeman, 1984). They do not have legitimacy in the early stages and must expend a great deal of effort to attract resources. Resources are spread thin. The firms are still developing new products; the products might not be ready for manufacture. The firms need specialized employees; they may have to encourage local universities to create programs to educate people for jobs in their fledgling industry. The firms need funding and people with connections

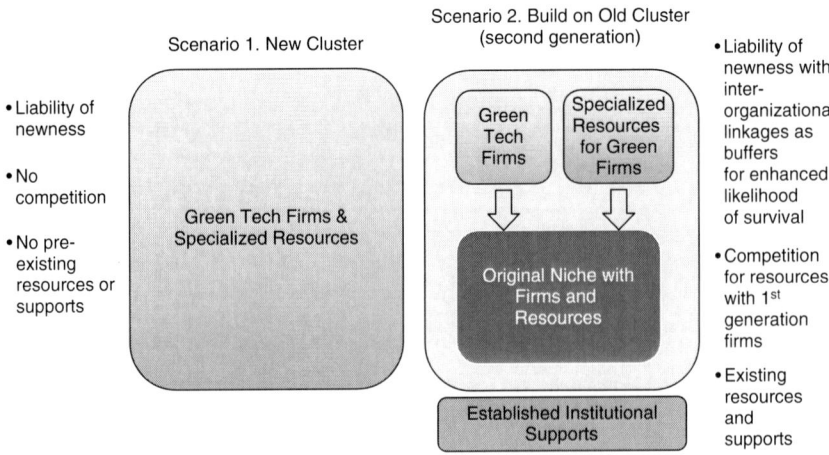

Figure 8.1 Two scenarios for green technology clusters. In which case are new firms more likely to survive and experience more rapid growth?

who can raise funds for them. The founders of new firms are stretched because they are involved in product development and fund-raising at the same time.

In an established cluster, some of these resource constraints may be alleviated. There are specialized employees, for instance, with transferable skills who can move to the new sector. Experienced investors are used to evaluating risky endeavors; they will be amenable to exploring the new firms' business prospects (Tam, 2010). The new firms then will benefit from the funding and management expertise that venture capitalists provide. Additionally, universities and local governments with multiple connections with cluster firms will be open to considering partnerships with new cluster firms as a result of their previous experience (Fromhold-Eisebith & Eisebith, 2005; Bramwell & Wolfe, 2008). However, there is also likely to be competition among the new and old firms for this existing pool of resources. They can compete over customers if the new firms develop substitute products. The incumbent firms have advantageous connections, which they may use to block the newcomers. So, while existing resources and institutional supports enhance the survival of new firms and reduce the liability of newness, the competition with old firms places them at a higher risk of failure.

The density dependence principle from population ecology predicts that the new firms will find less competition in a new cluster that is protective of their growth (Carroll & Swaminathan, 1991). The new firms are highly differentiated and have highly specialized workers (Iammarino & McCann, 2006). They encompass a wide range of possible products and services; in the green space they are likely to represent everything from home electricity monitoring equipment, to electric cars, to wind and tidal alternative energy production equipment, to solar powered high-tech equipment, to high-tech equipment that is recyclable, and the list continues. As these firms expand, they need additional workers and more funding, but they are unique enough and so preoccupied with their own development that they are not likely competing directly with other local fledgling firms, who are also engaged in a struggle for survival. Fatalities occur at this stage mainly because these start-up firms do not create enough value to attract customers and investors, not because another firm in the cluster does the same thing better. The new firms have unsure futures and therefore, they are not attractive enough to copy. Imitators are not common at this stage because success is too distant and uncertain.

With resource partitioning, specialists enter a cluster and use a different set of resources in which generalists are not interested (Carroll & Swaminithan, 2000). Generalists compete with each other as they consolidate, which leaves an opening for specialists; thus, resources are partitioned and the two types of firms coexist without being in vigorous competition (Carroll & Swaminithan, 2000). Similarly, second-generation population growth in a cluster may not directly threaten the existing first-generation firms' sales or markets if the industries are very different

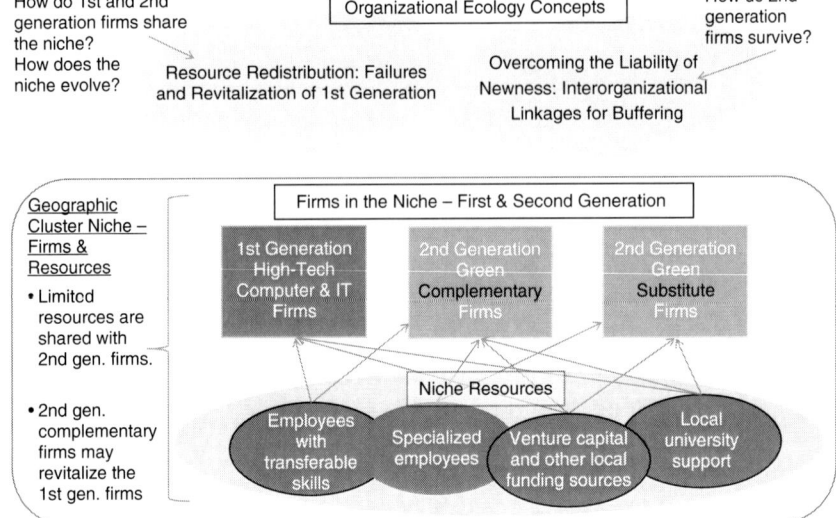

Figure 8.2 Organizational ecology concepts and niche conditions.

(see Figure 8.2). In this instance, the second-generation firms are not in the same industry. However, in factor markets, the two generations rely on an overlapping set of resources; they are not in competition for use of all of the same resources, though. The new entrants are attracted to the cluster, not because of the free space for them, but for other benefits.

Although the first-generation firms are more reliable because of a proven track record, it is likely that these firms are in a mature stage and are less interesting to investors because of lower growth opportunities. Investors are interested in the risky second generation because of the large potential gains, however far off; therefore, funding resources start to move to new firms, along with some employees. As a result, some first-generation firms will fail, but others will survive, possibly because they did not need the excess slack that went to the new firms. Also, the incumbents may benefit from linkages with new firms that help to revitalize them. These linkages also help the new firms. In resource partitioning, the specialists and generalists of the same industry do not affect each other as long as they stay in their respective domains. First- and second-generation cluster firms, on the other hand, compete and cooperate.

It is useful to understand under what circumstances a new industry may choose to locate in an existing cluster. Benefits for a second generation of firms in an existing cluster will depend upon reactions to the new generation by existing firms, customers, potential employees, and investors. I will analyze the interactions between first- and second-generation populations, categorizing their relationships as substitutes, complements, or unrelated; and as either competing, cooperating, or not interacting. These combinations generate six possibilities (see Figure 8.3).

		Sales Market		
		Compete	**Cooperate**	**No Interaction**
Factor Market	**Compete**	**Substitutes** • Factor market: thin resources distributed • Sales market: win-lose • Double liability of newness (reduced) and sales competition for new industry • Sales competition for incumbents and miss new technology opportunity (no revitalization) **Win (Old)-Lose (New) short term and Lose-Lose long term**	**Complements** • Factor market: thin resources distributed • Sales market: share technology, cross selling and marketing **Less Win – Win**	**Different offerings – no relation** • Factor market: thin resources distributed • Liability of newness (reduced) • No sales competition for new or old industries • Resource depletion for incumbents and no revitalization **Win (New) – Lose (Old)**
	Cooperate	**Substitutes** • Factor market: attract and share high tech employee pool, university resources, and venture capital • Sales market: win-lose • Much less liability of newness for new industry • Sales competition for both • Some revitalization of old industry **Win – Lose both ways short- and long-term**	**Complements** • Factor market: attract and share high tech employee pool, university resources, and venture capital • Sales market: share technology, cross selling and marketing • Revitalize incumbents • Reduce double liability for new industry **Max. Win – Win**	**Complements or Different offerings – no relation** • Factor market: attract and share high tech employee pool, university resources, and venture capital • Much less liability of newness for new industry • No Sales competition for either industry • Some revitalization of old industry **Less Win – Win**

Figure 8.3 First-generation (old) and second-generation (new) industry interactions.

In this analysis, the situation where new firms bring to the existing cluster additional funding and high-tech employees is not specifically considered because the amounts and types of resources and therefore the impact of the new resources can vary widely. This analysis considers existing resources in the niche of the original firms. In general, if the new firms bring resources with them, this will benefit them and will possibly benefit the entire cluster region, but more specific information about the resources is required to predict particular impacts.

The "double liability" of newness and competition, proposed and explained below, will be reduced for new firms in factor markets, but not necessarily in local sales markets. Overall, this reduction will improve their likelihood of survival. The generations may cooperate, compete, or have no interaction in sales markets since offerings may be complements, substitutes, or completely different products. The horizontal axis of the Figure 8.3 demarcates these competing and cooperative interactions. In the factor markets of land, labor, and capital, most new high-tech offerings, even if they are completely different, are able to draw upon some of the same niche resources the previous generation uses. This idea is represented in the matrix by the vertical axis with the two options of *cooperate* and *compete*. All firms, old and new, in any high-technology industry, can use capital, skilled employees, and help from universities, but use them to differing degrees. Older firms may not be as needy as newer ones for external capital, having generated organic success that new firms do not yet have. Also, specialized employees are transferable to some degree; certainly, their computer skills are useful and they tend to be technologically adaptable, but further specialization for developing new technologies may take some additional investment in them.

When both generations compete in sales and factor markets because their products are substitutes in sales markets, as in the upper left-hand box of Figure 8.3, factor market resources are distributed among the new and old firms. Because there is no sharing of resources or coordination, resources are not used as efficiently as possible. They are more thinly distributed compared to when they were used only by a smaller group of incumbent firms. Venture capital is interested in the greater growth opportunities that new firms offer; older firms appear less interesting and therefore they now attract less capital. However, they may not need the external capital as much. In fact, capital may be attracted to the region because of the new industry; this contributes to the revitalization of the entire cluster since there are positive spillovers for all types of businesses in the area as a result of the additional local investment and spending. Also, local high-tech workers may move to join the new firms; they may be of limited usefulness at first, as they require a learning curve. Local universities may supply well-prepared graduates for the new firms, if they educate them in accordance with the new technologies; also, research may move toward and support the new technologies because they are interesting to scientists. Incumbent high-tech firms may not receive the same attention and could lose out over the longer term in factor market competition.

In sales markets, the competition is challenging for new firms, because it comes on top of the liability of newness. New firms lack the resources that first-generation firms have for marketing and distribution. Also, incumbents have established customer, distributor, and other relationships and they may be able to use these to block new firms' activities. Thus, new firms face a double liability, one that decreases because of supportive factor markets and one that increases because of vigorous competition in sales markets. This may result in early failures of new firms such that the incumbents appear to win out over them; however, in the long term, without the first-generation firms, incumbent firms are not as motivated to innovate and changed factor conditions do not occur that would revitalize incumbent firms; for example, high-tech employees, having learned at new firms, may be enticed to work at larger, more stable old firms. They could breathe life, if allowed to, into the older firms, but this does not happen, because the new firms and their technologies are unable to survive. Also, without the survival of second generation firms, the first generation does not have the opportunity to build revitalizing alliances with them.

In the lower left hand box of Figure 3, the main difference with the previous discussion is that new and incumbent substitute industries cooperate in factor markets while still competing in sales markets. Improved, more efficient and coordinated use of factors helps more new firms survive, reducing the liability of newness to a greater extent. Also, incumbents are revitalized. For example, incumbents may engage in coopetition, allying to some extent with new firms, and providing capital and management expertise to them with expectations of a return. However, the sales competition still exists.

In the bottom middle box of Figure 8.3, conditions in the cluster are optimized when new and old industries are complements; they cooperate in both factor and sales markets. As complementary industries, they are not direct competitors, and the existence of each enhances sales for both. Thus, they form alliances, sharing factors of production, research, marketing and sales channels. For example, complementary offerings may be bundled or one may be part of each other. A solar cell may be incorporated into many types of electronic devices, for example. Consequently, the new industry blossoms because the liability of newness is reduced and competition is lessened. Incumbent firms' sales increase and they are encouraged to improve to meet the demands of the new complementary technologies. Employees are transferable among firms, and venture capital and universities are interested in both industries since they are complementary; when one grows and changes, so does the other.

In the upper middle box of Figure 8.3, a lesser win-win scenario is predicted because although the complementary industries are cooperating in the sales market, they are rivals in factor markets. The liability of newness is not as reduced when companies expend limited resources on competing in factor markets; incumbents try to block or do not cooperate with new firms. Though the new industry creates valuable new complementary products, it is not operating at the highest possible level. Similar benefits are also derived by incumbents as before, but not optimally, since they are expending resources on blocking new firms that are helpful by their very existence.

In the two right-hand side boxes of Figure 8.3, firms do not cooperate or compete in sales markets. If they are complements, their products are both chosen by consumers jointly, but the firms do nothing to promote each other. Alternatively, the new industry may be almost completely unrelated to the incumbent industry. For example, the incumbent is biotech and the new industry is wind power. They do not interact in the sales market, but in factor markets they will interact because venture capital, high-tech workers, and university research can adapt, although there are learning curves. If they compete in factor markets, as in the upper right-hand box of Figure 8.3, it is predicted that the new industry will win overall because factors are attracted by the new technologies and growth potential associated with them. For example, the biotech cluster in Cambridge, Massachuestts, has decayed and the state government is encouraging green cluster development to revitalize the area. The original decayed industry, attempting to compete for factors of production, already has lost and will continue to lose. Since established resources exist for the new industry in the cluster, the liability of newness is reduced. There is also no competition in the sales market. The new industry will win and the incumbent industry will lose.

On the other hand, if the two unrelated industries cooperate in factors of production (see the lower right-hand box in Figure 8.3), both could win, but to a lesser extent than in the scenario when their offerings are complementary (lower middle box). For example, incumbents invest in the new industry and provide management expertise which provides a return to incumbents.

Overall, this systematic analysis of potential interactions in the cluster offers a decision-making framework for a new generation considering its options. For example, if a second generation can predict that its offerings are going to directly compete with those of incumbents, it may think twice about locating in that particular cluster. However, the other situations in Figure 8.3, aside from the directly competitive ones, suggest that locating in a cluster for a second generation can be beneficial and may even revitalize the first generation when it is cooperative. Thus, a new generation may consider a different cluster if the one under consideration is too competitive. It can decrease its liability of newness in another existing cluster without facing high competitive threats. However, if alternative cluster options are limited, the new industry may prefer to set up its own new cluster, which would take a lot of development. The liability of newness would be high; this situation would match the theoretical case of density dependence and the realistic conditions that many existing clusters grew under originally. In a new cluster, competition would be nonexistent in the early stages; this situation would provide a higher likelihood of survival for a new industry, compared to an existing cluster offering a highly competitive environment.

Industrial clusters are often motivated and planned by governments, as mentioned, and the previous discussion assumes the location choice to be that of the new generation. However, government planning could be helpful and supportive while limiting in this location choice. Some governments, like that of the Japan, have decided already where industries will locate. Government has motivations such as economic development for encouraging cluster growth; this could mean directing firms to locate in regions that have little to offer. Although a new cluster is potentially beneficial to a poorer region, business conditions may hamper new firms and they may be much better off locating elsewhere—in existing clusters, for example. Although a government is likely interested in seeing and supporting the success of new industries, this involvement could become too dictatorial and could unintentionally prove to create negative conditions for innovation; we have yet to see it, but governments getting too much involved in business could be problematic. This latter statement represents interesting future research.

Conclusions

This chapter examined how industries are changing at a macro population level in the most promising hotbeds for revolutionary technological change—industrial clusters. This type of change will likely repeat itself in the future, only next time it will be much improved for research studies that this study prompts. The view in this chapter is that, on balance, a new industry in an old cluster is benefited by incubation more so than hurt by competition. Also, a revitalized first generation thrives; both populations coexist, better for the initial struggle. An existing cluster that does not invite challenging new entrants may experience a downward spiral of decay, leaving the community around it in trouble; Cambridge's biotech sector has experienced this. However, if economic development of an otherwise needy area is the desired goal, then starting a new cluster may be advised. This is not the best scenario for the industry population though; policymakers may understand and heed this tradeoff through this chapter's research.

Additionally, this chapter has theoretically developed an organizational ecology foundation for a new theory of the first- and second-generation industry interaction. This theory predicts that a first generation will be diminished when a second generation arrives. However, incumbent survivors will be revitalized through cooperative behavior in factor and sales markets with the second generation. The second generation will face different levels of struggle with incumbents in factor and sales markets, depending on whether their products and/or services are substitutes, complements, or unrelated; the greatest threats to survival exist when they are direct substitutes. However, these new direct substitutes may be much more valuable innovative offerings than existing ones; thus, in the long run, the cluster decays if the superior new firms are not allowed to overtake weak incumbents. For second-generation firms that are complementary, the cluster incumbents could actually provide a boost. Thus, the choice to locate in an existing cluster rather than starting anew is unequivocally preferred in the case of complements. In general, it is better for the geographical cluster if a second generation arrives, survives, and transforms it.

Empirical work has to be done now to support the theory. Research may examine specific interorganizational relationships, such as those between new firms and universities, think tanks, and incumbent firms. In the alliance literature, interfirm relationships often fail (Li & Guisinger 1991; Park & Ungson, 2001), yet in this context of first–second generation industries, alliances are beneficial; research evidence demonstrates that alliances support cluster growth (Casper, 2007; Eisingerich et al., 2010). Also, future research may investigate the benefits of government linkages, like the work of Japan's Ministry of Economy, Trade and Industry (Casper, 2007). Fromhold-Eisebith & Eisebith (2005) have investigated institutionalized support. Research may also consider when too much government involvement in new industries has negative effects. Complex relationships among firms, the scientific community, and government agencies in industrial clusters seem to be beneficial. Although future research can investigate these existing relationships, new developments such as when a new green generation arrives are intriguing and important to consider also. Answering the question, "How do industrial clusters change and grow?" will be no less important in the future than it is now as we face the impacts of accelerating climate change.

Notes

1. ETIC is an Environmental Technology Industry Cluster growing in Arizona, USA (http://www.az-etic.com/index.cfm), accessed January, 30, 2010.
2. See the Industrial Cluster Project website (http://www.az-etic.com/index.cfm), accessed January, 30, 2010.

References

Amburgey, T. L., Kelly, D., & Barnett, W. P. (1993). Resetting the clock: The dynamics of organizational change and failure. *Administrative Science Quarterly, 38*, 51–73.

Bramwell, A., & Wolfe, D. A. (2008). Universities and regional economic development: The entrepreneurial University of Waterloo. *Research Policy, 37*, 1175–1187.

Bresnahan, T., & Gambardella, A. (2007). *Building high-tech clusters.* New York: Cambridge University Press.

Burns, T., & Stalker, G. M. (1961). *The management of innovation.* London: Tavistock.

Carroll, G. R. (1984). Organizational ecology. *Annual Review of Sociology, 10*, 71–93.

Carroll, G. R., & Delacroix, J. (1982). Organizational mortality in the newspaper industries of Argentina and Ireland: An ecological approach. *Administrative Science Quarterly, 27*, 169–198.

Carroll, G. R., & Swaminathan, A. (1991). Density dependent organizational evolution in the American brewing industry from 1633 to 1988. *Acta Sociologica, 34*, 155–175.

Carroll, G. R., & Swaminathan, A. (2000). Why the microbrewery movement? Organizational dynamics of resource partitioning in the U.S. brewing industry. *American Journal of Sociology, 106*, 715–762.

Casper, S. (2007). How do technology clusters emerge and become sustainable? Social network formation and inter-firm mobility within the San Diego biotechnology cluster. *Research Policy, 36*, 438–455.

Chandler, A. D. (1962). *Strategy and structure: Chapters in the history of the American industrial enterprise.* Cambridge, MA: MIT Press.

Cyert, R. M., & March, J. G. (1963). *A behavioral theory of the firm.* New Jersey: Prentice-Hall Inc.

Eisingerich, A. B., Bell, S. J., & Tracey, P. (2010). How can clusters sustain performance? The role of network strength, network openness, and environmental uncertainty. *Research Policy, 39*, 239–253.

Freeman, J., Carroll, G. R., & Hannan, M. T. (1983). The liability of newness: Age dependence in organizational death rates. *American Sociological Review, 48*, 692–710.

Fromhold-Eisebith, M., & Eisebith, G. (2005). How to institutionalize innovative clusters? Comparing explicit top-down and implicit bottom-up approaches. *Research Policy, 34*, 1250–1268.

Gersick, C. J. G. (1991). Revolutionary change theories: A multilevel exploration of the punctuated equilibrium paradigm. *Academy of Management Review, 16*, 10–36.

Granovetter, M. (1973). The strength of weak ties. *American Journal of Sociology, 78*, 1360–1380.

Hannan, M. T. (1986). *Competitive and institutional processes in organizational ecology.* Technical Report 86–13, Department of Sociology, Cornell University.

Hannan M. T., & Freeman, J. (1977). The population ecology of organizations. *American Journal of Sociology 82,* 929–964.

Hannan M. T., & Freeman, J. (1984). Structural inertia and organizational change. *American Sociological Review,* 49, 149–164.

Hannan M. T., & Freeman, J. (1989). *Organizational ecology.* Cambridge: Harvard University Press.

Hockerts, K., & Wüstenhagen, R. (2010). Greening Goliaths versus emerging Davids — Theorizing about the role of incumbents and new entrants in sustainable entrepreneurship. *Journal of Business Venturing, 25*(5), 481–492.

Iammarino, S., & McCann, P. (2006). The structure and evolution of industrial clusters: transactions, technology and knowledge spillovers. *Research Policy, 35,* 1018–1036.

Knutson, R. (2009). Oregon looks to clean tech for revival. *Wall Street Journal,* New York, N.Y., Aug 28, A.5.

LaMonica, M. (2009). Massachusetts goes green to relive tech glory. CNET news. [Online access http://news.cnet.com/8301-11128_3-10265420-54.html January, 30, 2010.]

Li, J. T., & Guisinger, S. (1991). Comparative business failures of foreign-controlled firms in the United States. *Journal of International Business Studies, 22,* 209–224.

Marshall, A. (1920). *Principles of economics,* 8th ed. London: Macmillan.

Miner, A. S., Amburgey, T. L., & Stearns, T. M. (1990). Interorganizational linkages and population dynamics: Buffering and transformational shields. *Administrative Science Quarterly 35,* 689–713.

Ministry of Economy, Trade, and Industry (METI; Japan) (2004). Industrial cluster project. [Online access http://www.cluster.gr.jp/en/about/index.html March, 21, 2011]

Myint, Y. M., Vyakarnam, S., & New, M. J. (2005). The effect of social capital in new venture creation: the Cambridge high-technology cluster. *Strategic Change, 14,* 165–177.

Nair, A., D., Ahlstrom, D., & Filer, L. (2007). Localized advantage in a global economy: The case of Bangalore. *Thunderbird International Business Review, 49,* 591–618.

O'Reilly, C.A.I., & Tushman, M. (2004). The ambidextrous organization. *Harvard Business Review,* April, 74–81.

Park, S. H., & Ungson, G. R. (2001). Interfirm rivalry and managerial complexity: a conceptual framework of alliance failure. *Organization Science, 12,* 37–53.

Porter, M. E. (1998). Clusters and the new economics of competition. *Harvard Business Review 76,* 77–90.

Roughgarden, J. (1979). *Theory of population genetics and evolutionary ecology: An introduction.* New York: Macmillan.

Saxenian, A. (1994). *Regional advantage: Culture and competition in Silicon Valley and Route 128.* Cambridge, MA: Harvard University Press.

Stinchcombe, A. L. (1965). Organizations and social structure. In J. G. March (Ed.), *Handbook of organizations* (pp. 153–193). Chicago: Rand McNally.

Tam, P. W. (2010). Painting Silicon Valley a new shade of green. *Wall Street Journal* (online*),* New York, N.Y., May 12.

Tushman, M. L., & Anderson, P. (1986). Technological discontinuities and organizational environments. *Administrative Science Quarterly, 31,* 439–465.

CHAPTER NINE

Shecopreneuring: Stitching Global Ecosystems in the Ethical Fashion Industry

KIM POLDNER, OANA BRANZEI, AND
CHRIS STEYAERT

This chapter extends the literature on socially and ecologically minded entrepreneurship (Nicholls 2008, p. xix)—to ask how individuals can (re) imagine and realize more sustainable global ecosystems. Human action can create landscapes that are "at least as rich and as stable, occasionally as beautiful as those shaped by nature" (Lyle, 1999; Campbell, 2006). Taking responsibility for the environment begins with individual transformation and practices (Ruether, 1992). As individuals grow, experiment, and change (Bronfenbrenner, 1979; Pardeck, 1988), they may influence their own ecosystem (Paolucci, Hall, & Axinn, 1977; Slocombe, 1993) and change how others perceive and interact within that ecosystem (Lustermann, 1985). Some individuals can develop intricate systems of practices to sustain their ecologically embedded livelihoods (Whiteman & Cooper, 2000), yet in our increasingly global ecosystems, ecological embeddedness risks becoming the exception rather than the rule. Irresponsible choices prevail; against their backdrop, responsible practices deserve further study.

Traditional fashion has come under harsh scrutiny for its harmful use of pesticides to produce cotton, for water waste and chemical pollution in leather tanneries, for child labor in the global supply chains, and for unfair wages to workers overseas. Protesting against the negative footprint of traditional fashion, entrepreneurs with a passion for fashion and an ecological conscience began experimenting with more sustainable products and production processes. As of 2010, more than 500 ventures have been established since 2005. The emerging industry, known as ethical fashion, had been particularly appealing to women: over 90 percent of the ethical fashion entrepreneurs are female fashionistas, fondly described as *shecopreneurs*. Most of these shecopreneurs went into business to change the world—often driven by a feminin ethics of care, at times exacerbated by motherhood, which heightened their sense of duty to preserve the Earth for their newborn children (Thopte, 2009).

The number of shecopreneurs and their economic impact is still small, but their influence is being felt globally as growing visibility of ethical fashion motivates demand for organic fabrics, which in turn drives experimentation with different crops, sourcing approaches, and more ecologically minded production methods; colocation of production and consumption; and direct and fair engagement of local and remote communities (Ferrigno, 2008).

This chapter examines how shecopreneurs experience and influence global ecosystems. Ecosystems vary in scope and size. A fair-trade cooperative producing alpaca fiber sweaters, for example, brings together only a handful of people, who tend the alpaca using very limited local resources (for grazing, breeding, and shearing) and simple, often traditional, artifacts (for processing, dying, and knitting). An ecosystem that produces organic cotton outfits, on the other hand, requires a global supply chain, including new methods of cultivation (to prevent cross-contamination), cross-pollination, harvesting, and dedicated manufacturing lines.

The literature on ecosystem design has identified three stages: romance, precision and generalization. The romance stage is characterized by "[A] spirit of boundless anticipation" (Lyle 1999, p. 136). The precision stage narrows down a designer's choice set by focusing on "landscapes small enough to be perceived and understood in their entireties" (ibid., p. 146). The generalization stage presents the designer with "a task of assembly" (ibid., p. 162). In the romance stage, the imaginary comes to the foreground, the real fades to the background; the precision stage reverses the two dimensions while the generalization stage (re)balances them. This chapter focuses on the entrepreneurial micro-practices which enable shecopreneurs to (re)imagine and (re)build more sustainable ecosystems in the ethical fashion industry.

This chapter leverages the notion that tension between the real and the imaginary can inspire entrepreneurial action that changes the world (Hjorth & Steyaert, 2004; Carlson, 2000, p. 228). Our intended contribution is to explain how changes in individual self-understanding may lead to new sets and new meanings of practices, and, vice versa, how experimenting with and adopting new practices creates a different self-understanding and cultivates new relationships to the ecosystem. We focus on how individuals reconceptualize ecological and social constraints and work *with* rather than against these constraints to develop and model more sustainable practices. Specifically, we are interested in how the duty of care permeates and transcends shecopreneurs' self-understanding to expose the unfit underpinnings of global ecosystems in the traditional fashion industry and helps them replace such underpinnings with more ecologically and socially responsible alternatives in the emerging ethical fashion industry.

Social Imaginaries

The interface between the real and the imaginary is not fixed, but rather is a dynamic, socially negotiated frontier of collective action, which shifts as given groups of society start employing new practices that carry new understandings. Taylor (2005, p. 23) describes it as the *social imaginary*—"the ways people imagine their social existence, how they fit together with others, how things go on between them and their fellows, the expectations that are normally met, and the deeper normative notions and images that underlie these expectations." A social imaginary is less abstract than theory and more accessible to ordinary people through images, stories, and legends. Albeit complex, it is inherently functional and "makes possible common practices and a widely shared sense of legitimacy" (ibid.).

Changes in practices (or the meaning of practices) among a given group diffuses across other groups to usher in individual transformations and societal transitions. A social imaginary itself changes through individual practicing (ibid., p. 29–30). Most changes are gradual, but add up and eventually become profoundly influential (ibid., p. 43). As people reimagine their social existence, they develop a new kind of self-consciousness and redefine the very practices that underpin their way of life.

Such changes require imagination—but cannot be reduced to dreams (ibid., p. 183). They inspire individual action, which leads to altered self-understanding. At first, such self-understandings transform individuals; their actions make apparent new sets of practices and give new meanings to old ways. As changes in practices diffuse among individuals, a new kind of historical consciousness permeates society.

The core premise of social imaginaries is that a handful of protagonists who invent, adopt, and set out to enact new realities are sufficient to prototype social change. The entrepreneurship literature concurs that individuals can and often do transform themselves, their ventures, and their community (Rindova, Barry & Ketchen, 2009).

By qualitatively exploring the interface between the real and the imaginary, this chapter deepens our understanding of the connection between individual transformation and broader social change. We rely on the practice-based literature, which suggests that practices change—both in their content and their meaning (Sonenshein, 2010). We also make use of the resource-based view literature's related argument that incremental changes in daily routines can add up and can provoke radical change (Plowman, Baker, Beck, Kulkarni, Thomas, & Villareal, 2007). Finally, we draw from the entrepreneurship literature that emphasizes the mundaneness underpinning entrepreneuring (Steyaert, 2004, p. 19; Steyaert, 2007), especially in the case of socially marginalized actors (Calás, Smircich, & Bourne, 2009). The intersection of these three literatures enriches our understanding of varied yet compatible understandings of real practices.

Feminist perspectives draw attention to the coexistence of the real and the imaginary in everyday entrepreneuring practices. A feminist lens on entrepreneuring (Calás et al., 2009) suggests that gender destabilizes interpretative categories and encourages a plurality of meaning. More broadly, in her review of French feminism, Lattas (1991) emphasizes the transition from the first generation of feminists' desire to be the same as men to the second generation's striving for difference. French feminists celebrate the fluidity between the imaginary and the real as uniquely feminine and feminizing. They underscore that one is not born, but rather becomes a woman, through processes of reimagining and realizing womanhood (Lattas, 1991, p. 102). Cixous (1976), in particular, calls for a female imaginary, which would celebrate new feminine symbols (nonlinearity, multisensorial perception, and sexual difference). More broadly, the emphasis on voice emphasizes women's roles in the figurative realm of images, subjectivity, and emotion. Yet little work has so far explored the processes of entrepreneurial becoming in general; accounts of gendered practices of entrepreneurial becoming are even scarcer. This chapter explores how such gendered practices of entrepreneurial becoming might help shecopreneurs change the world.

Methods

We rely on multisensorial, multisource ethnographies of three shecopreneurs who started in London, Toronto, and Vancouver roughly at the same time. Because all three have achieved a high degree of local and international recognition in high-fashion fairs and reviews, we were able to develop rich narrative and visual accounts, which helped us to triangulate accounts from multiple sources for each collection, as well as observe the evolution of their collections over time. We collected data both prospectively (designs, materials, and news about forthcoming collections) and retrospectively (interviews, observations, and reactions after their launch). To preserve the confidentiality of their accounts, we refer to the three shecopreneurs by their start-up location (London, Toronto, and Vancouver). We have omitted specific details that would easily give away their identity, used archival details sparingly in our narrative, and have shared our manuscript drafts with the protagonists to guard against exposing or creating unintended vulnerabilities—still, anonymity remains a challenge, given the uniqueness of their collections.

London is a young, independent fashion designer. Design comes first, but the idea of contributing to a healthier planet through using ecofabrics, minimizing waste, and recycling adds value to her designs. London is also a keen student of ecosystems, which inspire her designs with biomimicry. Her initial inspiration came from a West African organic cotton farmer who was invited to speak at a sustainability event at her school. Soon after, London launched her first collection in organic cotton. She kept looking for sustainable fabrics and discovered that most companies have a green range but few exhibit in the preassigned green corner due to their still limited assortment. London nonetheless felt that it was important to buy those fabrics. Not everything London makes is 100 percent sustainable, but she tries her best. For example, based on the idea of creating as little trash as possible, one of London's first dresses featuring sculptural pockets was designed to be composted after its lifecycle; even the back zipper could go in the iron-recycling bin.

Toronto grew up sewing things; design has been her language, her way of communicating. Her first collection was not sustainable, but her venture is now 100 percent ethical. Toronto gradually tackled first her lifestyle, then her venture also, in a green direction. Her choice to use only ecofriendly fabrics required some difficult ecosystem tradeoffs, but the most important aspect for Toronto is "that experience I like to give people, like, the experience of putting something on and it's not wearing them; it's enhancing them and making them feel really good and confident and sharing my language with them, which is . . . I consider when I'm designing that's my language. So, that's me communicating with people and allowing them to communicate in their language with some of the tools that I'm giving them to do that with."

Vancouver grew up in a family that valued healthy nutrition, recycling, and treating others in a respectful way. She sought out world-class mentors, who influenced her journey of becoming a designer and a businesswoman without straying from her own values. Vancouver is passionate about community building—locally, among ecofashion designers.

For each shecopreneur, we developed a rich narrative that supplemented their own stories (first-hand interviews, press interviews, blogs, and reflections) with our own notes, observations, and understandings. We contextualized these narratives, because narratives are social practices that reflect and constitute their own context (Ewick & Silbey, 1995, p. 211). We went back and forth through archival accounts, photos, fashion collections, and video footage to identify key actors, resources, and artifacts and to understand how these came together as the protagonists (re)claimed different ecosystems. The multisensorial nature of the data is hard to describe in words at times, but is possible to convey textually.

The first and second author jointly developed and independently coded the narratives for our three protagonists. The first author was also an industry insider; her own experience as a shecopreneur offered a rich understanding of the industry and personal connections with the three protagonists. The second and third authors were industry outsiders; the second author worked closely with the data (but not directly with the protagonists), while the third author maintained a distance from both the data and the protagonists throughout the analyses. All three authors were sensitive to gendered and gendering practices and mindful to preserve the voices, nuances, and meanings intended by our protagonists in our text. Our intent, however, was not to describe (and certainly not to evaluate) their unique journeys, but rather to develop a conceptual framework that can shed new light to the broader research question of how the practices of shecopreneurs enable the interplay between the real and the imaginary to (re)claim responsibility for global ecosystems.

Findings

As feminist theories of entrepreneuring suggest, shecopreneurs reconstruct their femininity by offsetting the real and the imaginary. We also expected the duty of care to factor prominently in shecopreneurs' journeys of entrepreneurial becoming, in part because the duty to care is considered a key part of women's becoming (Lattas, 1991). We adopted an inclusive definition of the duty to care, which included looking after and nurturing others as well as a sense of stewardship towards the environment, towards resources and towards nonhuman actors. Looking after the ecosystem is neither exclusively nor necessarily the duty of women (Whiteman & Cooper, 2000), but the earth and the environment have been traditionally conceptualized as feminine energy, acutely perceived and often defended by women (Gnanadason, 2005).

Our three protagonists honed their duty of care in specific directions. London's passion for biomimicry grew out of compassion for bees—their complex role and vulnerability to colony collapse disorder. Learning to think like the bees helped London reimagine herself. She rethought her own identity. Vancouver first focused on helping others—mostly other women—feel happier. Then, because Vancouver felt strongly about fair labor practices, she produced all her clothes ten minutes away from her house, with a single exception: "one sweater that was knit instead of cut from fabric and made in Peru." For Toronto, it was all about restoring a sense of self and maintaining a sense of community. Later, her designs explicitly encouraged self-expression. Her clothes spoke about the things customers should care about, like the tar sands or nomadic cultures in the tundra threatened by climate change. Toronto's recent collections were inspired by the idea of moving around—being displaced, like the disappearing animals. She "keeps money in [her] community"; although nothing is entirely local anymore, she advocates producing locally in order to sustain the community.

To understand how shecopreneurs reclaimed responsibility for global ecosystems, we extracted all instances of ecosystem design (actors, resources, and artifacts); for each ecosystem, we first reordered these instances sequentially. This replicated the three-stage pattern previously discussed in the ecosystem design literature: romance, precision, and generalization. Because we focused on the corresponding sets of practices, we use verbs—*romanticizing, being precise,* and *generalizing*—to emphasize their deliberate and dynamic nature. These three sets are macro-practices, because they are not isolated acts but rather recurrent acts, which transcend time and space to convey a holistic way of interpreting and influencing the ecosystem.

Romanticizing. Our protagonists initially *romanticize* the ecosystem—they all become enchanted with nature and disenchanted with our negative impact on it; they call for new and different ways of living our lives. Clothes speak about our good or poor choices; changing how we dress calls attention to our place in the world and creates an occasion for restoring nature (London), feeling (Vancouver), and community (Toronto). Romanticizing macro-practices recasts the relationship between individual and ecosystem in a less confrontational and more integrative way. For our protagonists, organic fabrics protect our skin and the Earth; timeless trends bring out our essence, our individual expression, and also save resources. Design gives us strength, protection, and happiness but also requires us to take a stance, do the right things.

Being Precise. Each of the three shecopreneurs created unique landscapes that reflected their needs and their means. They gradually evolved these ecosystems—either to address additional needs or to take advantage of additional means. Precise practices were experimental and somewhat transient; yet all three shecopreneurs looked for better and different gestalts—ways to relate to the ecosystems they were experimenting with.

Generalizing. A common macro-practice among our three protagonists was the search, find, and assemble of ecosustainable fabrics, which was still challenging, although the availability of fabrics had come a long way since the early 2000s, when sourcing sustainable fabrics was either prohibitively expensive for small start-ups like theirs or unsustainable due to the distance they had to source these fabrics from. Sourcing sustainably remained, however, a double-edged sword. First, shecopreneurs need more and more varied and more local and more ethical fabrics. Second, everyone needs fewer fabrics that do not meet sustainability criteria.

All three shecopreneurs reflected on their own becoming. The three shecopreneurs spoke about how they felt, how they accomplished their work, and their aesthetic motivations and reactions. We first sorted their accounts into nine first-order themes; then we regrouped them into three second-order themes—*affect, effect,* and *art.* Table 9.1 provides definitions and examples for the three second-order themes (and the nine first-order themes underpinning them). These themes recurred for all our three protagonists, across multiple accounts (data available from the authors). In Table 9.1. we illustrate each first-order theme with a single example.

Affect

Feelings inspired the designs, were sewn into the clothes; they were tried on, rearranged, and passed on from shecopreneurs to their customers, through three sets of first-order affective practices: *awake, attract* and *attend.*

Awake. Shecopreneurs worked to become and remain self-aware; they (re)attuned to their own emotions. They paid constant attention to their emotions and (re)adjusted course to stay true to their feelings.

Table 9.1 Micro- and macro-practices of ecosystem design: definitions and illustrations.

Micro-Practices: Toronto			Dis(connect)		Macro-Practices: Toronto
Examples	First-order Themes	Second-order Themes	Function	Categories	Examples
"Fashion design has been a compulsion since childhood. [...] I am inspired by so much, my surroundings, far away places, music, architecture, Japanese and Scandinavian design, other designers, art, nature, people who have a very individual style, and the fantastical. [...] sometimes it is a friend, sometimes it is an idea, or an imagined muse." (Interview with Toronto, April 13, 2009)	**Awake:** to become self-aware, attuned to one's own emotions.	**Affect:** Expressions and interpretations of one's duty of care by fostering emotional connections.	**Reimagining the real:** Rediscovering the self (which had been lost or corrupted or distracted by unacceptable constraints) enables shecopreneurs to "see anew" the obstacles and the potential of global ecosystems.	"A spirit of boundless anticipation. Fragments of images of what might be light up all around us and myriad pathways flicker into the haze. It is all confusing, challenging, stimulating, intriguing, daunting, enormously exciting. [...] this is a time for letting impressions sink in [...] for questions not answers" (Lyle 1999, p. 136).	"I have a more limited choice of fabrics, so I have to take into consideration what is available to me to translate into specific designs. [...] They are more breathable and many feel so silky smooth against the skin, you can definitely feel the quality difference when you touch or wear them (ecofabrics)." (Interview with Toronto, April 22, 2011)
[Toronto] has hammered out some pretty serious-looking gear to protect "urban nomads against the elements." The all-black palette was inspired by black-clad anime warriors who stick to themselves but somehow manage to get pulled into trouble anyways. (Write-up about Toronto's Fall/Winter 2010 Collection, March 27, 2010).	**Attract:** to inspire others in ways that simultaneously share and reinforce one's inspiration.				"Rather than using color to add interest, [Toronto] chose a black-on-black theme laced with shape and texture. We applaud her gutsy decision and think it makes a lot of sense to design timeless clothes that real people will actually put on. After all, it isn't very sustainable to

"I want my pieces to be unique, but something someone can have in their wardrobe and pull out at different times in their life and it won't be dated. So many designers are continually looking to the past, but I want to design for today's woman. She has a lot of things to do in a day, and I want her to look polished while doing it, but with an edge." (Toronto, 2010).

produce super-trendy garments that won't be worn. And hey, black will never go out of style." (Write-up about Toronto's Fall/Winter 2010 Collection, February 17, 2010)

Attend: to sense, capture, and respond to others' emotions.

"By choosing to use organic, eco-friendly fabrics, [Toronto] wants not only to save the Earth, but to benefit her shoppers as well: fabrics made from bamboo, soy, and lyocell lend great comfort to the garments as they are soft, breathable and excellent in absorbing moisture. [. . .] Moreover, organic and ecofibres are all natural and do not contain irritating chemicals, making them a boon for those with sensitive skin." (Write-up about Toronto, April 7, 2010).

Continued

Table 9.1 Continued

Micro-Practices: Vancouver			Dis(connect)	Macro-Practices: Vancouver	
Examples	First-order Themes	Second-order Themes	Function	Categories	Examples
"When I finished school I knew I wanted to come home… being one of the first eco-designers out of Vancouver, we're really helping pave the way." (Vancouver, 2010).	**Anchor:** to recognize, respect and (re)appraise one's economic constraints, goals and abilities.	**Effect:** Expressions and interpretations of one's duty of care by taking pragmatic steps.	**Realizing the imaginary:** One's economic constraints, goals and abilities become guideposts for redesigning an ideal ecosystem across all aspects of operations.	**Precise:** During the precision stage, designers "deal with landscapes small enough to be perceived and understood in their entireties [and piece together] a gestalt resource inventory." (Lyle, 1999 p. 146).	"When local womenswear designer [Vancouver] was ready to set up a boutique to house her [. . .] label, there was no better place in her mind than Kitsilano, a neighbourhood known for its free-spirited and laidback attitude, devoted to yoga, coffee, and above all, ecoliving." (January 21, 2010, Fashion Critic Blog about Vancouver)
"The renovations and the things that we'll be doing will all be green [. . .] Whether it's the paint that we use or the fixtures that we use, it's just keeping a conscious mindset on that." (Vancouver, January 21, 2010, Interview with the Straight).	**Align:** to arrange one's economic activities in ways that are coherent and consistent with relevant economic constraints, goals and abilities.				"All of the pieces are made locally, and the materials are sourced locally when available. Naturally, [Vancouver] plans on translating this ecofriendly mindset to her retail space." (January 21, 2010, Interview with the Straight).
"It has to be all done very sustainably for the earth as well, which is exciting because the more and more you go the more and more sustainable you can get." (Vancouver, 2010).	**Amplify:** to expand one's activities to enable the refinement and implementation of future goals and abilities.				"Help rebuild communities and educate them and give them a way to live on their own. [. . .] really far down the road, but to be able to eventually build [. . .] a co-op." (Vancouver, 2010)

Table 9.1 Continued

Micro-Practices: London			Dis(connect)		Macro-Practices: London	
Examples	First-order Themes	Second-order Themes	Function	Categories		Examples

Examples	First-order Themes	Second-order Themes	Function	Categories	Examples
"The first collection, the spring emergence, was really about doing something that was, kind of, like, totally different to what people were doing with sustainable design to that point, really putting something new out there and seeing how people responded to that. And then, the second collection was about, [. . .] can I actually make a really wearable and elegant collection that still looks like me? [. . .] I think it's important [. . .] to know yourself, as the designer, what you're able to do at each stage of when you're doing it." (London, 2009).	**Authenticate**: to stay true to one's (growing) aesthetic appreciation of ecological dimensions as one (re)imagines products and production.	**Art**: Expressions and interpretations of one's duty of care by honing one's sense of aesthetics.	**Juxtaposing the real and the imaginary**: One's aesthetic expressions and interpretations create occasions for comparing and (re) calibrating the real and the imaginary. Products and production embody ecological dimensions in order to both emphasize their vulnerability to human action and inspire more responsible human action (including redesigning the artifacts to achieve a better balance between the real and the imaginary).	**Generalize**: "A task of assembly [. . .] by looking at each of the problems and sub-problems from differing points of view, the designer develops a kit of alternatives that could be combined in various ways to shape the final plan" (Lyle, 1999. p. 162).	"Like, what percentage of 30- to 35- to 40-year-olds, you know, care about ethical fashion or want to wear it or buy it or how much disposable income those people have to spend on fashion and how much of them are interested in actually, you know, ethical and sustainable products? [. . .] So, if the design's fantastic and then it comes with the added value of being sustainable, then you've got to see that a hundred percent of those people could be your customer; it's just about how do you reach those people?" (London, 2009).

Continued

Table 9.1 Continued

	Micro-Practices: London		Dis(connect)		Macro-Practices: London
Examples	First-order Themes	Second-order Themes	Function	Categories	Examples
"I take inspiration from nature. I heard this word recently that really fits: bio-mimicry. It has to do with looking at the way nature works and questioning why it works so well. What I try to do is apply the concepts that work in nature, and think about them in terms of a high-end fashion label." (Article about London's Autumn/Winter 2009 collection, August 1, 2009).	**Adopt**: to (re)design products and production in ways that respect and replicate one's aesthetic appreciation of ecological dimensions.				"I also used silk Jacquard [. . .] woven in England [. . .] for more than two hundred years [. . .] all vertical manufacturing [. . .] all in one building on one floor. [. . .] they're very happy to walk you around the mill and you can see everything from spinning to dying all the silk. [. . .] an hour away from London on a train and here I am with all these people making this amazing fabric." (London, 2009).
"I wanted to make a connection that really inspired people and really informed people about the amazing qualities that bees have, but in a way that was very, like, directional and fashion forward. [. . .] I started to look at the way that bees think. [. . .] bees construct everything using hexagons. And, so, I started to construct garments using hexagons and also to insert [them into] body conscious kind of garments as well. [. . .], that's how the whole collection evolved from that shape and from the idea of [. . .] building the [collection] the way that bees build their colony." (London, 2009).	**Associate**: to stimulate more ecologically mindful patterns of consumption that reflect one's aesthetic appreciation for ecological dimensions.				"So, I kind of used hexagons. [. . .] It's a fantastic shape because they all just slot together. [. . .] it's brilliant because you waste very little. [. . .] our shapes just happened to be so ergonomic as well and almost by accident. [. . .] We also used this one as a bag handle as well as a necklace. [. . .]. And then, we also had made all of our buttons so that all our buttons in the collection match the outside fabrics. And, these were all made in London as well." (London, 2009).

Attract. Inspiring others was important to all three protagonists. They often shared their inspiration with their customers through education. They frequently reflected on the opportunity for everyone—including themselves and society at large—to become more conscious about their choices. Toronto "gets essences and feelings from everything [. . .] is constantly researching all of it and taking it all in. And then it's coming out of me the way I communicate, which is with the clothing." She echoes Vancouver's point that evolution is gradual, but also emphasizes the need to accelerate this progression.

Attend. Shecopreneurs also sensed, captured, and responded to others' emotions, typically those of their customers. "You can see it in someone's eyes when they have something on it's, like, that makes them feel the way they want to be feeling [. . .]. And, that's a really awesome experience."

Effect

Being effective was important to our three shecopreneurs. They all spoke about commercial viability and overcoming setbacks and shortages; they sought seed money and money for growth and patient funding; they shared expansion plans, online and retail, and eventual ambitions for completing global supply chains that would connect producers of organic textiles in developing countries with ethical fashionistas in Europe and North America. Across their narratives we identified three sets of first-order effective practices: *anchor, align,* and *amplify.*

Anchor. Our protagonists constantly (re)appraised changes in economic constraints, goals, and abilities. Each had a reference point. Vancouver, for example, explained that "we are a little bit in between, like, the eco, which is a little bit more casual and a lower price point, and then the fashion, which can handle the higher price point, but doesn't understand the eco. So, we're somewhere in between."

Align. Shecopreneurs rearranged their economic activities around relevant economic constraints, goals, and abilities. For example, Toronto—a staunch advocate of local sourcing whenever possible—considered working "with some people in Ethiopia and have them making stuff and coming up with a product that will be sellable here because they're disenfranchised from their community, which would be great to be able to help people and kind of spread the wealth here around a bit more."

Amplify. Our protagonists also expanded and on occasion shrank their activities to keep pace with their economic constraints, goals, and abilities. Toronto, for example, constantly weighed "a multitude of a billion things that cause me to make a decision, whether it's conscious or subconscious."

Art

All three protagonists developed a growing aesthetic appreciation of ecological dimensions; as they did, they reimagined their products and production processes. Toronto initially practiced traditional fashion, but quickly moved to 100 percent sustainable fabrics. Vancouver evolved her aesthetics starting with sustainable fabrics. We identified three sets of first-order aesthetic practices: *authenticate, adopt* and *associate*.

Authenticate. Expressing oneself artistically is quintessential—"when you know what you want in your head and you have no restrictions on that, it makes it really easy to flow something out," Toronto explains. All our protagonists developed a deep sense of authenticity, which imbues their products and production processes—they designed for themselves.

Adopt. Our protagonists redesigned products and production processes in ways that respect and replicate their evolving aesthetic appreciation of ecological dimensions. Vancouver patterned life—at different stages. London adopted the hexagon shape, which she reinterpreted in blue, red, and white. Toronto borrowed the darkness of tar sands to signal our vulnerability and need for protection.

Associate. Art is a means to a greater end. Through unique aesthetics, our protagonists sought to stimulate more ecologically mindful patterns of consumption that reflect their—and their customers'—growing aesthetic appreciation for specific ecological dimensions.

Last, we iterated back and forth between the theory, the macro-practices of ecosystem design, and the micro-practices of entrepreneurial becoming to understand how shecopreneurs reclaimed responsibility for global ecosystems. We observed a strong link between self-transformation and ecosystem design. Our framework, shown in Figure 9.1, suggests that by reimagining the real and realizing the imaginary, shecopreneurs organically designed and developed more responsible ecosystems.

All our protagonists identified and actively managed who they were becoming as women/ethical designers/entrepreneurs; these micro-practices

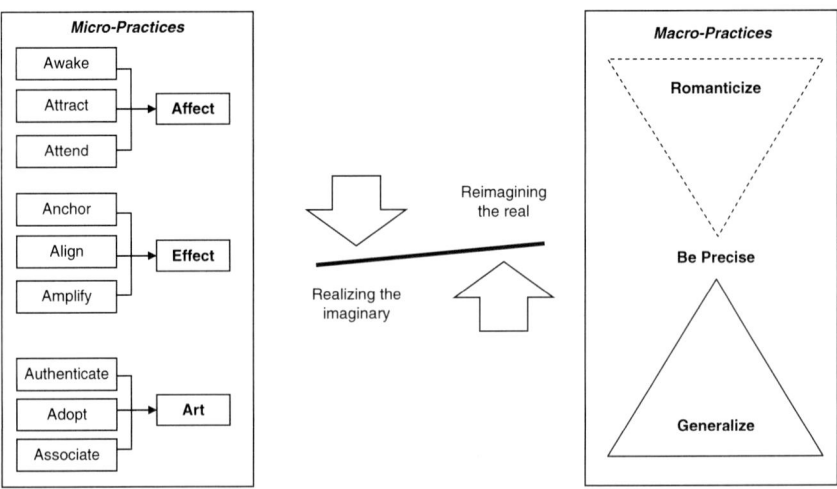

Figure 9.1 Practices of ecosystem design.

of entrepreneurial becoming shaped their products or production processes. Vancouver, for example, was keen to grow her social ecosystem to mills abroad, but couldn't source organic cotton from India because this would require a trade-off in her commitment to fair labor practices. Overcoming these disconnects sometimes required a leap of faith. The transition from their individual becoming to ecosystem design was often deliberate; all three shecopreneurs reflected on the transition as it happened. Our protagonists often saw these transitions as organic: they grew into the design of ecosystems as a natural extension of their individual becoming. Sometimes this link was also recursive: reclaiming more responsibility for local and global ecosystems prompted further self-transformation. Recursive links tied the imaginary and the real together—either by reimagining the real or by realizing the imaginary; nonrecursive links were either imaginary or real, but never both.

Vancouver developed her Healing Heart collection during her divorce; she had just had a baby, moved out of her home, and was working from her parents' house. Keeping the business running part-time was important to her, but she didn't have a lot of time nor energy for putting together a new collection. So she simply revamped some of her best-selling pieces. She already looked forward to her next collection, which helped her reimagine her next stage of becoming.

Toronto's Fall 2010 collection was inspired by a movie about the Alberta Tar Sands. She deliberately chose fabrics that were "all black and dark: it was this organic cotton with beeswax that looked like the tar sands oils. The dark shades captured her feeling that "the world's in this really dark place right now." She imagined herself as a warrior and designed her Fall 2010 for this imagined warrior.

Implications

Our conceptual framework offers three important insights. First and foremost, we propose and show that ecosystems are neither separate nor separable from individuals. Ecosystems are constantly reimagined and realized through practices. Reconceptualizing them as social imaginaries draws attention to the shifting and uneven interface between individuals and ecosystems. Humankind's damaging effects notwithstanding, subgroups of individuals, like the three shecopreneurs we study in this chapter, pursue their duty of care to gradually reclaim responsibility. They commit and create more sustainable ecosystems.

Second, we suggest that individuals (re)claim responsibility for global ecosystems through a gendered and gendering set of practices. Not only is the duty of care typically associated with the feminine energy of the Earth, but it is exercised through combinations of affect, effect, and art that require a broader palette of theories and theorizing that organizational studies currently afford.

Third, the individual strides we discuss in this chapter have not yet been fully realized. All social change starts small. The notion of social imaginaries makes us collectively alert to exemplary practices, like those showcased by the three shecopreneurs whose stories helped us articulate and illustrate our conceptual framework. The more attention we pay to such exemplary practices through engaged research, the quicker and broader their impact on society may become.

Acknowledgments: Financial support from the oikos Foundation, the Social Sciences and Humanities Research Council Grant 865-2008-0020 (Designing Eco-Social Organizations) and the Richard Ivey School of Business, The University of Western Ontario, is gratefully acknowledged.

References

Bronfenbrenner, U. (1979). *The ecology of human development: Experiments by nature and design.* Cambridge: Harvard University Press.

Calás, M., Smircich, L., & Bourne, K. (2009). Extending the boundaries: Reframing 'entrepreneurship as social change' through feminist perspectives. *Academy of Management Review, 34*(3), 552–569.

Campbell, K. (2006). Women, Mother Earth and the business of living. In D. Hjorth & C. Steyaert (Eds.), *Entrepreneurship as social change* (Vol. 3, pp. 165–187). Cheltenham: Edward Elgar Publishing.

Carlson, A. (2000). *Aesthetics and the environment: The appreciation of nature, art and architecture.* Routledge: London & New York.

Cixous, H. (1976). *The exile of James Joyce.* John Calder: London.

Ewick, P., & Silbey, S. S. (1995). Subversive stories and hegemonic tales: Toward a sociology of the narrative. *Law & Society Review, 29,*197–226.

Ferrigno, S. (2008). *Farm development report.* San Francisco: Organic Exchange.

Gnanadason, A. (2005). *Listen to the Women!* Geneva, Switzerland: World Council of Churches.

Hjorth, D., & Steyaert, C., eds. (2004). *Narrative and Discursive Approaches in Entrepreneurship.* Cheltenham, UK: Edward ElgarLimited.

Lattas, J. (1991). French feminisms. In P. Beilharz (Ed.), *Social theory: A guide to central thinkers.* North Sydney: Allen & Unwin Pty Ltd.

Lustermann, D. (1985). An ecosystematic approach to family-school problems. *American Journal of Family Therapy 13*(1), 22–30.

Lyle, J. T. (1999). *Design for human eco-systems: Landscape, land use and natural resources.* Washington, DC: Island Press.

Nicholls, A., ed. (2008). Social entrepreneurship: New models of sustainable social change. Paperback Edition (with new preface), Oxford, UK: Oxford University Press.

Paolucci, B., Hall, O., & Axinn, N. (1977). *Family decision-making: An ecosystem approach.* New York: John Wiley and Sons.

Pardeck, J. (1988). Social treatment through an ecological approach. *Clinical Social Work Journal, 16*(1), 1573–3343.

Plowman, D., Baker, L., Beck, T., Kulkarni, M., Thomas, S., & Villareal, D. (2007). Radical change accidentally: The emergence and amplification of small change. *Academy of Management Journal, 50*(3), 515–543.

Rindova, V., Barry, D., & Ketchen, D. (2009). Entrepreneuring as emancipation. *Academy of Management Review, 34*(3), 477–491.

Ruether, R.R. (1992). *Gaia & God: An ecofeminist theology of earth healing,* San Francisco: HarperSanFrancisco.

Slocombe, D. (1993). Environmental planning, ecosystem science and ecosystem approaches for integrating environment and development. *Environmental Management,* 17(3), 289–303.

Sonenshein, S. (2010). We're changing—or are we? Untangling the role of progressive, regressive, and stability narratives during strategic change implementation. *Academy of Management Journal,* 53(3), 477–512.

Steyaert, C. (2004). The prosaics of entrepreneurship. In D. Hjorth and C. Steyaert (Eds.), *Narrative and discursive approaches in entrepreneurship* (Vol. 2, pp. 8–21). Cheltenham, UK: Edward Elgar.

Steyaert, C. (2007). Entrepreneuring as a conceptual attractor? A review of process theories in 20 years of entrepreneurship literature. *Entrepreneurship and Regional Development,* 19(6), 453–477.

Taylor, C. (2005). *Modern social imaginaries.* Duke University Press, Durham North Carolina, US

Thopte, I. (2009). A profile of ethical fashion entrepreneurs. London College of Fashion, London, UK

Whiteman, G., & Cooper, W. (2000). Ecological embeddedness. *Academy of Management Journal,* 43(6), 1265–1282.

PART III

Global Supply Chains Matter

In a globalized world, where information is easy to access, supply chain management has acquired growing importance as a key driver of companies' success in many different industries. Consumers and nongovernmental organizations (NGOs) are becoming more and more interested in the origins of products and the provenance of raw materials, asking companies for additional transparency. Managers have started to understand that their environmental and social responsibilities include not only the direct impact generated by their own manufacturing activities, but also the indirect effects generated upstream and downstream by different tier suppliers, retailers, and consumers. To thrive in the green economy, companies must transform their business models. This dramatically affects the way in which supply chains are designed, managed, and controlled. Moreover, this issue also influences firm internationalization strategies and export decisions.

Using different theoretical perspectives and methodological approaches, the chapters in this section address the topic of how and why sustainability standards are becoming relevant for companies in managing their supply chains. One chapter looks at endogenous factors, exploring the relationship between firm capabilities and the institutionalization of supply chain sustainability standards. Another chapter investigates how exogenous features related to the supply chain structure influence firms to engage and adopt specific socially responsible supplier practices. Another perspective is taken by the other two chapters of the section, which focus on the relationship between internationalization strategies and firm environmental performance. Can environmental performance influence the firm's export activity? What are the relationships between environmental practices, innovation, and the firm's internationalization processes? These chapters provide clues that enhance our understanding of the multidimensional challenges that firms face in the transition towards a more sustainable economy.

CHAPTER TEN

Institutionalizing Proactive Sustainability Standards in Supply Chains: Which Institutional Entrepreneurship Capabilities Matter?

Jörg H. Grimm, Joerg S. Hofstetter,
Martina Müggler, and Nils J. Peters

External stakeholders have built up sustainability consciousness and expectations, putting companies under regular surveillance by nongovernmental organizations (NGOs) and by the media (Doh & Guay, 2006). Stakeholders often do not differentiate between a company's operations and its suppliers' operations; they hold the company responsible for all practices involved in the making of the product, including any potential sustainability concerns (Rao, 2002). Thus, suppliers not complying with the company's promised values are likely to damage corporate reputation or harm customer confidence. Levi's, Nike, and Mattel are prominent examples that show how brands can suffer as a result of using noncompliant suppliers (Wagner, Lutz, & Weitz, 2009). A proactive supply chain sustainability strategy is therefore vital (Handfield, Scroufe, & Walton 2005; Rao & Holt, 2005). To implement such a strategy, companies provide specific sustainability standards for their supply chains (Bansal & Hunter, 2003; Luo & Bhattacharya, 2006). These standards, known as proactive supply chain sustainability standards (PSCSS) may provide statements to comply with legal requirements and may add elements that go beyond the law. Having introduced these PSCSS, companies still face the challenge of ensuring that their supply chain partners comply. Monitoring supplier processes and assessing the quality of procured products is challenging, because global supply chains have become more complex (Matten & Moon, 2008; Roth, Tsay, Pullman, & Gray, 2008). The large number of suppliers, as well as the organizational and geographical distance between the company and its direct and indirect suppliers, hinder a company from controlling its suppliers' sustainability practices (Bremer & Udovich,

2001). Further problems may arise when suppliers are located in developing countries, as local legal standards may not conform to the main company's requirements (Detomasi, 2007).

Institutional entrepreneurs "create technical and cognitive norms, models, scripts and patterns of behavior consistent with their identity and interests, and establish them as standard and legitimate to others" (Dejean, Gond, & Leca, 2004, p. 743). What role do institutional entrepreneurs play in establishing PSCSS (Peters, Hofstetter, & Hoffmann, 2011)? They act on institutional fields composed of "diffused practices, technologies, or rules that have become entrenched" (Lawrence, Hardy, & Philips, 2002, p. 282). The institutional field on which companies operate includes both direct and indirect supply chain partners (Hargrave & Van de Ven, 2006). Companies can create "proto-institutions" that have normative, mimetic, and coercive dimensions (King & Lenox, 2000; Matten & Moon, 2008). Yet, according to Lawrence et al. (2002, p. 283), the PSCSS must become more than a proto-institution, it must become a "full-fledged" one that is entrenched and diffused throughout the field. Once inherent norms, cognitive schemata, and rules are accepted by the field, full institutionalization takes place (Matten & Moon, 2008), which is reflected by the compliant behavior of supply chain partners (Peters, 2010; Peters, Hofstetter & Hoffmann, 2011). In this chapter, we examine the institutional entrepreneur's (the focal company's) attempt to achieve the compliance of supply chain partners.

Wright, Filatototchev, Hoskisson and Peng (2005, p. 25), as well as Hamprecht and Sharma (2006), called for research to investigate the capabilities that enable an institutional entrepreneur to change the existing institution or to create a new institution successfully. Following this call, Peters, Hofstetter, and Hoffmann (2011) examined which key capabilities are specifically needed to develop "voluntary sustainability initiatives" in supply chains. Their work mainly covers the first phases of the institutional entrepreneur's endeavor, such as the design phase and the collective action plan phase (Hargrave & Van de Ven, 2006). It remains unclear what further key capabilities are required in order to institutionalize and maintain the PSCSS (cf. Battilana & Leca, 2009). Lawrence (1999, p. 168) explains that the success of an institutional entrepreneur is reflected by an ability to ultimately "influence legislative or regulatory frameworks, affect cultural norms or values, or establish some structures or processes as "taken-for-granted". Consequently, we argue that institutional entrepreneurs are only successful in institutionalizing the PSCSS if they can change the institutional field, with the optimal end result that supply chain partners comply with the PSCSS. This leads us to the question: What capabilities do institutional entrepreneurs require to institutionalize "proactive supply chain sustainability standards" (PSCSS) effectively, ultimately leading to their supply chain partners' compliance with the PSCSS?

The *resource-based view* (RBV) outlines several criteria for key capabilities that contribute to achieving long-term competitive advantages.

These capabilities must be valuable, rare, inimitable, and nonsubstitutable (Barney, 1991; Peteraf, 1993). The capabilities enable a company "to conceive and implement strategies that improve its efficiency and effectiveness" (Barney, 1991, p. 101). The value of a capability is in its effectiveness in achieving targeted institutional change (George, Chattopadhyay, Sitkin, & Barden, 2006). Thus, the higher the value of a capability, the greater is its contribution to institutional change (Hargrave & Van de Ven, 2006). A number of capabilities are described in the literature that enable a company to successfully introduce sustainability strategies in supply chains. Some examples are: "supply chain environmental management" (Rao, 2002; Zhu & Sarkis, 2004), "total quality environmental management" (Shrivastava, 1995), "environmental collaboration" (Vachon & Klassen, 2008), and "supply chain implementation" (Peters, Hofstetter & Hoffmann, 2011). However, these capabilities do not fully correspond with the above-described challenge of ensuring supply chain partners' compliance with PSCSS.

Compliance might be achieved by measures such as (1) contracts, (2) assessments, (3) supplier development, and (4) monitoring (e.g., Carter & Jennings, 2002; Frenkel & Scott, 2002; Rao, 2002; Vachon, 2007). However, the "mere" execution of these compliance management activities and measures appears to produce only limited results (Locke, Qin, & Brause, 2007, p. 2). A focal company also requires organizational capabilities (cf. Christmann, 2000) to ensure compliance with its PSCSS. For this reason, our case study research sought to identify the capabilities that act as enablers of the above-mentioned activities and measures to achieve supply chain partner compliance.

Case Selection and Analysis

We applied the following criteria: First, we selected cases where there was an established proactive supply chain sustainability initiative with implemented PSCSS. Second, we chose companies with average or above-average supply chain complexity. Further, we looked for cases in which there had been institutionalization problems that ultimately were resolved. This helped us identify capabilities that lead to success. We selected four cases upon which to base our analysis: two cases in the food/retail industry and two in the paper industry. We conducted two case studies (Musgrave and SKG) by ourselves; the other two (Migros and Axel Springer) were based on secondary data from previous projects (Hamprecht, 2006; Peters & Schaupp, 2009; Peters, 2010; Peters, Hofstetter, & Hoffmann, 2011). The secondary data were complemented by interviews with key persons to include missing information with respect to our research question.

We applied a three-step data collection process for construct validity (cf. Gibbert, Ruigrok, & Wicki, 2008). First, we collected secondary data from each company and its suppliers. Second, we conducted interviews

Table 10.1 Theme analysis of intraorganizational and interorganizational capabilities needed for compliance to PSCSS

First-Order Themes	Second-Order Themes	Final Themes*
Ability to clearly formulate sustainability values and visions	Ability to communicate with supply chain partners about corporate sustainability	Interfirm dialogue (Black and Härtel, 2004, adapted from "dialogue" subsection)
Ability to explain sustainability standards and policies towards supply chain partners		
Ability to outline expectations towards suppliers	Ability to gain mutual understanding concerning sustainability	
Ability to identify gaps between own and suppliers' understanding concerning sustainability factors	Ability to explain relevance of corporate sustainability program and persuade suppliers	
Ability to make suppliers understand the purpose of PSCSS and the sustainability standard itself		
Ability to understand suppliers' behavior and practices	Ability to provide and receive feedback	
Ability to demonstrate relevance of sustainability standards and persuade suppliers to comply with them	Ability to adapt communication to specific cultural and local needs	
Ability to sensitize suppliers to sustainability factors		
Ability to communicate findings from auditing activities to suppliers, leading to specific actions concerning the development of supplier capabilities		
Ability to share information about sustainable practices with suppliers		
Ability to take cultural context and local specificities into consideration during interactions		
Ability to recognize sustainability issues regarding direct suppliers	Ability to identify sustainability risks	Risk management (Foerstl, Reuter, Hartmann and Blome, 2010, adapted from 'Supplier Sustainability Risk Management')
Ability to anticipate sustainability issues in upstream supply chain processes	Ability to assess impact of sustainability issues	
Ability to map out entire supply chain	Ability to prioritize sustainability risks	
Ability to identify root causes of sustainability issues in supply chain	Ability to mitigate sustainability risks through the application of appropriate mechanisms and resources	
Ability to assess supply base concerning business risk and the impact resulting from sustainability issues in the supply chain		
Ability to assess sustainability threads in upstream supply chain processes		
Ability to preassess critical suppliers or components who might hide noncompliant business practices		
Ability to prioritize impact of identified sustainability issues		
Ability to transfer sustainability requirements into supplier selection criteria		
Ability to streamline supply chain (reduce supply base and focus on most capable suppliers)		
Ability to anticipate potential sustainability issues in supply chain		
Ability to develop proactive solutions for foreseen/upcoming sustainability issues		

Ability to select best-fitting stakeholders/partners	Ability to build relationships with strategic stakeholders	External stakeholder collaboration (Sharma and Vredenburg, 1998, adapted from 'Stakeholder Integration')
Ability to select and build up relationships with strategic stakeholders	Ability to share tacit knowledge with strategic stakeholders	
Ability to maintain frequent dialogue with stakeholders	Ability to integrate stakeholders to solve sustainability issues	
Ability to continuously exchange experiences and share knowledge with stakeholders (e.g. suppliers, NGOs)	Trust of strategic stakeholders	
Ability to analyze chain-of-custody by including supply chain partners to increase insights		
Ability to solve sustainability issues collaboratively with stakeholders		
Stakeholders' trust in company's competence to approach sustainability factors		
Stakeholders' trust in company's "sustainability vision"		
Ability to provide fair supplier treatment		
Ability to form project teams working on sustainability factors with representatives from various affected corporate functions	Ability to coordinate affected corporate functions for the implementation of sustainability standards in the supply chain	Cross-functional integration (Verona, 1999)
Ability to exchange experiences on sustainability factors from different functional perspectives	Ability to bundle competencies of affected corporate functions to approach sustainability issues	
Ability to perform the evaluation of sustainability factors jointly with affected corporate functions		
Ability to integrate affected corporate functions for solving sustainability issues in supply chains (e.g., integration of sourcing experts into environmental teams)		
Ability to integrate the competencies of affected corporate functions for the implementation of sustainability standards in the supply chain (e.g., supplier training)		
Ability to exploit feedback from stakeholders concerning sustainability practices	Ability to exploit feedback and lessons learned	Continuous improvement (Hart, 1995; Benner and Tushman, 2003)
Ability to identify best practices and improve sustainability policies accordingly	Ability to assess current supply chain processes with respect to their social and environmental performance	
Ability to incorporate experiences from previous "sustainability projects"	Ability to adapt policies and standards to identified sustainability issues	
Ability to modify supply chain processes according to findings in supply chains concerning sustainability issues	Ability to improve supply chain processes with respect to social and environmental performance	
Ability to adopt purchasing practices by incorporating sustainability factors		
Ability to consider sustainability factors in new product development		
Ability to improve compliance management activities (e.g., supplier audits) to increase the likelihood of detecting potential noncompliance to sustainability standards		

*Final themes represent the linkage to existing literature.

with key personnel responsible for the institutionalization of the PSCSS, such as senior sustainability and purchasing managers. Further key informants were identified by following the snowball principle (Sharma & Vredenburg, 1998). Interviews were transcribed, verified by interviewees, and subsequently analyzed to enable early identification of emerging results (Yin, 2003). Then, narrative accounts were explicitly analyzed for discrepancies and if any were identified, further data were consulted to obtain the "true story" (Pentland, 1999). Emerging concepts with respect to the targeted identification of capabilities were categorized and were constantly compared during these data collection steps (Eisenhardt, 1989). We consolidated key quotes and crafted structured maps, studied theories that could explain emerging concepts, and combined our empirical data with concepts in the literature (Gibbert, Ruigrok, & Wicki, 2008). Quotes retrieved from data collection comprised the first-order themes; second-order themes summarized the respective quotes; and the final themes represented the linkage to the existing literature (Sharma & Vredenburg, 1998), as illustrated in Table 10.1.

Institutionalization of Proactive Supply Chain Sustainability Standards

This section provides an overview of the four cases and their context, followed by the presentation of the organizational key capabilities that contributed to the institutionalization of the proactive sustainability standards in the companies' supply chains.

Migros. This company, a major Swiss retailer, worked with the International Roundtable on Sustainable Palm Oil (RSPO), the World Wildlife Fund (WWF), and other organizations to establish sustainable palm oil standards. Migros and its supply chain members, which are mainly based in Indonesia and Malaysia, were expected to improve their business practices according to principles and criteria defined by the RSPO. The criteria include defined indicators that make compliance verifiable. An underlying certification system guarantees sustainable production throughout the supply chain, which is made up of producers, processors, traders, consumer goods manufacturers, and retailers. Migros received the World Business Award in 2002 for its successful work toward sustainable palm oil production, which includes the avoidance of slash-and-burn farming.

Musgrave. This major privately owned Irish company partners with independent food retailers operating under its brand in Ireland, the United Kingdom, and Spain. Musgrave was the first Irish company that ratified the UN Global Compact, and it emphasized its commitment to sustainable business practices in its supply chains. To respect this commitment, Musgrave established the Musgrave Ethical Trading Policy, which is also binding on its suppliers. Musgrave's supplier base can be classified into three types: global brand manufacturers (e.g., Coca-Cola), nonexclusive suppliers, and exclusive suppliers. Various suppliers of all three categories deliver products for Musgrave's own-brand product assortment. The trading policy outlines sustainability standards and seeks to especially shield its own-brand products from potential damage through noncompliant supply chain partners.

Smurfit Kappa Group (SKG). This company is one of the biggest paper-based packaging producers in Europe and Latin America as well as an operator of various paper mills. The industry in which it participates is one of the largest users of industrial water and fossil fuels. Thus, sustainabilty is an integral part of SKG's business and supply chain strategy. SKG seeks to avoid any purchases of wood from controversial sources and simultaneously respects international treaties and domestic law. The company established sustainability standards, rooted in its Sustainable Forestry Policy Statement, which are applicable to its direct suppliers and also to indirect suppliers. For example, its suppliers must guarantee that pulp is sourced only from mills that can prove chain-of-custody certification or other credible certification schemes. These international forest certification schemes are monitored by independent accredited organizations. In following this path, SKG became a role model for other members in the paper industry.

Axel Springer. As a major German publisher, Axel Springer recognized the many environmental, social, and reputational risks resulting from non-sustainable paper production, since this could be directly linked to Axel Springer as a customer of the paper industry. The optimization of supply chain processes to be as sustainable as possible became a company priority. A main focus of Axel Springer's efforts was Russian logging practices. Together with one of its main paper suppliers, Axel Springer established guidelines and standards for sustainable supply chain practices. The objective was to sensitize suppliers to sustainability factors, to change suppliers' mindsets, and to make processes within the paper supply chain transparent, from the raw material supplier to the publisher.

In all cases, companies performed compliance management activities and measures such as supplier monitoring or supplier development to achieve supply chain partners' compliance with the respective PSCSS.

In line with institutional entrepreneurship (IE) theory and the RBV, we identified a set of capabilities that companies used to achieve compliance (cf. Dacin, Goodstein, & Scott, 2002; Lawrence, 1999; Barney, 1991; Dyer & Singh, 1998). Table 10.1 summarizes the identified capabilities that resulted from the analysis.

Interfirm Dialogue. The interviewees emphasized the importance of open dialogue between their company and supply chain partners. The companies communicated PSCSS-related objectives and made sure that requirements were understood. For example, Musgrave checked via interaction and discussions with suppliers on potential gaps between Musgrave's own understanding and suppliers' understanding of the different sustainability factors, instead of just informing the suppliers about requirements. The explicit consideration of the local and cultural context is considered key for effective communication and eventually, for a common understanding. The Migros case highlighted that a trustful dialogue allows information-sharing and is the basis for any improvement in sustainability practices. Proper interfirm dialogue was not only considered key for presenting, explaining, and demonstrating the importance of sustainability standards when introducing the PSCSS, but also for later phases when assessing and auditing supply chain practices at suppliers' sites. In all cases, gaps and poor conditions that were identified in supplier assessments and audits lead to improved practices when appropriately reported afterwards. Interfirm dialogue creates a common understanding of sustainability standards and factors, motivates suppliers to follow sustainable practices, enables the development of suppliers' societal and environmental capabilities, and increases the probability that suppliers will adapt their business practices according to defined requirements.

The rareness and inimitability of interfirm dialogue in this context can be illustrated by the limited availability of personnel who have experience in the field of corporate sustainability and supply chain management to perform these dialogues effectively. Interfirm dialogue is a two-way process that breaks down existing assumptions, uncovers shared meanings, and facilitates collective learning in the field of corporate sustainability by exchanging arguments and experiences (cf. Burchell & Cook, 2006). In a different context, it has also been argued that a company's ability to engage in dialogue with its stakeholders contributes to its success (Black & Härtel, 2004). This pattern finds support in IE literature where institutional entrepreneurs' discursive skills (Maguire & Hardy, 2006; Munir & Phillips, 2005) and communication skills (Bansal & Clelland, 2004; Suchman, 1995) have been acknowledged.

Risk Management. Nearly all interviewed managers reported that they faced the challenge of deciding how they should efficiently control for suppliers' compliance with their company's sustainability standards. They aimed for a guarantee throughout the entire supply chain, which requires that all supply chain partners be audited and monitored regularly. However, the case study companies do not have sufficient financial and human resources to audit all direct suppliers. Only when they decided to differentiate among their suppliers, and apply compliance-management activities of different scales as needed, did control become economically feasible. For instance, SKG initially relied on undifferentiated compliance management activities to ensure sustainability compliance, but then changed its approach when it reevaluated its supplier base according to sustainability risks it identified. The other cases also illustrate how companies started to follow approaches that structured their supply chains and prioritized suppliers that should be audited and monitored regularly. Companies had to increase their knowledge of practices within their supply chain, and had to increase the awareness of potential sustainability issues in order to preassess critical suppliers, which might hide noncompliant business practices. Musgrave, for instance, identified critical paths by mapping its supply chains. Consequently, it categorized its suppliers in accordance with the risk associated with a sourced product and the potential business impact if noncompliant behavior were revealed. The categorization ranged from "very high risk" to "very low risk". High-risk suppliers were audited on a regular basis with optional supplier development programs; low-risk suppliers were only asked to fill in self-assessment questionnaires as a monitoring tool. Musgrave's approach enabled the efficient use of limited resources to maximize control over suppliers' sustainability compliance. To further reduce risk of contracting with noncompliant suppliers, Migros began very early to include sustainability aspects into its supplier selection criteria, which helped it to select strongly sustainability-oriented suppliers and to minimize later monitoring efforts (cf. Tang, 2006). As a positive side-effect, purchasing managers reported that suppliers that were not selected stated their intention to improve their sustainability practices in the future in order to be considered during upcoming tenders. Literature on RBV and IE hardly refers to the concept of risk management (cf. Battilana, Leca & Boxenbaum, 2009; Foerstl, Reuter, Hartmann, & Blome, 2010; Peters, 2010). However, we observed in our cases that the capability of risk management permitted a rigorous prioritization of sustainability risks in the supply chain, which made it possible to implement auditing and supplier development programs of different scales, in accordance with the prioritized risks. This enabled an efficient use of limited resources to institutionalize the PSCSS (cf. Neilson & Pritchard, 2007). The rareness and path-dependence of risk management in the context of sustainable supply chain practices is reflected by the limited availability of relevant experience from which the companies could draw (Millington, 2008). Inimitability is shown by the respective supply chain specificities that must be considered and by the comprehensive adaptive learning routines of risk-management-related activities. All companies followed similar stages in performing risk management: identification, analysis, and response (cf. Borge, 2001; Kutsch & Hall, 2009; Raftery, 1994). Risk management in our context is the identification, assessment, and prioritization of sustainability-related risks, followed by the aligned and efficient application of resources to examine and to minimize the probability of and/or impact of unwanted noncompliance (cf. Foerstl et al., 2010).

External Stakeholder Collaboration. The companies repeatedly highlighted their collaboration with various strategic stakeholders such as NGOs, auditors, and suppliers, as positively contributing to the institutionalization of proactive sustainability standards in their supply chains. Migros's ability to identify and build up strong relationships with credible and competent NGOs enabled it to acquire further knowledge in sustainability practices, which was essential for the joint development of environmental management and certification systems and their subsequent transfer into suppliers' supply chain practices. This included the ability to integrate external auditing bodies, since from an early phase on the necessary competence to audit in accordance with the PSCSS was not available in house. Similarly, Axel Springer built up strong partnerships with key suppliers, which were considered a prerequisite for successfully institutionalizing its PSCSS. This collaborative approach enabled both sides to combine different aspects of the supply chain and to explore the counterpart's competencies during joint activities. Together, they approached indirect suppliers in the upstream supply chain with "one voice", leading to an increased credibility with respect to sustainability practices.

In all cases, we observed that the integration of competent stakeholders concerned with sustainability allowed companies to explore missing internal knowledge about sustainability issues and let them bundle forces for joint efforts in the supply chain. However, building relationships usually followed long and intensive interactions. The rareness of the capability of external stakeholder collaboration is reflected by the limited availability of stakeholders (e.g., NGOs) that are willing to build such relationships and that also can provide targeted sustainability competencies. Established trust with external stakeholders is furthermore path-dependent and "cannot be easily imitated by competitors" (Sharma & Vredenburg, 1998, p. 740). We define "external stakeholder collaboration" as the cooperation of strategic stakeholders in which sustainability-concerned solutions are jointly developed and implemented in supply chains by making use of each other's knowledge, resources, and competencies (cf. Olden, 2003). The capability of external stakeholder collaboration enables the identification of competent partners and thus effective cooperation with those (Sharma & Henriques, 2005). Further, it facilitates the establishment of trusted relationships (Oliver & Holzinger, 2008) as a basis for the subsequent exploration of external stakeholders' tacit knowledge and competencies (Lavie & Rosenkopf, 2006; Rothaermel & Deeds, 2004).

Cross-Functional Integration. In all cases, interview partners considered it crucial that the institutionalization of PSCSS be run by cross-functional teams, integrating the different perspectives of corporate functions. All studied companies had included experts with different backgrounds, such as purchasing, law, communications, and quality, in "sustainability task forces". SKG emphasized that its sustainability group first struggled with the challenges of understanding the existing supply chain configurations, since it lacked related knowledge. After the integration of dedicated purchasing and supply chain personnel, SKG was much more effective in solving sustainability issues that had their origin in the upstream supply chain.

Migros also noted that it was only able to solve certain issues by the systematic approach of its cross-functional sustainability team. The team was able to bundle necessary knowledge relevant to specific problems that was not exclusively available in one single corporate function. These cases support the trend for cross-functional teams to become common management practice. Nevertheless, the majority of interviewed managers stressed that underlying processes and management systems were specifically adopted in given interorganizational configurations. The capability of "cross-functional integration" was described as being causally ambiguous and socially complex, since various corporate functions with different personnel, objectives, and tasks were involved (Peters, Hofstetter, & Hoffmann, 2011). The RBV literature mainly discusses "cross-functional integration" in the context of new product development processes. It can be defined as a capability that serves "as adhesive by absorbing critical knowledge from external sources and by blending the different technical competencies developed in various company departments" (Verona, 1999, p. 134). We observed that in most cases, "cross-functional integration" is able to include and to coordinate representatives from various sustainability-concerned corporate functions, facilitating the bundling of different areas of expertise (Brown & Eisenhardt, 1995; Eisenhardt & Martin, 2000), which effectively supports the institutionalization of PSCSS.

Continuous Improvement. The companies studied continuously improved their supply chain processes as well as the applied methods and tools that subsequently contributed to their suppliers' compliance with the respective PSCSS. Musgrave emphasized that it makes use of the input and feedback it receives from its trading managers and from various external stakeholders. Musgrave's sustainability team was thus able to incorporate recognized best practices into its sustainability policy.

At Axel Springer, we observed how policies and principles were improved by incorporating experiences from former projects. Frequently, interviewees mentioned that Axel Springer learned from project collaborations with suppliers and vice versa—for example, identified and subsequently analyzed sustainability issues were incorporated into improved management systems, thus closing gaps. Axel Springer and Migros reported their ambitions for continuous improvements to the methods and tools that they used for supply chain assessment and auditing. Axel Springer tried to accelerate its learning curve and incorporated past experiences into better methods for risk assessment with respect to legality and health, safety, and environmental (HSE) risks. Similarly, Migros sought continuing innovation in its auditing mechanisms to reveal suppliers' potential noncompliant business practices. We observed how Migros analyzed findings that it gained in one supply chain and applied them successfully in other supply chains. This, for example, lead to better purchasing practices through adopted selection processes and influenced new product development. The rarity of approaches to improve sustainability performance was reflected by the limited availability of experts and partners with valuable experience in sustainability practices who could contribute to improvements. The adaptive learning routines we observed that lead to improvements underline the inimitability of each company's solutions. The capability for continuous improvements that was enabled by intraorganizational and interorganizational routines means that valuable tacit knowledge was made explicit (Brown & Duguid, 1991), resulting in improved compliance management activities and measures, which consequently improved the sustainability performance in the supply chains. We define "continuous improvement" as the ongoing effort to improve processes, policies, and products in terms of social and environmental performance by the evaluation of current practices and the incorporation of feedback and lessons learned (cf. Hart, 1995; Benner & Tushman, 2003).

Summary and Conclusions

This research focused on the institutionalization of a company's proactive sustainability standards in supply chains after those standards had been defined. Although the importance of the institutional entrepreneur's capability to drive institutional change is widely acknowledged (DiMaggio, 1988; Oliver, 1991; Phillips, Lawrence, & Hardy, 2004; Powell, 1988; Suchman, 1995; Zimmerman & Zeitz, 2002), researchers explicitly called for a more systematic approach to examine such capabilities (Battilana & Leca, 2009; Hamprecht & Sharma, 2006; Wright et al., 2005) for which the resource-based view (RBV) provides appropriate concepts. Specifically, by conducting four comparative case studies, we sought to examine capabilities that enable the institutional entrepreneur to implement previously defined proactive sustainability standards within its supply chain. We were able to identify five key capabilities which contribute to the institutional change, namely (1) interfirm dialogue, (2) risk management, (3) external stakeholder collaboration, (4) cross-functional integration, and (5) continuous improvement. We propose that these capabilities are positively related to the institutionalization of PSCSS, which is reflected by supply chain partners' compliance. The study contributes to institutional entrepreneurship (IE) literature by putting a stronger focus on the factors that facilitate the final institutional change (Battilana & Leca, 2009; Battilana, Leca, & Boxenbaum 2009). Although our case study research notes key capabilities in a company that support the effective institutionalization of proactive sustainability standards in its supply chains, further research is required. While we observed influential linkages between identified capabilities within the case studies (e.g., the capability for "continuous improvement" was often associated with "crossfunctional integration" or "external stakeholder collaboration"), the analysis of collected data did not permit the development of propositions about the complementarities of identified capabilities, and our case studies have not addressed yet how external contingencies (e.g., Aragón-Correa & Sharma, 2003; Sharma, Aragòn-Correa, & Rueda-Manzanares, 2007) influence the value of capabilities in our research context. Furthermore, our cases were limited to the retail and paper industry; thus, studying other industries may reveal additional insights. Testing the propositions against a large set of data that include other industries would allow us to draw generalized conclusions. As we relied mainly on data provided by informants from companies and their direct suppliers, future research may put a stronger focus on including indirect upstream suppliers beyond tier-1 suppliers (Lee, 2008; Millington, 2008; Vermeulen & Ras, 2006).

As global supply chains grow longer and more complex, companies face the problem of how to control supply chain partners' compliance with introduced PSCSS. In addition to execution of supply-chain compliance management measures and activities, such as supplier monitoring or auditing, companies should consider their organizational capabilities in relation to our findings.

Interfirm dialogue is essential for gaining a common understanding about visions, values, and requirements, and therefore it opposes the common practice of simply informing suppliers about the requested PSCSS. Constant dialogue between a company and its suppliers, for example during auditing processes or during respective follow-ups, reminds suppliers of the company's expectations and facilitates guidance about how to correct deficiencies. Supplier training, workshops, "supplier days", awards, etc. are appropriate platforms for addressing relevant sustainability factors. Furthermore, companies benefit by motivating their tier-1 suppliers to be in close dialogue with key subsuppliers (tier-2) to push PSCSS up the supply chain. Within the company, interfirm dialogue must be aligned among the various concernedcorporate functions (i.e., procurement, legal, HSE, etc.) to speak with one voice to supply chain partners.

Only rigorous risk management can enable efficient usage of a company's limited resources so it can maximize control over suppliers' and subsuppliers' compliance with PSCSS. Mapping and visualizing a company's supply chain lets the company identify any hidden potential sustainability risks. Supply chain categorization should be done in accordance with the risk associated with a sourced product and the potential business impact that a publicly known breach in sustainability practices would have. Factors such as characteristics of products and production processes, geographical regions, and supply chain partners' track records should be considered for this risk assessment. High-risk supply chain partners should be audited on a frequent basis, with optional supplier development programs offered, in order to reduce the risk of any noncompliance; low-risk suppliers might to a greater extent be controlled by less costly assessments (e.g., supplier self-assessments).

External stakeholder collaboration plays an important role in acquiring missing knowledge and integrating the stakeholder's forces for the successful implementation of PSCSS. Companies need to identify capable stakeholders, such as those who, for example, possess wide-ranging expertise in the fields of social and environmental sustainability issues and relevant cultural and legal issues. Relevant stakeholders could range from strategic suppliers with important contribution to NGOs or certification bodies, to specialized consulting companies. Partnering with such stakeholders typically improves companies' own sustainability performance, as well as performance of of supply chain partners' in compliance with PSCSS.

Similarly, *cross-functional integration* allows for the integration of specialists and the alignment of various corporate interfaces. People with corporate functions such as procurement or supply chain management can understand certain requirements regarding sustainability, which might be outlined by other departments and they in turn may have relevant information to share. The establishment of dedicated sustainability task forces with members from all relevant corporate functions is an especially appropriate means to holistically address sustainability issues within supply chains and to enable the joint development of solutions.

Finally, *continuous improvement* techniques contribute to the ongoing optimization of processes and policies in terms of social and environmental performance in supply chains. Companies must analyze where sustainability issues are located within a supply chain and how these issues could be positively influenced in the future—for example, the advancement of a certain product or production characteristic may positively influence the supply chain partners' practices. Many existing supplier management programs still lack the explicit consideration of sustainability factors and still need to be adapted to existing sustainability issues within supply chains. Appropriate principles and indicators must be integrated into assessment, selection, and auditing processes and must be explicitly addressed in supplier development programs.

References

Aragón-Correa, J. A., & Sharma, S. (2003). A contingent resource-based view of proactive corporate environmental strategy. *Academy of Management Review, 28*(1): 71–88.

Bansal, P., & Clelland, I. (2004). Talking trash: Legitimacy, impression management, and unsystematic risk in the context of the natural environment. *Academy of Management Journal, 47*(1), 93–103.

Bansal, P., & Hunter, T. (2003). Strategic explanations for the early adoption of ISO 14001. *Journal of Business Ethics, 46*(3), 289–299.

Barney, J. (1991). Firm resources and sustained competitive advantage. *Journal of Management, 17*, 99–120.

Battilana, J., & Leca, B. (2009). The role of resources in institutional entrepreneurship: Insights for an approach to strategic management combining agency and institutions. In L.A. Costanzo & R.B. MacKay (Eds.), *Handbook of research on strategy and foresight* (pp. 260–274). Norwell, MA: Kluwer.

Battilana, J., Leca, B., & Boxenbaum, E. (2009). How actors change institutions: Towards a theory of institutional entrepreneurship. *The Academy of Management Annals, 3*(1), 65–107.

Benner, M. J., & Tushman, M. L. (2003). Exploitation, exploration, and process management. The productivity dilemma revisited. *Academy of Management Review, 28*(2), 238–256.

Black, L. D., & Härtel, C. E. J. (2004). The five capabilities of socially responsible companies. *Journal of Public Affairs, 4*(2), 125–144.

Borge, D. (2001). *The book of risk*. New York, NY: Wiley.

Bremer, J., & Udovich, J. (2001). Alternative approaches to supply chain compliance monitoring. *Journal of Fashion Marketing and Management, 5*(4), 333–352.

Brown, J. S., & Duguid, P. (1991). Organizational learning and communities of practice. *Organizational Science, 2*(1), 40–47.

Brown, S. L., & Eisenhardt, K. M. (1995). Product development: Past research, present findings, and future directions. *Academy of Management Review, 20*(2), 343–378.

Burchell, J., & Cook, J. (2006). Assessing the impact of stakeholder dialogue: Changing relationships between ngos and companies. *Journal of Public Affairs, 6*(3), 210–227.

Carter, C. R., & Jennings, M. M. (2002). Logistics social responsibility: An integrative framework. *Journal of Business Logistics, 23*(1), 145–180.

Christmann, P. (2000). Effects of "best practices' of environmental management on cost advantage: The role of complementary assets. *Academy of Management Journal, 43*(4), 663–682.

Dacin, M. T., Goodstein, J., & Scott, R. W. (2002). Institutional theory and institutional change: Introduction to the special research forum. *Academy of Management Journal, 45*(1), 45–56.

Dejean, F., Gond, J.-P., & Leca, B. (2004). Measuring the unmeasured: an institutional entrepreneur strategy in an emerging industry. *Human Relations, 57*(6), 741–764.

Detomasi, D. A. (2007). The multinational corporation and global governance: modelling global public policy networks. *Journal of Business Ethics, 71*(3), 321–334.

DiMaggio, P. J. (1988). Interest and agency in institutional theory. In L. G. Zucker (Ed.), *Institutional patterns and organizations: Culture and environment* (pp. 3–23). Cambridge, MA: Ballinger.

Doh, J. P., & Guay, T. R. (2006). Corporate social responsibility, public policy, and NGO activism in Europe and the United States: An institutional-stakeholder perspective. *Journal of Management Studies, 43*(1), 47–73.

Dyer, J. H., & Singh, H. (1998). The relational view: Cooperative strategy and sources of interorganizational competitive advantage. *Academy of Management Review, 23*(4), 660–679.

Eisenhardt, K. M. (1989). Building theory from case study research. *Academy of Management Review, 14*(4), 532–550.

Eisenhardt, K. M., & Martin, J. A. (2000). Dynamic capabilities. What are they? *Strategic Management Journal, 21*(10/11), 1105–1121.

Foerstl, K., Reuter, C., Hartmann, E., & Blome, C. (2010). Managing supplier sustainability risks in a dynamically changing environment—sustainable supplier management in the chemical industry. *Journal of Purchasing and Supply Management, 16,* 118–130.

Frenkel, S. J., & Scott, D. (2002). Compliance, collaboration, and codes of labor practice. *California Management Review, 45*(1), 29–48.

George, E., Chattopadhyay, P., Sitkin, S. B., & Barden, J. (2006). Cognitive underpinnings of institutional persistence and change: A framing perspective. *Academy of Management Review, 31*(2), 347–385.

Gibbert, M., Ruigrok, W., & Wicki, B. (2008). What passes as a rigorous case study? *Strategic Management Journal, 29*(13), 1465–1474.

Hamprecht, J. (2006). *Sustainable purchasing strategy in the food industry*. Bamberg, Germany: Difo.

Hamprecht, J., & Sharma, S. (2006). *A resource-based view on institutional entrepreneurship*. Paper presented at the GRONEN Conference, July 10-12, 2006, St.Gallen, Switzerland. Further information: http://www.insme.org/files/1917

Handfield, R. B., Scroufe, R., & Walton, S. (2005). Integrating environmental management and supply chain strategies. *Business Strategy and the Environment, 14*(1), 1–19.

Hargrave, T. J., & Van de Ven, A. H. (2006). A collective action model of institutional innovation. *Academy of Management Review, 31*(4), 864–888.

Hart, S. L. (1995). A natural-resource based view of the firm. *Academy of Management Review, 20*(4), 986–1014.

King, A. A., & Lenox, M. J. (2000). Industry self-regulations without sanctions. The chemical industry's responsible care program. *Academy of Management Journal, 43*(4), 698–716.

Kutsch, E., & Hall, M. (2009). Deliberate ignorance in project risk management. *International Journal of Project Management, 28*(3), 245–255.

Lavie, D., & Rosenkopf, L. (2006). Balancing exploration and exploitation in alliance formations. *Academy of Management Journal, 49*(6), 797–818.

Lawrence, T. B. (1999). Institutional strategy. *Journal of Management, 25*(2), 161–187.

Lawrence, T. B., Hardy, C., & Phillips, N. (2002). Institutional effects of interorganizational collaboration. the emergence of proto-institutions. *Academy of Management Journal, 45*(1), 281–290.

Lee, S.-Y. (2008). Drivers for the participation of small and medium-sized suppliers in green supply chain initiatives. *Supply Chain Management: An International Journal, 13*(3), 185–198.

Locke, R. M., Qin, F., & Brause, A. (2007). Does monitoring improve labor standards? Lessons from Nike. *Industrial and Labor Relations Review, 61*(1), 3–31.

Luo, X., & Bhattacharya, C. B. (2006). Corporate social responsibility, customer satisfaction, and market value. *Journal of Marketing, 70*(4), 1–18.

Maguire, S., & Hardy, C. (2006). The emergence of new global institutions: A discursive perspective. *Organization Studies, 27*(1), 7–29.

Matten, D., & Moon, J. (2008). "Implicit" and "explicit" CSR: A conceptual framework for a comparative understanding of corporate social responsibility. *Academy of Management Review, 33*(2), 404–424.

Millington, A. (2008). Responsibility in the supply chain. In A. Crane, A. McWilliams, & D. Matten (Eds.), *The Oxford handbook of corporate social responsibility* (pp. 363–383). Oxford, UK: Oxford University Press.

Munir, K., & Phillips, N. (2005). The birth of the "Kodak moment": Institutional entrepreneurship and the adoption of new technologies. *Organization Studies, 26*(11), 1665–1687.

Neilson, J. & Pritchard, B. (2007). Green coffee? The contradictions of global sustainability initiatives from an Indian perspective. *Development Policy Review, 25*(3), 311–331.

Olden, P. C. (2003). Hospital and community health: Going from stakeholder management to stakeholder collaboration. *Journal of Health and Human Services Administration, 26*(1), 35–57.

Oliver, C. (1991). Strategic responses to institutional processes. *Academy of Management Review, 16*(1), 145–179.

Oliver, C., & Holzinger, I. (2008). The effectiveness of strategic political management: A dynamic capabilities framework. *Academy of Management Review, 33*(2), 496–520.

Pentland, B. T. (1999). Building process theory with narrative. From description to explanation. *Academy of Management Review, 24*(4), 711–724.

Peteraf, M. A. (1993). The cornerstones of competitive advantage. A resource-based view. *Strategic Management Journal, 14*(3), 179–191.

Peters, N., Hofstetter, J.S., & Hoffmann, V.H. (2011). Institutional entrepreneurship capabilities for interorganizational sustainable supply chain strategies. *International Journal of Logistics Management, 22*(1), 1–47.

Peters, N. (2010). *Inter-organisational design of voluntary sustainability initiatives: Increasing the legitimacy of sustainability strategies for supply chains*. Wiesbaden, Germany: Gabler.

Peters, N., Hofstetter, J. S., & Hoffmann, V. H. (2011). Institutional entrepreneurship capabilities for interorganizational sustainable supply chain strategies. *International Journal of Logistics Management, 22*(1), 52–86.

Peters, N., & Schaupp, J. (2009). Proactive sustainability in Russian wood supply chains: Just another CSR fad or institutional change in the making? Retrieved March 30, 2011, from http://www.alexandria.unisg.ch/export/DL/Nils_Peters/52133.pdf

Phillips, N., Lawrence, T. B., & Hardy, C. (2004). Discourse and institutions. *Academy of Management Review, 29*(4), 635–652.

Powell, W. W. (1988). Institutional effects on organizational structure and performance. In L. G. Zucker (Ed.), *Institutional patterns and organizations: Culture and environment* (pp. 115–139). Cambridge, MA: Ballinger.

Raftery, J. (1994). *Risk analysis in project management*. London, UK: Chapman and Hall.

Rao, P. (2002). Greening the supply chain: A new initiative in South East Asia. *International Journal of Operations and Production Management, 22*(5/6), 632–655.

Rao, P., & Holt, D. (2005). Do green supply chains lead to competitiveness and economic performance? *International Journal of Operations and Production Management, 25*(9), 898–916.

Roth, A. V., Tsay, A. A., Pullman, M. E., & Gray, J. V. (2008). Unraveling the food supply chain: Strategic insights from China and the 2007 recalls. *Journal of Supply Chain Management, 44*(1), 22–40.

Rothaermel, F. T., & Deeds, D. L. (2004). Exploration and exploitation alliances in biotechnology: A system of new product development. *Strategic Management Journal, 25*(3), 201–221.

Sharma, S., & Henriques, I. (2005). Stakeholder influences on sustainability practices in the Canadian forest products industry. *Strategic Management Journal, 26*(2), 159–180.

Sharma, S., & Vredenburg, H. (1998). Proactive corporate environmental strategy and the development of competitively valuable organizational capabilities. *Strategic Management Journal, 19*(8), 729–754.

Sharma, S., Aragón-Correa, J., & Rueda-Manzanares, A. (2007). The contingent influence of organizational capabilities on proactive environmental strategy in the service sector: An Analysis of North American and European ski resorts. *Canadian Journal of Administrative Sciences, 24*(4), 268–283.

Shrivastava, P. (1995): Environmental technologies and competitive advantage. *Strategic Management Journal, 16* (special Issue, Summer), 183–200.

Suchman, M. C. (1995). Managing legitimacy: Strategic and institutional approaches. *Academy of Management Review, 20*(3), 571–610.

Tang, C. S. (2006). Perspectives in supply chain risk management. *International Journal of Production Economics, 103*(2), 451–488.

Vachon, S. (2007). Green supply chain practices and the selection of environmental technologies. *International Journal of Production Research, 45*(18/19), 4357–4379.

Vachon, S., & Klassen, R. D. (2008). Environmental management and manufacturing performance: The role of collaboration in the supply chain. *International Journal of Production Economics, 111*(2), 299–315.

Vermeulen, W. J., & Ras, P. (2006). The challenge of greening global product chains: Meeting both ends. *Sustainable Development, 14*(4), 245–256.

Verona, G. (1999). A resource-based view of product development. *Academy of Management Review, 24*(1), 132–142.

Wagner, T., Lutz, R. J., & Weitz, B. A. (2009). Corporate hypocrisy: Overcoming the threat of inconsistent corporate social responsibility perceptions. *Journal of Marketing, 73*(6), 77–91.

Wright, M., Filatotchev, I., Hoskisson, R. E., & Peng, M. W. (2005). Strategy research in emerging economies. Challenging the conventional wisdom. *Journal of Management Studies, 42*(1), 1–33.

Yin, R. K. (2003). *Case study research*. London: Sage.

Zhu, Q., & Sarkis, J. (2004). Relationships between operational practices and performance among early adopters of green supply chain management practices in Chinese manufacturing enterprises. *Journal of Operations Management, 22*(3), 265–289.

Zimmerman, M. A., & Zeitz, G. J. (2002). Beyond survival. Achieving new venture growth by building legitimacy. *Academy of Management Review, 27*(3), 414–431.

CHAPTER ELEVEN

Supply Chain Structure as a Critical Driver of Sustainable Supplier Practices

Amrou Awaysheh and Robert D. Klassen

As technology improves the transfer of information, a broader range of customers and stakeholders gain access to more information about what happens within supply chains. As a result, issues like poor worker conditions in suppliers' facilities are increasingly pushed into the limelight. What used to be hidden behind long distances and language differences is more visible (Lee, 2002; Van Der Zee & Van Der Vorst, 2005). As a result, consumers, governments, and nongovernmental organizations (NGOs) are demanding that companies be held more accountable for what happens. Concerns include the use of sweatshop labor, the provision of safe working conditions, and the payment of a living wage to their employees. In response, a growing number of firms are exploring how to identify, assess, and monitor supplier-related social issues and practices. They can monitor their suppliers to ensure adherence to social expectations, conduct audits, or use a certification provided by an independent third-party. Fairtrade (Fairtrade, 2007) is one such third-party certification for agricultural commodities such as coffee and cocoa beans. Following an audit, certification is granted to cooperative farms in developing countries that adhere to a number of sustainability-related principles, including safe working conditions for employees, payment of fair wages, and environmentally friendly cultivation techniques. In contrast, other firms choose to develop their own standards internally, for example Starbucks' system for assessing and working with farmers, termed Coffee and Farmer Equity (CAFE).

Unfortunately, it is not as simple as just dictating that a particular set of standards be employed by suppliers. Some firms might not have sufficient influence to drive change back through the supply chain to all suppliers. The cultural norms and expectations for improving human potential vary by industry, customer segment, and marketplace. As more manufacturing and supplier sourcing has shifted overseas, the geographic distance, and length of supply chains (i.e., tiers) between supply chain partners also has increased. In addition, the costs associated with adoption practices to deal with social issues can be prohibitive. So overall, much remains unclear about how the structure of the supply chain influences the management of social issues between a firm and its suppliers.

The objective of this chapter is to identify the factors that can influence firms to engage in specific socially responsible supplier practices and how these factors influence the adoption of these practices. We define and frame the construct of social issues within the broader debate on sustainable development and stakeholder management. We delineate social practices for supply chain management and empirically validate a set of scales for assessing the degree of development of socially responsible supplier practices. Finally, we examine the link between supply chain structure and the adoption of socially responsible supplier practices.

Defining Social Issues

Sustainable development, generally defined as "meeting the needs of the present without compromising the ability of future generations to meet their own needs" (Brundtland, 1987, p. 1) implies that a broad and complex range of issues must be actively managed by firms. One lens to operationalize the concept for operations and supply chain management is the *triple bottom line* or TBL (Elkington, 1997; Carter & Rogers, 2008), encompassing environmental and social performance, in addition to financial performance. Environmental management and performance has received a growing degree of attention in the operations literature, including such areas as green product design (Baumann, Boons, & Bragd, 2002), closed-loop supply chains, and green supplier development. A detailed literature review is provided by Seuring and Müller (2008).

Social issues and performance capture both individual-level human safety and welfare, and societal-level community development. Thus, by extension, social practices and performance in operations and the supply chain encompass all management practices that affect how a firm contributes to the development of human potential or protects people from harm, thereby capturing both positive and negative aspects. Examples include workforce policies for safety or diversity, and product safety. While corporate philanthropy might also be viewed as social performance (Porter & Kramer, 2006), it will not be considered further in this chapter, as that lies beyond the control of operations and supply chain managers.

Societal expectations for acceptable practices and standards of conduct continue to evolve and ratchet upwards as public opinion, NGOs, the popular press, and regulations uncover shortcomings in previously acceptable routines (Martin, 2002). Supply chain members, particularly customers, also apply pressure to firms to improve in areas where problems or inconsistent social practices become apparent, such as suppliers' treatment of their workers in developing economies (Carter, 2000). Thus, for operations, a mid-range definition of socially responsible supplier practices is critical to structuring research. Further research is needed to identify tools, systems and programs that operations can put in place to assess and monitor the degree to which social issues are managed by their suppliers, and ideally, to improve performance.

Different streams of study help to organize our understanding of social issues in the supply chain: international labor practice standards; socially responsible purchasing; and parallels between environmental and social auditing. First, firms with international supply chains reaching into developing countries have increasingly adopted standards, such as SA8000, that set basic requirements for workforce practices in internal operations and in multiple tiers of suppliers (Social Accountability International, 2008). Nine areas are explicitly examined, including child labor; forced labor; health and safety; freedom of association and collective bargaining; discrimination; disciplinary practices; working hours; compensation; and related management systems. A related standard, ISO 26000, remains in development, and is expected to parallel environmental management standard, ISO 14001 (Castka & Balzarova, 2008). In contrast, reporting frameworks like the Global Reporting Initiative (GRI) provide a means for companies to identify, structure, and communicate their triple bottom line performance to stakeholders both inside and outside the firm (Global Reporting Initiative, 2006).

In addition to required key indicators, others are developed in consultation with stakeholders. However, the objectives can vary quite significantly between different standards and frameworks, with some focusing on societal expectations, e.g., not using child labor, and others on improved working conditions, e.g., paying overtime for additional work (Jantzi, 2008). Socially responsible purchasing has been based on the deployment of firm- or industry-specific codes of conduct (COC). COC dictate specific guidelines, behaviors, and buying criteria that employees in the purchasing department or supply chain group must follow during their interactions with potential and current suppliers (Mamic, 2005). In addition, COC can be used as leverage to encourage change in, or in the worst cases termination of, contracts with suppliers that were found to violate social expectations (Emmelhainz & Adams, 1999). More recently, Carter & Jennings (2002) have extended this work and proposed a multidimensional index to operationalize socially responsible purchasing, including environment, diversity, safety, human rights, and philanthropy. Although the index is a much-needed step in the right direction, it captures more than just social practices, including such aspects as environmental management and philanthropy. In contrast, focusing on COC provides a clearer view of specific measurable elements within the firm. The third stream focuses on social auditing, and borrows heavily from research in environmental management. Operations can either monitor or collaborate with their suppliers to assess and improve performance, respectively (Zhao,

Flynn, & Roth., 2007). With monitoring, either the firm itself or a third party physically inspects the operations of first- or second-tier suppliers to ensure that specific practices are being used. These audits go well beyond quality and other supplier performance criteria that have been in place for many years, and thus represent a significant, incremental investment and commitment (Krause, 1999). Subsequent to inspection and auditing, a firm may undertake collaboration to work with supply chain partners to enable the development of new skills and capabilities (Klassen & Vachon, 2003). Collectively, these streams point toward four broad categories of socially responsible supplier practices:1) international standards that seek to ensure supplier human rights are being protected; 2) extended frameworks that capture supplier labor practices; 3) supplier Codes of Conduct (COC) that structure how a firm should interact with its supply chain partners, and finally, 4) supplier social audits that seek to ensure adherence to human rights and labor practices.

Operations managers must consider many factors when designing, developing and restructuring their supply chains as markets evolve. Moreover, social practices for suppliers do not develop in isolation, but instead must be connected with, and take into account, the nature of both the upstream and downstream portions of the supply chain. And at a minimum, the location of suppliers and the forms of interaction between supply chain members have important social implications, and cannot be ignored. Three dimensions relating to the structure of the supply chain were identified as potentially having a significant impact on the tools, systems, and programs that are in place to address social issues with suppliers: transparency, dependency, and distance.

Transparency

When considered within the context of supply chains, transparency captures the extent to which information is readily available to end users and other firms in the supply chain. Transparency has become increasingly important for social issues, including the origins of commodities (i.e., provenance) and product safety (Lee, 2002; Van Der Zee & Van Der Vorst, 2005). For example, supply chain provenance tracks products from raw material sourcing to supply chain partners, to manufacturing, to distribution, and to end-consumer use (and beyond, if needed). Organic foods and "conflict" diamonds are but a few labels that require all supply chain partners to ensure traceability of goods (New, 2004).

Legitimacy is enhanced if third-party standards and independent auditors are employed, such as SA8000 or Fairtrade certification (Bansal & Hunter, 2003). An alternative is prequalification or direct audit of suppliers against a specific set of social standards established internally by the firm itself (Settings, 2004). Nike, for example, has developed its own standards for workforce and environmental practices using input from a variety of stakeholder groups. Third parties might then assist with monitoring suppliers for compliance with these standards, although fulfillment is not necessarily assured, because information remains difficult to verify (Nike, 2004).

Large firms with highly visible brand names might be expected to actively work to guard against unexpected criticism of social performance in their supply chain, which in turn can harm the value of their brands. Moreover, the Internet encourages rapid dissemination of negative information, and customers will react quickly after questionable behavior is identified (Tapscott & Ticoll, 2003). Thus, the degree to which supply chains are transparent and subject to scrutiny by NGOs, the media, and the public is likely to influence the extent to which a firm actively develops socially responsible supplier practices (Graafland, 2002), which leads to Hypothesis H1.

H1: As the level of transparency in the supply chain increases, the firm's use of socially responsible supplier practices increases.

Dependency

The dependency dimension represents the degree to which a firm relies on other members of the supply chain for critical resources, components, or capabilities. For supply chains, several factors affect the degree of dependency, including concentration, vertical integration, and credible commitment (Cool & Henderson, 1998). Each of these factors facilitates a firm's ability to control and influence change in the operations of its suppliers, and they extend beyond simple pricing power (i.e., the continual drive to force prices down) to encompass multiple aspects of the buyer-supplier relationship. Moreover, a firm can potentially transfer responsibilities for social issues, and pressures to improve social issues, away from itself to either upstream or downstream partners in the supply chain through rewards, coercion, or legal instruments, to name several means (Maloni & Benton, 2000). Naturally, having power to transfer responsibility implies a corresponding imbalance in dependency, and it can be a factor either upstream or downstream in the supply chain, i.e., in the firm's dependency on its suppliers or on customers, respectively.

Greater competition and fragmentation in markets increase a firm's dependency on supply chain partners. Dependency theory (Pfeffer & Salancik, 1978) also suggests that power, and ultimately profitability, decrease for a buying firm as the number of potential suppliers falls. Furthermore, credible commitment also is likely to decrease. Credible commitment, a concept borrowed from game theory, captures the believability of the firm's assurances (positive or negative) about the actions that the firm will take in response to specific behaviors or outcomes of a supplier. Thus, signaling by a firm is less likely to alter suppliers' priorities and actions if the firm is highly dependent on suppliers (Dixit & Nalebuff,

1991). Collectively, these outcomes arise because few alternative supply arrangements are available for a firm.

In contrast, firms that are vertically integrated tend to have greater information about products, processes, and markets, which provides leverage in negotiations with their supply base (Harrigan, 1985) and affects supplier involvement in product development (Carr, Kaynak, Hartley, & Ross, 2008). Applied to social issues, a vertically integrated firm knows the workings of the industry (i.e., what is practically achievable), can lead by example, and can force suppliers to report about such aspects as workforce practices. However, such reporting is unlikely to occur if the firm is highly dependent; the ability to adopt and enforce socially responsible supplier practices is very limited, at best, as expressed in Hypothesis 2a.

H2a: As the dependency of the firm on its suppliers increases, the firm's use of socially responsible supplier practices decreases.

Turning upstream, in an industry where a firm confronts intense competition, such a firm might attempt to differentiate itself by engaging in socially responsible practices, possibly as a means of offering enhanced value with its products. Furthermore, firms that are highly dependent on customers also might be viewed as an easy target by NGOs, downstream customers, and the general public to be pushed to improve workforce conditions in their suppliers – thereby becoming an example for competitors (see, e.g., Grow, 2005), as expressed in Hypothesis 2b. The effects of dependency are not expected to be symmetrical, with customer dependency stimulating the use of socially responsible supplier practices, and supplier dependency dampening their development.

H2b: As the dependency of the firm on its customers increases, the firm's use of socially responsible supplier practices increases.

Distance

If a firm uses local suppliers to serve local customers, the need to use a variety of socially responsible supplier practices is dramatically reduced, as all three (supplier, firm, and customers) can be expected in move in tandem. However, as the distance increases, managers confront problems in data gathering, assessment and implementation (Klassen & Vachon, 2003). Distance encompasses three subdimensions: geographical, cultural, and organizational distance. First, as geographical separation expands, firms have more difficulty interacting frequently with their suppliers (Choy & Lee, 2003), and by extension, have more difficulty in ensuring good working conditions are maintained. In addition, either the supplier or the distant firm may view the other party as having less commitment. As a result, firms facing the challenge of geographic distance may feel compelled to establish monitoring or auditing systems to help mitigate shortcomings derived from limited access to information and uncertain commitment (Koplin, Seuring, & Mesterharm, 2007).

Second, cultural distance reflects the differences that exist between the cultures of the societies in which the firms are based (Hofstede, 1980). Cultural distance influences the approaches that managers choose when confronting new or challenging problems (Joynt & Warner, 1996), as well as relationships between firms (Reynolds, Simintiras, & Diamantopoulos, 2003). If firms and their suppliers are based in societies with similar cultures, discussions about expectations and possible changes to operations are straightforward as managers in both organizations draw from similar cultural experiences. Moreover, regulatory frameworks and expectations, including enforcement, in each jurisdiction are likely to be similar. Thus, potential misunderstandings and problems can be avoided (Hofstede, 2001). In contrast, if the firm is located in a developed country and sources are from developing countries, the firm must conform to cultural expectations (and regulations) in its own local market (O'Grady & Lane, 1996), which may not understood by suppliers. Finally, managers of firms operating in countries with democratic institutions and a free press are aware that improper or deficient practices are more likely to become public knowledge. However, this may be of little concern to particular suppliers.

Third, organizational distance is defined by the number of tiers that exist between the firm and suppliers or customers, and the length of the supply chain (Banet, 1976). Increasing organizational distance tends to increase complexity with more frequent handoffs of information in such areas as product tracing, material specification, and operational procedures. Longer supply chains also potentially increase the number of relationships that might be managed. Thus, increased organizational distance between firms necessitates that additional mechanisms must be put in place, generating higher total transaction costs across the supply chain (Williamson, 1979; Grover & Malhotra, 2003). It should be stressed that the three forms of distance are not necessarily highly correlated. For example, a firm in Europe might source from either Brazil or Thailand, both roughly equivalent in geographic distance, but very different in cultural distance.

Collectively, as the distance in the supply chain increases, a firm is more likely to develop a stronger set of socially responsible supplier practices to manage the distance, differences in culture and inter-organizational complexity, as expressed in Hypothesis 3.

H3: As distance increases in a supply chain, a firm's use of socially responsible supplier practices increases.

The basic construct dimensions of supply chain structure and socially responsible supplier practices are depicted in Figure 11.1.

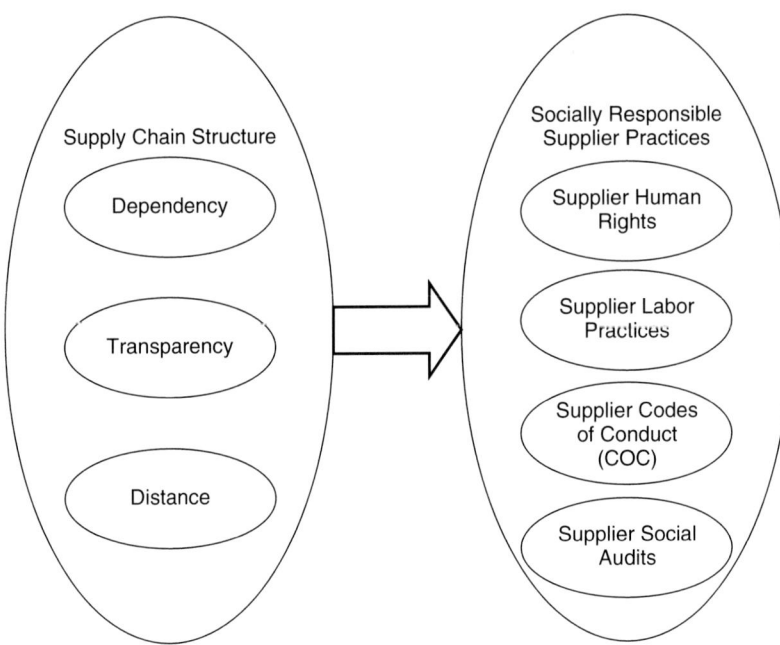

Figure 11.1 Conceptual model of supply chain structure.

Research Methods

The relationship between the structure of the supply chain and socially responsible supplier practices was explored using a plant-level survey. An ideal set of industries for this research has several characteristics: high degree of variation in approaches to managing social issues, a competitive marketplace to ensure at least some responsiveness to multiple stakeholders, and a multiplicity of different structures for their supply chains. The last criterion points to the need to capture supply chains that reach into both developing and developed economies, encompass a range of social concerns, and have different production technologies (e.g., process vs. discrete product). In addition, if an industry is facing some pressure for change in its management of social issues, whether driven by regulation, public pressure, or customer demands, some firms are likely to lead the industry with innovative or proactive initiatives, while others wait and see, or even resist changes. Based on these criteria, three industries were selected: food (North American Industrial Classification System [NAICS] code 311), chemicals (NAICS 325), and transportation equipment (NAICS 336). Each of these industries is very competitive, is facing evolving social pressures, and has supply chains that span from local to global networks.

The targeted respondent for the survey also was carefully considered. First, it was important to focus on one well-defined individual in a firm, as it was impractical to expect multiple managers across many firms to respond to a lengthy survey instrument. Second, this person must be intimately involved in managing internal operations, the supply chain, and ideally, the surrounding community. Rather than targeting either a senior supply chain manager (who often focuses only on either the upstream or downstream portion of the supply chain) or a corporate staff person (who can be rather disconnected from day-to-day operations), this research targeted the plant manager. In many ways, this individual becomes the nexus for many social issues in the supply chain. Internal operations clearly are under her or his responsibility, and the plant is intimately connected to both upstream suppliers for incoming parts and downstream customers for outgoing products. Moreover, approaches to social issues can vary between plants, even within the same firm, depending on the network of suppliers used and customers served. One concern could be raised as to how many of the respondents were actually plant managers and how many delegated the survey to someone else who is more junior. To address this concern, a survey question that asked for the title of the survey respondent was examined in depth: 79 percent of the respondents had a senior title of plant or operations manager; 11 percent had a more junior title such as production manager; and 10 percent of the respondents did not provide information about their title.

Scott's Directory of Manufacturing (Group, 2007) was used to identify plants from each of the three industries in Canada. This research followed Dillman's (2000) five-point contact protocol: an initial introductory telephone contact; then two waves of postal surveys separated by a fax reminder; and a final telephone call to encourage participation. The survey was available in both English and French, and either a paper or online version could be completed and returned. This practice has been used in recent operations management research to help increase response rate (Johnson, Klassen, Leenders, & Awaysheh, 2007).

The data for this survey was collected over a six-month period ending in early 2008. To increase the likelihood that the plant would have at least some systems in place to manage social issues, plants with a minimum of 100 employees were targeted. A total of 1209 surveys were distributed (574 plants in the food industry, 300 in chemicals, and 335 in transportation industries). In total, 42 were removed because the survey was undeliverable, or the facility was no longer in business or not a manufacturing plant. Another 42 surveys were returned with virtually no responses; and these surveys were removed from further analysis. Three hundred and seven completed surveys were received, yielding an effective response rate of 25 percent. To examine possible nonresponse bias, the respondent plants were compared to the pool of nonrespondents in terms of sales, industry, and number of employees (Lessler & Kalsbeek, 1992). Early and late respondents were compared using the same criteria (Armstrong & Overton, 1977), as well as the survey technology (i.e., mail versus Internet). No evidence was found that the respondents were not representative of the target sample. List-wise deletion was used for responses that had missing data.

A key objective of this research was the development of a new set of scales to characterize the use of socially responsible practices in the supply chain. A five-step approach was employed. First, the literature base was reviewed to identify the items that would be necessary to form the constructs in question. Five major sources provided the items used in the survey (Klassen & Vachon, 2003; Carter & Jennings, 2004; Fairtrade, 2007; Jantzi, 2008; Social Accountability International, 2008). Second, the survey was pretested with five business managers and seven academics to assess face and content validity of the items and improve readability. Within the questionnaire, the items were presented by scale rather than mixing the items for the various measures (Forza, 2009). Third, reliability was assessed for each individual construct, i.e., Cronbach's alpha (Segars, 1997). Fourth, Confirmatory Factor Analysis (CFA) was conducted across the related constructs for supply chain structure, and then for supplier social responsibility practices (O'Leary-Kelly & J. Vokurka, 1998). Finally, discriminant validity was assessed both for the items within a set of constructs (i.e., items within a construct are expected to be more highly correlated than those between constructs), and between pairs of constructs (constrained versus unconstrained CFA models).

Four dimensions of socially responsible supplier practices were examined: supplier human rights, supplier labor practices, supplier codes of conduct (COC), and supplier social audits. *Supplier human rights* measures the extent to which practices are in place to reduce the possibility that suppliers employ vulnerable groups of people, such as children. *Supplier labor practices* assesses the conditions in which employees perform their duties, and how an employer contributes to the overall welfare of employees. *Supplier codes of conduct (COC)* measures the extent to which specific procedures are explicitly spelled out by the firm to ensure that suppliers adhere to ethical expectations. Finally, the *supplier social audits* measures the degree to which suppliers are monitored by the firm to ensure their adherence to social expectations.

The values of Cronbach's Alpha for the scales were .96, .95, .82, and .92 for supplier human rights, supplier labor practices, supplier COC, and supplier social audits scales, respectively. Appendix 1 presents the estimates and the model fit statistics for the CFA estimates for the four final socially responsible supplier practices constructs that were used in the analysis; all were within acceptable ranges.

As previously discussed, three dimensions for supply chain structure were conceptualized as being critical for socially responsible supplier practices: transparency, dependency, and distance. Cool and Henderson (1998) developed a set of scales that examined supply chain dependency by examining two related aspects of buyer–supplier dependency. Drawing from this work, two multi-item scales were developed to assess the degree to which the firm was dependent on its suppliers and customers (Appendix 2).

The transparency construct was divided into two subdimensions that attempted to capture the degree to which the plant perceived that an end user (i.e., not necessarily the immediate, next-tier customer) was aware of both the firm's product and process. One subdimension dealt with the *end user knowledge of the supply chain*, and the other dimension dealt with *product visibility* as perceived by the firm (Appendix 2). As with the previous set of constructs, the literature provided a basis for all of the items that were used in the scales and their formation (Cool & Henderson, 1998; Tapscott & Ticoll, 2003; New, 2004). Furthermore, these items were tested and refined to help ensure consistency and validity.

Cronbach's alpha values for these scales were acceptable, at .81, .87, .85 and .77, for supplier dependency, customer dependency, end user knowledge of supply chain and product visibility, respectively. All general-fit statistics for the CFA models also were within acceptable limits. Third, individual items were more highly correlated within their respective scales than between scales, providing evidence of discriminant validity. Finally, the fit of the unconstrained model was significantly better than for the constrained model, further supporting discriminant validity in the CFA model.

Finally, measures for distance were operationalized across three subdimensions: geographical, cultural, and organizational distance. To measure the first two dimensions of distance, respondents provided the extent to which their primary suppliers were located in six different regions: Canada; United States; Latin America, including Mexico; Europe; Asia, including Russia, India, and China; and Africa. The items for each plant were scaled to equal one; these six metrics then served as weights applied to the different regions to estimate the weighted distance (either geographical or cultural) between the plant in Canada and its supply base. An analogous set of weights was also generated for the plant's customer base. Finally, to estimate the overall weighted geographical distance for the supply base, the regional weights were multiplied by the haversine distance (i.e., distance between two points on a sphere) between Canada and the geographic center of that particular region (Vincenty, 1975). While this metric has potential weaknesses, it provided a reasonable sense of the distances that must be managed between suppliers and customers. Additional sensitivity analysis using different geographic centers for a region did not reveal any significant differences in the parameter estimates of the regression models.

For cultural distance, the difference between Canada and each of six regions was estimated using a weighted average of the difference for each of four dimensions of culture (Hofstede, 1980), as done by others (Kaufmann & Carter, 2006). Finally, the overall weighted cultural distance (CD) for the supply base was estimated using the regional weights described earlier:

$$CD_j = \sum_{i=1}^{4} \left\{ \left(I_{ij} - I_{ip}\right)^2 / V_i \right\} / 4$$

where I_{ij} = the index for the ith cultural dimension and the jth region, V_i = the variance of the index of the ith dimension, p = Canada's score on the dimension, CD_j = the cultural distance of the jth region from Canada. An analogous cultural distance measure was estimated for the customer base.

Lastly, organizational distance was assessed based on the number of firms in the supply chain between the plant and the final end user, and between the plant and the primary basic raw material supplier. These two items were combined into a single overall measure of the total length of the supply chain (i.e., number of tiers). A related metric was estimated on the plant's relative upstream positioning in the supply chain, i.e., a higher value means further upstream, ranging from 0 (i.e., plant is the end user) to 1 (i.e., plant is the raw material supplier). The correlation between the overall and relative measures was low, at .097; providing evidence that these two metrics assessed two distinct constructs.

Social desirability refers to a potential bias that might be introduced if respondents answer questions consistent with perceived social expectations (i.e., political correctness) or a desire to please the survey administrator. To address this issue, respondents were asked to give their responses to the questions about the practices at their plant, not their personal practices or beliefs. This subliminal technique has been used in the past to help reduce social desirability bias (Rudelius & Buchholz, 1979).

In addition, an accepted scale was used to assess the degree of social desirability and its impact on responses. A shortened form of the Crowne-Marlow Social Desirability Scale (Crowne & Marlowe, 1960), X2 (Strahan & Gerbasi, 1972); further validated by Fischer and Fick (1993) was tested against scales measuring the four dimensions of socially responsible supplier practices. Ideally, a low, nonsignificant correlation is expected, as was found here; thus no evidence of social desirability bias was found.

Results

A series of linear models was used to examine the relationship between the structure of the supply chain and the use of the four socially responsible supplier practices. Table 11.1 presents the descriptive statistics and correlation table of all the variables used in the regressions. Each regression model also controlled for industry, firm sales, unionization level, and the degree of plant-level responsibility for supplier policies. The results are reported in Table 11.2.

Support for Hypothesis H1 was found in all four of the models; higher levels of transparency were related to higher levels of socially responsible supplier practices. More specifically, as product visibility increased, the use of supplier human rights ($p < .01$), supplier labor practices ($p < .05$), supplier codes of conduct ($p < .05$), and supplier social audits ($p < .01$) increased. In contrast, support for Hypothesis H2b was very limited, with some evidence of greater use of supplier human rights ($p < .1$) as customer dependency

Table 11.1 Descriptive statistics and correlation table

Variable	Mean	s.d.	1	2	3	4	5	6	7	8	9	10	11	12	13	14	15	16	17	18
1 Chemical industry	.312	.464																		
2 Transportation industry	.272	.446	−.412																	
3 Firm sales	3.438	1.253	.106	.022																
4 Unionization level	.456	.461	.016	−.149	.017															
5 Plant-level responsibility for supplier policies	.2553	1.168	−.169	.047	.009	.110														
6 End-user knowledge of supply chain	3.403	1.441	−.151	.063	.011	−.019	.042													
7 Product visibility	4.748	1.845	−.022	−.043	−.024	−.027	.020	.469												
8 Dependency on suppliers	4.901	1.128	.065	.049	.133	−.139	.044	.035	.040											
9 Customer dependency on plant	4.408	1.337	−.024	.268	.122	−.128	−.006	.091	−.018	.281										
10 Supplier geographical distance	.317	.098	.126	.001	.158	.000	−.134	.038	.105	.069	.019									
11 Supplier cultural distance	.294	.273	.147	.177	.174	.014	−.105	−.057	.000	.059	.109	.635								
12 Total length of the supply chain	3.921	2.950	−.109	.134	−.027	−.065	−.072	−.128	−.187	.053	.049	.048	.086							
13 Customer geographical distance	.291	.092	.059	−.172	.093	.034	−.056	.061	.088	.161	−.081	.130	−.019	−.099						
14 Customer cultural distance	.261	.274	.094	.200	.190	−.012	−.049	−.064	−.040	.175	.145	.192	.291	.015	.376					
15 Relative upstream positioning in the supply chain	.570	.291	−.014	−.139	.024	.016	.057	−.088	−.136	.037	.011	−.134	−.148	.122	−.003	−.049				
16 Supplier human rights	3.414	2.207	.029	.008	.211	.121	−.063	−.031	.109	.015	.087	.074	.075	.048	−.012	.063	.046			
17 Supplier labour practices	.2941	1.838	.046	.051	.273	.097	−.068	.004	.083	.059	.111	.064	.068	.093	.007	.091	.035	.844		
18 Supplier codes of conduct	4.110	1.563	.084	−.070	.326	.134	.058	.026	.087	.063	.061	−.045	.003	.028	.031	.014	.112	.548	.601	
19 Supplier social audits	.2430	1.657	−.044	.015	.209	.139	−.032	.039	.123	−.014	.058	.062	.041	.089	−.012	.006	−.008	.748	.803	.609

|ρ| ≥ .154 are statistically significant at $p < .01$; and |ρ| ≥ .116, $p < .05$.

Table 11.2 Regressions results

	Supplier Human Rights		Supplier Labor Practices		Supplier Codes of Conduct		Supplier Social Audits	
Control								
Chemical industry	.090	(.326)	.261	(.267)	.269	(.223)	−.065	(.247)
Transportation industry	−.071	(.354)	.062	(.290)	−.221	(.241)	−.083	(.268)
Firm sales	.382	(.107)**	.407	(.088)**	.404	(.073)**	.340	(.081)**
Unionization level	−.382	(.285)	−.324	(.234)	.163	(.195)	−.193	(.216)
Plant-level responsibility for supplier policies	.255	(.114)*	.181	(.094)*	.187	(.078)*	.208	(.087)*
Supply Chain Structure								
Transparency								
End-user knowledge of supply chain	−.158	(.107)	−.044	(.088)	.029	(.073)	−.052	(.081)
Product visibility	.246	(.079)**	.151	(.065)*	.112	(.054)*	.154	(.060)**
Dependency								
Dependency on suppliers	−.005	(.101)	.093	(.083)	.010	(.069)	−.053	(.076)
Customer dependency on plant	.171	(.096)†	.061	(.079)	.059	(.065)	.045	(.073)
Distance								
Supplier geographical distance	−1.129	(1.967)	−1.512	(1.613)	−3.051	(1.340)*	.132	(1.490)
Supplier cultural distance	−.166	(.689)	−.073	(.565)	.258	(.470)	−.171	(.522)
Total length of the supply chain	.055	(.044)	.083	(.036)*	.066	(.030)*	.092	(.034)**
Customer geographical distance	−.930	(1.834)	−1.027	(1.504)	−.022	(1.251)	−.308	(1.389)
Customer cultural distance	.427	(.600)	.379	(.492)	.169	(.410)	−.008	(.455)
Relative upstream positioning in supply chain	.379	(.460)	.346	(.377)	.731	(.315)*	.105	(.349)
Constant	.059	(.877)	.059	(.377)	1.404	(.731)*	.040	(.811)
N	277		277		276		277	
R-squared	.116		.144		.194		.121	

† ≤ .10, * ≤ .05, ** ≤.01   Standard errors noted in paranthesis

increases. No evidence of a relationship between supplier dependency and the use of supplier social practices was found (Hypothesis H2a).

Finally, multiple dimensions of distance also were significantly related to the use of socially responsible supplier practices (Hypothesis H3). Several important relationships emerged. First, as the organizational length of the supply chain increased, the use of supplier labor practices, COC, and social audits increased. Moreover, being positioned further upstream in the supply chain (i.e., closer to raw material sourcing) also was related to greater use of supplier COC. However, contrary to expectations, as the weighted geographical distance for suppliers increased, the use of supplier COC decreased. Finally, no evidence was found that customer geographic or cultural distance was related to the use of socially responsible supplier practices.

Additional diagnostics were considered to assess any potential impact of collinearity. Only two independent variables had a correlation coefficient greater than 0.3, suggesting few concerns. Further modeling assessed the impact of deleting individual variables, and the regression results changed little.

Discussion

The chapter identified a number of drivers that increased the adoption of socially responsible practices. First, one form of transparency, namely product visibility, was significant across all models. Thus, the use of multiple socially responsible supplier practices is linked to greater product visibility. Potentially, firms might engage in these practices to proactively protect their firm reputation and brands. For example, firms that make greater use of a COC have a more formalized process in place for the manner in which they interact with suppliers. This formalized process helps to educate both managers in the firm and its suppliers about socially acceptable (but culturally bound) procedures. Furthermore, with greater use of supplier social audits, firms follow up with their suppliers to ensure adherence. Collectively, these practices decrease the probability that suppliers are engaged in improper practices, such as using child labor, which could harm the firm and the value of its brands.

It is important to consider how this study might be interpreted in light of prior literature. Tapscott and Ticoll (2003) argued that firms would no longer be able to hide from their customers in this new technology age. Our research provides empirical evidence to support this perspective; firms engaged in socially responsible supplier practices when customers were aware of the firm and its brand. However, little evidence was found to support the notion that the structure of the supply chain significantly impacted the adoption of socially responsible supplier practices. Perhaps this occurred because customers might not yet fully comprehend the notion of supply chains, or might have little interest in understanding the flow of products through many firms on its way to the end user. Naturally, this might evolve and change over time as communication technology further advances and customers come to expect clear reporting and traceability of products through multiple firms in complex supply chains.

Only limited support was identified for the relationships between dependency and the various socially responsible supplier practices. In particular, for human rights, greater customer dependency did encourage the greater use of supplier practices. To some extent, one might argue that this is the most basic form of socially responsible supplier practices, and the first hurdle that end users are likely to worry about in a developed market, such as Canada. This concern is passed backwards up the supply chain—at least to firms that are highly dependent on specific customers. There are several factors that might explain the limited findings for the other practices. First, dependency may have much less influence on other socially responsible practices than do related constructs, such as the degree of power or control exercised by a firm over supply chain partners. Second, the costs of implementing these practices were not measured, and likely vary by supply chain context, which may outweigh the impact of dependency.

On the other hand, distance in the supply chain did impact the adoption of socially responsible supplier practices. Organizational distance was a dimension of distance that was important in the majority of the models. First, the relationship between the total length of the supply chain and socially responsible supplier practices was positive. All other aspects being equal, having more tiers in the supply chain translates into greater complexity and greater uncertainty. Establishing systems to formally develop socially responsible supplier practices can attenuate the range and number of concerns that must be managed. For example, social audits of suppliers are a clear mechanism to actively monitor a global supply base and ensure that far-flung suppliers are acting in ways that are consistent with the firm's own standards.

This finding can be linked to prior research in operations management that has studied supplier monitoring. The practices that make up the social auditing dimension of socially responsible supplier practices were heavily borrowed from the environmental management field (Klassen & Vachon, 2003; Zhao et al., 2007), where some work has considered the impact of the length of the supply chain on the adoption of supplier audits (Kovács, 2008). Thus, the findings of this research can potentially be transferred to environmental management research. More generally, firms might be expected to establish increasingly sophisticated audit systems for their suppliers' practices and performance (be they social, environmental, quality, etc.) as the supply chain increases in length.

At a superficial level, one might argue that stronger collaboration with suppliers might remove the need for auditing and monitoring of suppliers (Vereecke & Muylle, 2006; de Leeuw & Fransoo, 2009). However, even in the presence of collaboration, a firm must clearly signal both its concern and commitment to particular standards of performance, and not simply "trust" that its expectations are met. Thus, while auditing is undoubtedly important in relatively short supply chains, its importance grows as the supply chain lengthens and increases in complexity. Both trust and open communication are likely to decline as the number of tiers between the firm and a supplier increases. Thus, auditing provides a tangible indicator of the importance of appropriate supplier behavior, while also exercising greater control over an extended supply chain. Future research might more clearly explore the relationship between collaboration, auditing, and supply chain length to better disentangle the role and importance of each factor. For example, collaboration might attenuate, to some degree, the need for auditing; alternatively, auditing might encourage collaboration as both the firm and the supplier become more aware of the other party's needs, priorities, and capabilities.

The empirical analysis clearly indicated that firms that are closer to raw material extraction are more likely to put supplier COC in place. Several possible explanations support this outcome. First, as a firm gets closer to raw material sourcing, it is conceivable that the firm is more likely to be directly buying from plants and operations in developing countries.

Depending on the degree of legal enforcement present, local suppliers may be tempted to cut corners, or in fact, see little reason to meet vague expectations imposed by foreign buyers. Thus, COC ensure that the supplier is well aware of what is expected, and of the consequences of not meeting those expectations. Second, several certifications gaining widespread adoption, e.g., Fairtrade, are related to raw material extraction or supply. These certifications can serve as proxies for COC. In contrast, certifications have not been applied to the same degree for exporters, distributors. and downstream processors. However, one might expect this to change over time as the management of social issues in the supply chain broadens and deepens; thus, new certifications might extend Fairtrade principles further downstream in the supply chain (Roberts, 2003).

However, at least one finding appeared to be counterintuitive: as geographical distance increased, the firm was less likely to establish supplier COC (despite the points raised in the previous paragraph). One possible reason is that as the geographical distance increases, the number of repeated interactions between a firm and individual suppliers might decrease; furthermore, any interactions might be more transactional in nature, and there are fewer opportunities for unethical dealings between the supplier and the firm. Thus, larger distances are traded off against position in the supply chain.

Managerial Implications

There are a number of reasons for operations managers to adopt socially responsible supplier practices and to invest time and financial resources in these practices. Overall, two possible rationales underpin these decisions: seeking new opportunities to position a firm's products and brands and minimizing the risks of criticisms and concerns from NGOs, the public, and customers. Moreover, developing socially responsible supplier practices can help mitigate the negative outcomes of unexpected events and/or revelations.

Managers with valuable, highly visible brands do not want their brand images tarnished by improper practices in the supply chain. Therefore, these managers are more likely to invest in practices that might help to protect their product's brand. By putting these socially responsible supplier practices in place, it is less likely that improper practices in the supply chain would occur. Furthermore, operations managers that currently have socially responsible supplier practices in place can promote these practices to their customers to further differentiate the firm and give it more of a competitive advantage. This research found that supply chain visibility drove the adoption of socially responsible supplier practices. By extension, firms with well-developed practices can launch educational programs for consumers to illustrate the tangential social benefits that have been embedded in their products and supply chains. As has happened with Starbucks, educational campaigns can both influence customer purchase behavior and create conditions that force competitors to match their actions. But not all competitors are likely to have the management capabilities or economies of scale essential to replicating socially responsible practices in a cost-effective manner. Thus, socially responsible practices simultaneously achieve social gains, blunt potential criticism, and erect barriers for potential competitors.

The number of tiers in the supply chain also influences what operations managers did to address social concerns. As a supply chain becomes longer, operations managers put more procedures and policies in place, such as audits, to manage supplier adherence to social expectations (Mamic, 2005). However, as a firm is positioned farther downstream, it becomes more difficult to first, identify specific suppliers (e.g., to try to name the particular third-tier supplier providing the iron ore for the steel casing of a computer), and second, to realistically track working conditions (e.g., labor) used for manufacturing particular materials and parts in organizationally distant suppliers. Moreover, some would argue that COC are less relevant as the firm's responsibility diminishes with more organizational handoffs (New, 2004). Additionally, firms that are closer to the commodity sourcing or extraction stage have demonstrated greater use of supplier COC, which in turn, can have a positive impact on the supply chain (Kovács, 2008).

It is interesting to look at the level of involvement of the four types of socially responsible supplier practices. Admittedly, while the data gathered were perceptual and subjective in nature, the relative use (i.e., rank order) of each practice can be assessed. The survey results indicate that the level of involvement in supplier COC was the highest, followed by supplier human rights, then supplier labor practices, and finally supplier social audits. Thus, we see evidence for progressive development from internal (i.e., COC involve both the buyer and supplier), to supplier-oriented (i.e., human rights and labor practices), to verification of practice (i.e., auditing). In essence, operations managers are beginning to address social issues by trying to get their own operational practices in order, as supplier COC ensure that buyer's employees have procedures and practices to deal with suppliers ethically. Additional practices then engage and push suppliers to improve their human rights and labor practices. Finally, operations managers would put supplier social audits in place to ensure that suppliers are adhering to these new social practices.

Limitations and Conclusions

There were several limitations to this research. First, the surveys were only administered to Canadian manufacturing plants. Although the general findings of this study are expected to be generalizable to firms in other developed countries, the extent and strength of particular relationships might vary from country to country. Also, costs of the four socially responsible practices were not assessed, and are likely to influence both the degree and rank order of the constructs. The resources to implement socially responsible supplier practices might also vary based on the overall degree of competitiveness and profitability of particular industries or regions. Thus, social auditing of suppliers in China might be much less expensive for a Japanese firm than for a Canadian firm, thereby influencing its adoption. Similar issues arise with the importance of particular practices to firms in different countries, as American customers might view their relative importance quite differently than, for example, European customers. While the research design tested for the potential of social desirability, it cannot be entirely ruled out. Further research along similar lines by others might provide additional insight into its relative impact on data collection.

One of the main contributions of this research is the development of an operations perspective of social issues in the supply chain. In contrast to earlier research, practices related to social issues were clearly delineated from the sustainable development literature, in general, and the environmental management literature, in particular. Prior work also has tended to employ proxies, such as accounting measures, which are difficult to translate to operationally relevant management levers. Based on prior literature, corporate best practice, and the foundational work of international standards organizations, a set of operations-based scales was developed and empirically validated, using a large-scale survey of three industries in Canada. Four dimensions of socially responsible supplier practices were identified: supplier human rights; supplier labor practices; supplier codes of conduct; and supplier social audits. These scales also represent an initial step toward quantifying the costs and benefits of developing socially responsible practices and assessing management attention and capabilities in this area.

Second, much of operations and strategy is concerned with the design of the supply chain. Thus, when considering a new area such as social issues, key supply chain variables were expected to influence the development and use of socially responsible management practices. To that end, relationships between supply chain structure and supplier practices were tested empirically. Evidence pointed to two factors in supply chain structure—transparency and distance—being related to socially responsible supplier practices. For transparency, the degree of product visibility to end users was significantly related to greater use of multiple dimensions of socially responsible supplier practices. Thus, firms can be construed to be using these practices to either enhance or protect the firm's brand and reputation. Also, organizational distance was related to socially responsible supplier practices, with more tiers in the supply chain being linked to greater use of these practices, which help better manage societal expectations in the supply chain. Finally, as the plant was positioned further upstream in the supply chain (closer to raw material extraction), it was more likely to have supplier codes of conduct in place.

The development of these scales provides an empirical basis for further scale refinement by other scholars characterizing socially responsible practices. This research focused very explicitly on supplier practices. However, suppliers capture only half of the supply chain. Customers too are expected to be instrumental, and the impact of customer expectations should be measured and modeled. Expectations might be captured directly, using surveys of representative customers, or indirectly, using panels of experts. Ideally, such survey work would be complemented by case studies that examine socially responsible practices across the entire supply chain. Both forms of empirical research would provide further details about new (or related) constructs, the nature of specific relationships, and the likelihood of different performance outcomes. For example, some practices such as social auditing might be strongly related to risk-reduction, while others, such as supplier labor practices, might be linked to opportunities, e.g., increased productivity.

Additionally, a survey could be conducted on the general population to explore how typical consumers assign responsibility for social issues in supply chains for common products, like food or manufactured durable goods. Possibly, operations managers need only take responsibility for some social issues a single tier back in the supply chain—for example, worker conditions—while other issues, such as product safety, encompass all tiers in the supply chain. Moreover, responses of consumers could be compared against those from plant managers to understand the degree of consistency or alignment. Collectively, understanding both the assignment and scope of responsibility can assist managers with identifying high-priority areas, and focusing the investment of limited resources.

Appendix 1. Estimates and model fit statistics for four socially responsible supplier practices

To what extent is your plant involved in the following policies and procedures related to your **primary suppliers**? In general, our plant: (*please circle a number for each item*)

Variable	Item		Estimate
Supplier Human Rights	HR1	ensures that our suppliers do not use child labor	.956
	HR2	ensures that our suppliers do not use forced labor (e.g., prison labor)	.975
	HR3	ensures that our suppliers pay their workers a living wage (i.e. a wage that is above minimum wage)	.868
	HR4	ensures that suppliers do not use sweatshop labor	.936
Supplier Labor Practices	SLC1	ensures that our suppliers have regulated over-time wage policies (e.g., employees are paid a higher wage for over-time work)	.890
	SLC2	ensures that our suppliers allow their employees to associate freely (e.g., join or create a union)	.859
	SLC3	ensures that our suppliers do not discriminate against its own workers	.925
	SLC4	ensures that our suppliers provide a healthy and safe working environment for their employees	.854
	SLC5	ensures that our suppliers provide their employees with protective equipment in hazardous areas	.864
	SLC6	ensures that our suppliers help improve the natural environment in which they operate	.884
Supplier Codes of Conduct	SCOC1	has a supplier code of conduct	.565
	SCOC2	has a management system to ensure that social expectations affecting our suppliers are identified	.526
	SCOC3	ends relationships with suppliers that do not adhere to our code of conduct	.834
	SCOC4	has a defined set of acceptable/ unacceptable behavior (e.g. ethics statement) our employees must abide by	.851
	SCOC5	has ethical sourcing training programs for the purchasing department	.747
Social Audits	SA1	monitors our suppliers to ensure adherence to our social expectations	.862
	SA2	conducts surprise visits to our suppliers to ensure adherence to our social expectations	.884
	SA3	has specific audit procedures to ensure that our suppliers adhere to our social expectations	.914

χ^2 = 56.597
p-value = .149
χ^2/d.f. = 1.204
NFI = .976
TLI = .994
CFI = .996
RMSEA = .030
RMSEA CI = .000, .055

Appendix 2. Assessment of firm's dependence on suppliers and customers

For the following items, please describe your **plant's** relationship with **your primary suppliers**. In general, our plant (*please circle a number for each item*)

			Estimate
Supplier Dependency	SD1	is greatly dependent on our suppliers	.536
	SD2	has difficulty changing suppliers	.807
	SD3	requires a long time to change to new suppliers	.868
	SD4	finds it very costly to change to new suppliers	.689
	SD5	will perform poorly if our suppliers don't perform well	.527

For the following items, please describe **your primary customer's** relationship with your **plant**. In general, our customers (*please circle a number for each item*)

Customer Dependency	CD1	are greatly dependent on our plant	.549
	CD2	have difficulty changing suppliers	.868
	CD3	require a long time to change to new suppliers	.929
	CD4	find it very costly to change to new suppliers	.806
	CD5	will perform poorly if our plant doesn't perform well	.599

The **end-user** is defined as the end customer that purchases and uses/consumes the product (the final link in the supply chain). For your primary product, please indicate the extent to which the **end-user** is aware of (*please circle a number for each item*)

End-User Knowledge of Supply Chain	SCS1	how our product is manufactured	.787
	SCS2	the type of raw materials that go into the product	.761
	SCS3	where the raw materials are sourced	.865
	SCS4	the structure of our supply chain	.817
Product Visibility	PV1	the name of the company that manufactures the product	.829
	PV2	our brand name (product name)	.818
	χ^2	= 53.130	
	p-value	= .219	
	χ^2/d.f.	= 1.155	
	NFI	= .965	
	TLI	= .993	
	CFI	= .995	
	RMSEA	= .026	
	RMSEA CI	= .000, .052	

References

Armstrong, J. S., & Overton, T. S. (1977). Estimating nonresponse bias in mail surveys. *Journal of Marketing Research, 14* (3), 396–402.

Banet, A. G. (1976). Organizational distance: A concept for the analysis and design of organizations. *Group & Organization Studies, 1*(4), 496–497.

Bansal, P., & Hunter, T. (2003). Strategic explanations for the early adoption of ISO 14001. *Journal of Business Ethics, 46*(3), 289–299.

Baumann, H., Boons, F., & Bragd, A. (2002). Mapping the green product development field: Engineering, policy and business perspectives. *Journal of Cleaner Production, 10*(5), 409–425.

Brundtland, G. H. (1987). *Our common future/World Commission on Environment and Development.* Oxford: Oxford Univ. Press.

Carr, A., Kaynak, H., Hartley, J., & Ross, A. (2008). Supplier dependence impact on supplier's participation and performance. *International Journal of Operations and Production Management, 28*(9), 899–916.

Carter, C. R. (2000). Ethical issues in international buyer–supplier relationships: A dyadic examination. *Journal of Operations Management, 18*(2), 191–208.

Carter, C. R., & Jennings, M. M. (2002). Logistics social responsibility: An integrative framework. *Journal of Business Logistics, 23*(1), 145–180.

Carter, C. R., & Jennings, M. M. (2004). The role of purchasing in corporate social responsibility: A structural equation analysis. *Journal of Business Logistics, 23*(1), 145–186.

Carter, C. R., & Rogers, D. S. (2008). A framework of sustainable supply chain management—moving toward new theory. *International Journal of Physical Distribution & Logistics Management, 38*(5), 360–387.

Castka, P., & Balzarova, M. A. (2008). The impact of ISO 9000 and ISO 14000 on standardisation of Social responsibility—An inside perspective. *International Journal of Production Economics, 113*(1), 74–87.

Choy, K. L., & Lee, W. B. (2003). A generic supplier management tool for outsourcing manufacturing. *Supply Chain Management: An International Journal, 8*(2), 140–154.

Cool, K., & Henderson, J. (1998). Power and firm profitability in supply chains: French manufacturing industry in 1993. *Strategic Management Journal, 19*(10), 909–926.

Crowne, D. P., & Marlowe, D. (1960). A new scale of social desirability independent of psychopathology. *Journal of Consulting Psychology, 24*(4), 349–354.

De Leeuw, S., & Fransoo, J. (2009). Drivers of close supply chain collaboration: One size fits all? *International Journal of Operations & Production Management, 29*(7), 780–739.

Dillman, D. A. (2000). *Mail and Internet surveys: The tailored design method.* New York: Wiley. .

Dixit, A. K., & Nalebuff, B. J. (1991). Thinking strategically: The competitive edge in business. In *Politics, and Everyday Life, New York:* WW Norton.

Elkington, J. (1997). *Cannibals with forks: The triple bottom line of 21st century business.* Oxford: Capston Publishing.

Emmelhainz, M. A., & Adams, R. J. (1999). The apparel industry response to "sweatshop" concerns: A review and analysis of codes of conduct. *Journal of Supply Chain Management, 35*(3), 51–57.

Fairtrade Labelling Organizations International (2007). Fairtrade Certification. Retrieved December 10, 2007, from http://www.fairtrade.net/

Fischer, D.G., & Fick, C. (1993). Measuring social desirability: Short forms of the Marlowe-Crowne Social Desirability Scale. *Educational and Psychological Measurement, 53*(2), 417–424.

Forza, C. (2009). Surveys. In Karlsson, C. (Ed.), *Researching operations management,*. New York: Routledge Taylor & Francis.

Graafland, J.J. (2002). Sourcing ethics in the textile sector: The case of C & A. *Business Ethics: A European Review, 11*(3), 282–294.

Global Reporting Initiative (GRI) (2006). Sustainability Reporting Guidelines. Boston: Global Reporting Initiative, 94.

Group, B. I. (2007). *Scott's national manufacturers: 20+ employees directory.* Toronto: Scott's.

Grover, V., & Malhotra, M. K. (2003). Transaction cost framework in operations and supply chain management research: Theory and measurement. *Journal of Operations Management, 21*(4), 457–473.

Grow, B. (2005). The debate over doing good. *Business Week.* 3947 (August 15): 76.

Harrigan, K. R. (1985). Vertical integration and corporate strategy. *The Academy of Management Journal, 28*(2), 397–425.

Hofstede, G. H. (1980). *Culture's consequences: International differences in work-related values.* Beverly Hills, CA: Sage.

Hofstede, G. H. (2001). *Culture's consequences: Comparing values, behaviors, institutions, and organizations across nations.* Thousand Oaks, CA: Sage Publications.

Jantzi. (2008). Jantzi research. from http://www.jantziresearch.com/index.asp?section=12.

Johnson, P., Klassen, R., Leenders, M., & Awaysheh, A. (2007). Selection of planned supply initiatives: The role of senior management expertise. *International Journal of Operations & Production Management, 27*(12), 1280–1302.

Joynt, P., & Warner, M. (1996). *Managing across cultures: Issues and perspectives.* London:International Thomson Business Press.

Kaufmann, L., & Carter, C. R. (2006). International supply relationships and non-financial performance—A Comparison of US and German Practices. *Journal of Operations Management, 24*(5), 653–675.

Klassen, R. D., & Vachon, S. (2003). Collaboration and evaluation in the supply chain: The impact on plant-level environmental investment. *Production and Operations Management, 12*(3), 336–352.

Koplin, J., Seuring, S., & Mesterharm, M. (2007). Incorporating sustainability into supply management in the automotive industry—The case of the Volkswagen Ag. *Journal of Cleaner Production 15*(11–12), 1053–1062.

Kovács, G. (2008). Corporate environmental responsibility in the supply chain. *Journal of Cleaner Production, 16*(15), 1571–1578.

Krause, D. R. (1999). The antecedents of buying firms' efforts to improve suppliers. *Journal of Operations Management, 17*(2), 205–224.

Lee, H. L. (2002). Aligning supply chain strategies with product uncertainties. *California Management Review, 44*(3), 106.

Lessler, J. T., & Kalsbeek, W. D. (1992). *Nonsampling error in surveys.* New York: John Wiley & Sons.

Maloni, M., & Benton, W.cC. (2000). Power influences in the supply chain. *Journal of Business Logistics, 21*(1), 49–73.

Mamic, I. (2005). Managing global supply chain: The sports footwear, apparel and retail sectors. *Journal of Business Ethics, 59*(1), 81–100.

Martin, R.L. (2002). The virtue matrix: Calculating the return on corporate responsibility. *Harvard Business Review, 80*(3), 68–75.

New, S. J. (2004). The ethical supply chain. In New, S., Westbrook, R. (Eds.) *Understanding supply chains: Concepts, critiques and futures.* Oxford: Oxford University Press, pp. 253–280.

Nike. (2004). Corporate responsibility report, fiscal year 2004. Beaverton, OR: Nike, Inc.

O'Grady, S., & Lane, H. W. (1996). The psychic distance paradox. *Journal of International Business Studies, 27*(2), 309–333.

O'Leary-Kelly, S. W., & Vokurka, R. J. (1998). The empirical assessment of construct validity. *Journal of Operations Management, 16*(4), 387–405.

Pfeffer, J., & Salancik, G. R. (1978). *The external control of organizations.* New York: Harper & Row.

Porter, M. E., & Kramer, M. R. (2006). Strategy and society: The link between competitive advantage and corporate social responsibility. *Harvard Business Review, 85*(12), 14.

Reynolds, N. L., Simintiras, A. C., & Diamantopoulos, A. (2003). Theoretical justification of sampling choices in international marketing research: Key issues and guidelines for researchers. *Journal of International Business Studies, 34*(1), 80–89.

Roberts, S. (2003). Supply chain specific? Understanding the patchy success of ethical sourcing initiatives. *Journal of Business Ethics, 44*(2), 159–170.

Rudelius, W., & Buchholz, R. A. (1979). What industrial purchasers see as key ethical dilemmas. *Journal of Purchasing and Materials Management, 15*(4), 2–10.

Segars, A.H. (1997). Assessing the unidimensionality of measurement: A paradigm and illustration within the context of information systems research. *Omega, 25*(1), 107–121.

Settings, M. (2004). Travail, transparency and trust: A case study of computer-supported collaborative supply chain planning in high-tech electronics. *European Journal of Operational Research, 153*(2), 445–456.

Seuring, S., & Müller, M. (2008). From a literature review to a conceptual framework for sustainable supply chain management. *Journal of Cleaner Production, 16*(15), 1699–1710.

Social Accountability International (2008). Social Accountability International SA8000: 2008. New York, NY.

Strahan, R., & Gerbasi, K.C. (1972). Short, homogeneous versions of the Marlowe-Crowne Social Desirability Scale. *Journal of Clinical Psychology, 28*(2), 191–193.

Tapscott, D., & Ticoll, D. (2003). *The naked corporation: How the age of openness will revolutionize business.* New York: Free Press.

Van Der Zee, D. J., & Van Der Vorst, J. (2005). A modeling framework for supply chain simulation: opportunities for improved decision making. *Decision Sciences, 36*(1), 65–95.

Vereecke, A., & Muylle, S. (2006). Performance improvement through supply chain collaboration in Europe. *International Journal of Operations & Production Management, 26*(11), 1176–1198.

Vincenty, T. (1975). Direct and inverse solutions of geodesics on the ellipsoid with application of nested equations. *Survey Review, 23*(176), 88–93.

Williamson, O.E. (1979). Transaction-cost economics: The governance of contractual relations. *Journal of Law and Economics, 22*(2), 233–261.

Zhao, X., Flynn, B. B., & Roth, A. V. (2007). Decision sciences research in China: Current status, opportunities, and propositions for research in supply chain management, logistics, and quality management. *Decision Sciences, 38*(1), 39–80.

CHAPTER TWELVE

Going Green by Exporting

EMILIO GALDEANO-GÓMEZ, EVA CARMONA-MORENO,
AND JOSÉ CÉSPEDES-LORENTE

The focus of this chapter is on the relationship between environmental performance and exports at country or industry level. The aim is to provide an analysis of these relationships at the firm level. Can environmental performance explain differences in export intensity between firms? Does a firm's environmental performance improve as its export activity improves? The approach followed is a composite equation model in which export intensity and productivity components, including environmental productivity, are jointly determined, taking as reference the exporting firms of the agro-food industry in southeast Spain.

Some studies point out that international trade may have the potential to contribute to overall improvements in the environmental performance of specific industries, or in regional environmental performance (e.g., Liddle, 2001; Frankel & Rose, 2002; Copeland & Taylor, 2003); others believe that export intensity may contribute to the risk of environmental damage (e.g., Managi & Karemera, 2005). In both cases, it may be argued that the relationship between regional export intensity and regional environmental performance (e.g., pollution) is bi-directional, as causal effects may go in both directions: regional export intensity may improve or worsen regional environmental performance. Also, environmental performance differences between regions can explain the differences in the regions' export intensity (e.g., Ederington & Minier, 2003; Cole & Elliot, 2003). Although these studies have provided useful insights into the interrelationship between environmental performance and export intensity at the regional or industry level, they have not explored the role of firm's specific factors (i.e., the firm level of analysis). In recent years, literature on international trade has paid attention to the role that firms play in mediating countries' imports and exports (e.g., Wagner, 2007). Several studies have highlighted the links between firm productivity and export activity, assuming that this relationship may be bi-directional (e.g., Bernard, Eaton, Jensen, & Kortum, 2003;

Helpman, Melitz, & Yeaple, 2004). Nevertheless, these studies have not considered environmental factors.

The analysis of bi-directional effects between environmental performance and export intensity at the firm level is important for several reasons. First, environmental performance has become a goal for firms, which are subjected to intense scrutiny from society and from policymakers (Tyteca, Carlen, Berkhout, Hertin, Wehrmeyer, & Wagner, 2002). In this context, environmental performance has to be considered as an additional component of total firm productivity. Second, it is important to note that the firm level may be more appropriate than the industrial or regional level to analyze this relationship, as the responses of firms to environmental policy is heterogeneous in terms of environmental strategies and the development of environmental capabilities (Aragón, 1998; Aragón & Sharma, 2003). For example, Madsen (2009) found that firm-specific environmental capabilities moderated the relationship between the stringency of a country's environmental regulations and the firm's investment in that country. Third, we would like to understand the effects of a firm's environmental performance on export intensity, an issue that has received scant attention (see Christmann & Taylor, 2001; Martín, Aragón, & Senise, 2008). Identifying these effects can shed light on the potential benefits of promoting environmental performance improvements in order to gain international competitiveness. Additionally, it is interesting to determine whether, and to what extent, export intensity can give rise to an environmental learning process that improves firm environmental performance. Finally, regardless of which direction of effect has been studied, it is necessary to account for the other direction of causal effect. In other words, environmental performance and export intensity are jointly determined. Hence, estimates of induced environmental performance effects that do not account for the joint endogeneity of export intensity and environmental performance are likely to be biased.

The objective of this chapter is to examine these interrelationships by means of productivity index estimation methods. Over the last two decades, research on productivity analysis has shown how firm productivity measurements are sensitive to environmental performance; firms frequently consider undesirable outputs (i.e., pollutant outcomes of production activity) in input–output relationships (see Tyteca, 1997, for an overview). Obtaining an environmental performance index, namely, environmental productivity (e.g., Ball, Lovell, Luu, & Nehring, 2004; Kaneko & Managi, 2004) as a component of a Total Factor Productivity (TFP) index, we try to provide evidence on issues of interest in international trade analysis: whether environmental performance can explain differences in export intensity between firms and whether firms' environmental performance improves (or worsens) as their export activity increases (or decreases).

To this end, we developed an empirical framework to analyze the relationship between exports and environmental performance in a sample of firms belonging to the agro-food sector. We also estimated the impact of exports on general productivity components (efficiency and technological change) and on environmental productivity. These variables were determined endogenously in our empirical model, following recent frameworks by Ederington and Minier (2003), and Managi and Karemera (2005).

Environmental Performance, Exports, and Productivity

Productivity has been one of the basic variables for measuring differences between firms in international trade analyses (e.g., Bernard et al., 2003; Helpman et al., 2004). Two alternative but not mutually exclusive hypotheses about why exporters can be expected to be more productive than nonexporting firms have been discussed and investigated empirically (International Study Group on Exports and Productivity, 2008). The first hypothesis points to self-selection of the more productive firms into export markets. If there exist additional costs of selling goods in foreign countries (transportation costs, distribution or marketing costs, personnel with skills to manage foreign networks, and production costs in modifying current domestic products for foreign consumption), these costs provide an entry barrier that less productive firms cannot overcome. The second hypothesis points to the role of learning by exporting. The pressure of international competition forces exporters to cut costs and improve efficiency by eliminating managerial and organizational inefficiencies (Egan & Mody, 1992; Clerides, Lauch, & Tybout, 1998). Also, the participation in a sophisticated network of international stakeholders favors knowledge flows and firm absorption of knowledge. Thus, exporting makes firms more productive.

Wagner (2007) summarized the results of a comprehensive survey of the empirical literature about the relationship between firm export intensity and productivity, and found that exporters are more productive than nonexporters, and that more productive firms self-select into export markets, although exporting does not necessarily improve productivity. However, this big picture hides a lot of heterogeneity, as the different empirical studies differ substantially in terms of type of data, analysis, and other details of the approach (International Study Group on Exports and Productivity, 2008).

The impact of environmental policy on international trade has frequently been analyzed based on assumption that environmental policy mainly affects standard factors, such as differences in resource endowments (e.g., skilled workers, technological innovation capabilities, efficiency in usage of resources) or technology, and determines trade patterns, commonly known as the "factor endowment hypothesis"[1] (see Jaffe, Peterson, Portney, & Stavins, 1995; Copeland & Taylor, 2003). Porter and Van der Linde (1995) argued that tighter environmental regulation spurs technological innovation, and hence tighter regulation could, in theory, raise exports or lower imports (this is known as the Porter hypothesis). The findings on this theoretical argument have often been contradictory. Van Beers and van den Bergh (1997) suggested a positive relationship between environmental policy and exports, while others found small and often insignificant relationships between these two variables (e.g., Tobey, 1990; Jaffe et al., 1995).

On the other hand, it is argued that competitive exports have contradictory impacts on environmental outputs, both increasing pollution and motivating reductions in it (Liddle, 2001). The direction and magnitude of these effects depend on the trade-induced changes in production patterns, the state of the environment, and the environmental policies in place to preserve and improve environmental quality (Shortle & Abler, 2001). Recently, analyses of this issue have been conducted. Ederington and Minier (2003), and Levinson and Taylor (2008) both found evidence that by treating pollution regulations as endogenous, such that pollution regulations affected international trade and vice versa, pollution abatement costs were a significant determinant of trade competitiveness at an aggregate level. Frankel and Rose (2002) treated income and trade as endogenous and found an impact of trade on air pollution. Managi and Karemera (2005) used environmental productivity as a measure of abatement pollution effort, obtaining interrelationships between exports and this productivity indicator at an aggregate level (United States) in a simultaneous estimation model.

At the firm level, these relationships have scarcely been debated, although in recent years international economics literature has paid increasing attention to the role that firms' environmental strategies play in mediating countries' imports and exports. Some studies have analyzed the relationship between firms' environmental performance and their export activity, finding a positive relationship between them (Christmann & Taylor, 2001; Bellesi, Lehrer, & Tal, 2005; Martín et al., 2008). These authors assume that the firms' adoption of advanced environmental management practices improves their international competitiveness, showing findings in favor of the Porter hypothesis. Using the resource-based theory (see, e.g., Barney, 1991), some authors argue that the adoption of advanced environmental strategies that go beyond regulatory requirements by dealing with waste reduction at the source and pollution prevention influences the development of firm capabilities and favors an improvement in its financial (and export) performance (e.g., Hart, 1995; Russo & Fouts, 1997; Sharma & Vredenburg, 1998; Martín et al., 2008). The adoption of advanced environmental practices increases the firm's environmental productivity and consequently it: (1) encourages innovation of both products and processes (Beise & Rennings, 2005), which make the firm more productive; (2) promotes better firm relationships with international stakeholders, including customers and governments; (3) facilitates organizational learning processes (Sharma & Vredenburg, 1998), which are essential in the international context; and (4) improves the firm's reputation, which is a crucial asset in accessing international markets (Martín et al., 2008).

When environmental regulations imposed by importing countries function as protective trade barriers, firms can address such problems by developing practices that allow them to comply with the strictest environmental regulations prevailing in the largest export market (Christmann & Taylor, 2001). Thus, only firms with the highest environmental productivity can export to these markets.

The foregoing observations and arguments lead to the following hypothesis:

Hypothesis 1. Improvements in a firm's environmental productivity will increase the firm's export intensity.

However, although the previous arguments suggest a process of self-selection in firms' export activity, other factors may explain to what extent greater export intensity reinforces improvement in environmental performance, as firms learn by exporting (Grossman & Helpman, 1991). Exporting firms are exposed to knowledge (e.g., environmental knowledge) that is unavailable to domestic firms. This knowledge can be of two types (Salomon & Jin, 2010): market knowledge and technological knowledge. Exporting firms receive information related to tastes and preferences from consumers in the target markets. This information allows them to tailor products to meet these customers' specific needs, including environmental attributes. For instance, since consumers in the country of destination cannot control environmental performance of the exporting firms, it is highly likely that they require these firms to obtain some specific environmental quality certificates (e.g., Bellesi et al., 2005). In the same vein, as the pressure for environmental quality increases in the country of destination, and it imposes more environmental requirements for products, exporting firms (or countries) also feel obliged to improve their environmental performance by using greener processes. Thus, increasing stringency of a country's environmental policy would increase the exporting firms' environmental investments, which, in turn, would lead to an increase in the firms' environmental productivity, if they wish to continue exporting to this country.

In both cases, exporting firms' environmental learning is extended by the acquisition and application of technological knowledge. Exporters can access international knowledge networks in which customers, other firms, associations, research centers, et cetera, from different countries participate. Intensive involvement in such networks can facilitate the acquisition of knowledge on operating processes, methods, and techniques that enhance the firm's environmental capabilities, and by extension their environmental performance. These arguments reflect the fact that the intensity of the firm's exports is related to the possibility of acquiring knowledge about green technologies and practices. This knowledge is a fundamental part of the learning processes that facilitate the improvement of the firm's environmental productivity. Thus,

Hypothesis 2. The increasing export intensity of a firm will increase the firm's environmental productivity.

Methods

The empirical analysis has been based on balanced panel data using annual financial reports and surveys of 65 firms belonging to the agri-food industry in southeast Spain over the 1994–2006 period. This sample represents 81 percent of the value of total regional exports, mainly destined for the EU market (which receives over 90 percent of total exports). The agro-food industry in southeast Spain centers around a series of products: fresh and processed vegetables, olive and olive oil, grapes and wine, and fresh and processed citrus produce. Our project focuses on fresh fruit and vegetables and minimally processed produce (cut and pre-prepared), since these products are those that sell most on the international market (Agricultural Council of Andalusia, 2007) and have similar production processes. The firms included in our empirical study are farming–marketing firms, which consist mainly of cooperatives. Such entities do sales and marketing on behalf of the farmer producers, who in this case total more than 9000 farmers. We define the production process as the farming, handling, processing, and packaging of produce. Farming production utilizes greenhouses, the main technological system of production since the 1970s, because of the climatological conditions of this area (intense heat and scarcity of water).

However, this system has generated considerable amounts of waste and residues (fertilizers, packaging materials, waste produce, and so on) and makes an intensive use of resources. As far as water consumption is concerned, considerable advances have been made, and currently almost 100 percent of farming activity is carried out using modern systems of irrigation and water-saving techniques (hydroponic planting, drip irrigation, etc.). The intensive use of fertilizers and pesticides has persisted up to recent years. Since the 1990s, however, as a result of the increasing relevance of environmental quality components to consumers in the EU markets, the managerial activity of the firms has changed considerably.

Environmental performance in these firms has been primarily aimed at promoting more ecological produce for marketing: the application of integrated pest management; investment in irrigation systems to save water and avoid pollution; the application of methods that avoid soil pollution; improvements in residue and waste management; and more frequent analysis of soil, water, plants, and waste (Galdeano-Gómez, Céspedes-Lorente, & Rodríguez-Rodríguez, 2006). In many cases these measures are included in certified environmental management systems (EMS), such as the EUREP-GAP code, the ISO 14001, and the Integrated Production Certificates. In most cases, these practices deal with waste reduction and pollution prevention at the source.

In general, the current intensification of environmental performance implies major changes in the managerial activity of these firms, but it may also be an essential factor in making their exports competitive (Céspedes-Lorente & Galdeano-Gómez, 2004).

Total Factor and Environmental Productivity Measurement. Studies on productivity have become more popular, thanks to the application of Malmquist indices. The Malmquist Total Factor Productivity (*TFP*) is a specific output-based measure of *TFP*. It measures the *TFP* change between two data points by calculating the ratio of two associated distance functions (Caves, Christensen, & Diewert, 1982). Chung, Färe, and Grosskopf (1997) defined an output-oriented Malmquist-Luenberger index of productivity, which included productivity changes with respect to both desirable and undesirable outputs.

The Malmquist-Luenberger *TFP* index and its components, efficiency change (*EFC*) and technological change (*TEC*) are defined in terms of the ratios of distance functions (Färe, Grosskopf, Norris, & Zang, 1994; Chung et al., 1997). In order to estimate the environmental productivity, we calculate two productivity indices (Kaneko & Managi, 2004) by comparing distance functions in two different time periods (t and $t+1$) using Data Envelopment Analysis (DEA) technique. First, a basic model is used to calculate total productivity of market output, TFP_{market}, using the usual production input and output. This measure can be operationally divided into two major components, namely *TEC* and *EFC*. Second, a joint model, TFP_{total}, which measures the total effect of increases in productivity due to improvements in technology and efficiency for the multi-production of marketable and undesirable outputs, is used. Increases in market output, and/or reduction in undesirable output, at a given input level, will increase TFP_{total}. Thus, the residual effects of two factors explain the changes in productivity related to the environmental output, i.e., environmental productivity (*EP*), given by Equation 1,

$$EP = TFP_{total} / TFP_{market} \qquad 1$$

where an increase in *EP* implies productivity improvements related to environmental performance. Given the same market productivity level, reduction in undesirable output increases *EP*, whereas an increase in undesirable output decreases *EP* (Managi, Opaluch, Jin, & Grigalunas, 2004). Thus, the growth or reduction of *EP* can be considered a positive or negative effect of the firm's abatement effort or the firm's environmental performance.

Export Performance Model. The empirical model of export performance proposed in this chapter is based on the assumption that export activity depends on firm-specific resources (Martín et al., 2008). Considering specialization differences, in this framework, export performance is measured as intensity in exporting (EX) and is considered a dependent variable of productivity components, including TEC, EFC, and EP, and the other company characteristics, such as the management's international experience and firm's size (Sterlacchini, 2001; Wagner, 2001). Thus, the model is specified as shown in Equation 2:

$$\ln EX_{it} = a_0 + a_1 \ln TEC_{it} + a_2 \ln EFC_{it} + a_3 \ln EP_{it} + a_4 \ln IE_{it} + a_4 \ln PM_{it} + a_5 \ln FS_{it} + e_{it} \qquad 2$$

where EX_{it} is the export intensity ratio of firm i at time t, which depends on: technological change, TEC; efficiency change, EFC; environmental productivity, EP; average price of marketed output, PM; international experience, IE (i.e., number of years the firm has been exporting); and firm size (i.e., firm market share), FS; and where e_{it} is the error term. This model allows us to estimate the effect of environmental productivity on export intensity (Hypothesis 1), once we have taken into account the effects of the other components of total productivity.

Technological and Efficiency Change Models. Exporting may contribute to productivity growth via efficiency improvement by cross-efficiency promotion and resource reallocation, and via technical progress by technological spillovers and encouragement of investment in research and development, or R&D (e.g., Wagner, 2007). Due to the nature of small or medium-sized agro-food firms in the sample (low investment in R&D activities), we consider the capital intensity (i.e., annual investment in machinery and equipment over total assets) and the percentage of intermediate imported inputs (raw materials) over total inputs as alternative explanatory variables (Harmse & Abuka, 2005). In addition, we consider that higher wages are expected to accompany higher productivity levels (Galor & Moav, 2000; Bernard et al., 2003) and that the concentration of exporting firms and firm size have a positive effect on productivity (Fu, 2005; Greenaway & Kneller, 2008). Disaggregating the productivity, we consider the two following models, Equation 3 and Equation 4:

$$\ln TEC_{it} = b_0 + b_1 \ln CI_{it} + b_2 \ln IS_{it} + b_3 \ln EX_{it-1} + b_4 \ln WG_{it} + b_4 \ln FS_{it} + b_6 \ln CR_t + e_{it} \qquad 3$$

$$\ln EFC_{it} = l_0 + l_1 \ln CI_{it} + l_2 \ln IS_{it} + l_3 \ln EX_{it-1} + l_4 \ln WG_{it} + l_4 \ln FS_{it} + l_6 \ln CR_t + v_{it} \qquad 4$$

where the productivity measures of firm i at time t depend on: the capital intensity, CI; the intermediate imported inputs, IS; the export intensity of previous year, EX_{t-1}; the wage rate, WG; the firm size, FS; and the concentration ratio, CR. The error terms are represented by e and v. It should be noted that the foregoing equations are necessary as we are interested in jointly estimating the relationships between each component of productivity and firm's export intensity.

Environmental Productivity Model. The effects of export intensity may be extended to the environmental productivity of firms. In our case, we use the reduction of waste (undesirable output) as a measure of abatement effort, this being a form of such productivity. We consider that some determinants of this environmental indicator may be different from the productivity equations set out above. Thus, in the equation set out below, instead of capital intensity (*CI*), an environmental investment intensity variable, *EI* (Anton, Deltas, & Khanna, 2004) is used (environmental investment over sales). Also included as explanatory variables are the environmental practices spillovers in the sector (measured by total investment in environmental practices by the other firms) and the firm's experience in environmental practices (Mazzanti & Zoboli, 2006; Galdeano-Gómez et al., 2006). With such modifications, the environmental productivity, *EP*, model is specified as shown in Equation 5:

$$\ln EP_{it} = f_0 + f_1 \ln EI_{it} + f_2 \ln ES_{it} + f_3 \ln EX_{it-1} \\ + f_4 \ln WG_{it} + f_4 \ln FS_{it} + f_6 \ln AGE_t + w_{it} \qquad 5$$

where the *EP* of firm i at time t depends on: the environmental investment intensity, *EI*; the environmental spillovers, *ES*; the export intensity, EX_{t-1}; the wage rate, *WG*; the firm size, *FS*; and the number of years that environmental practices have been adopted, *AGE*. The error term is represented by *w*.

For the estimation of *TFP*, we have used one market output, one undesirable output, waste, and two inputs, labor and capital. Market output is measured by the firm's value added (sales output minus intermediate inputs; Tyteca et al., 2002). The labor factor was obtained from the number of employees; the capital factor was obtained from the depreciation expenditures (accountable replacement value of fixed assets, including buildings, equipment and machinery; Martínez, Díaz, Navarro, & Ravelo, 1999). The environmental output (undesirable output) was measured by the waste and residues generated by the firm that was detected in environmental quality tests by the Agricultural Council of Andalusia (see Galdeano-Gómez, et al. 2006). (Until 2008 there was no plant for treating waste produce as compost or fertilizer in the region), We used an average quantity of wasted divided by marketed output.

Several indicators can be used to measure firms' export performance, but the most common ones refer to an export sales ratio or export intensity, *EX*, which may offer an adequate measurement of export activity relevance (e.g., Verwaal & Donkers, 2002). As regards the determinants of export intensity not previously defined, the average price of marketed output (*PM*) is calculated by a weighted average price index, and firm size (*FS*) is measured by the ratio of firm's sales to the total sales of the sample.

Among the determinants of productivity indicators of the firm, Equations 3 and 4, we considered the proportion of total inputs represented by intermediate imported inputs *(IS)*, mainly packaging materials.[2] The wage rate *(WG)* is calculated by the annual labor costs divided by the number of workers, and the concentration ratio *(CR)* in the sector is estimated by the Hirschman-Herfindahl ratio.

In Equation 5, we considered that the firm's environmental productivity *(EP)* must have a direct relationship with environmental investment intensity *(EI)*, as measured by the annual expenditure on environmental practices (which includes management of waste, contracting engineers and technicians and, in general, application of certified environmental systems) over sales. Also, the *EP* of each firm may be related to the number of years these environmental practices have been applied *(AGE)* and the existence of an environmental spillover effect *(ES)* in the sector, measured by the environmental expenditures of the other firms in the sample (see Galdeano-Gómez & Céspedes-Lorente, 2008).[3] It is assumed that geographical proximity increases the probability of environmental knowledge diffusion between firms (Mazzanti & Zoboli, 2006).

The monetary variables have been corrected for inflation (using the consumer price index, 1994 = 100) and are expressed in real terms. Table 12.1 shows the definition of variables, their expected signs, and data sources. Table 12.2 shows the descriptive statistics and the correlation matrix.

Results

The accumulated growth of technological change *(TEC)*, efffficiency change *(EFC)* and environmental productivity *(EP)* are shown in Figure 12.1.

The productivity results of the total sample indicate that the productivity improvement is mainly due to technological change, *TEC*, which increases from 1.009 in 1995 to 1.021 in 2006, showing an average annual growth of 1.35 percent over the period studied. Efficiency change *(EFC)* reaches 1.013 and environmental productivity *(EP)* reaches 1.022 by 2006, showing an average annual growth from 1995 to 2006 of 0.5 percent and 0.85 percent respectively. Nevertheless, the increase in the later *EP* is particularly noteworthy. For example, taking the unity as reference, inefficiencies in environmental performance *(EP < 1)* are observed for the three first years; however, from 1997 onwards these are turned into efficiency *(EP > 1)*.

The estimation of our composite equation model (equations 2, 3, 4 and 5) is carried out using the Generalized Method of Moment (GMM). This approach is usual when heteroskedasticity may be present in the simultaneous estimations (Arellano & Bond, 1991). Preliminary tests confirmed the presence of heteroskedasticity (White and Breusch-Pagan tests) and endogeneity (Wu-Hausman test). A model of common fixed effects for all the firms was considered, treating the observations as a pool of data

Table 12.1 Definitions of variables, and their sources

Variables	Independent variable in Equations	Predicted sign	Definitions (and data sources)
Export intensity (EX)[a]	3, 4, 5	?	Export sales ratio (firm's annual financial reports)
Technological change (TEC)[b]	2	+	Technological change index estimated as a component of total factor productivity (TFP_{market})
Efficiency change (EFC)[c]	2	+	Efficiency change index estimated as a component of total factor productivity (TFP_{market})
Environmental productivity (EP)[d]	2	?	Environmental productivity index estimated as the ratio of two total factor productivity indices (TFP_{total} / TFP_{market}). Indicator of firm environmental performance
International experience (IE)	2	+	Number of years in exporting (firms survey)
Average market price (PM)	2	?	Weighted average price index of exports, in euros per ton (Agricultural Council of Andalusia)
Firm size (FS)	2, 3, 4, 5	+	Ratio of firm's sales to the total sales of the sample (annual financial reports)
Capital intensity (CI)	3, 4	+	Annual investment in machinery and equipment over total assets (annual financial reports)
Imported inputs (IS)	3, 4	−	Amount of intermediate imported inputs over total inputs (firm's survey)
Wage rate (WG)	3, 4, 5	+	Annual labor costs (thousands of euros) divided by the number of workers (annual financial reports)
Concentration ratio (CR)	3, 4	+	Hirschman-Herfindahl ratio (annual financial reports)
Environmental intensity (EI)	5	+	Annual expenditure on environmental practices divided by sales (annual financial reports and Agricultural Council of Andalusia)
Environmental spillovers (ES)	5	+	Annual expenditures on environmental practices of the other firms in the sample (firm's survey and Agricultural Council of Andalusia). See Galdeano-Gómez and Céspedes-Lorente (2008)
Years of application (AGE)	5	+	Number of years applying environmental practices (firms survey)
Outputs and inputs			
Value added			Accountable added value = sales output minus intermediate inputs in thousands of euros (annual financial reports)
Waste			Waste and residues divided by marketed output, in kilograms per ton (Agricultural Council of Andalusia and firms survey). See Galdeano-Gómez et al. (2006)
Capital			Accountable replacement of fixed assets in thousands of euros (annual financial reports). See Galdeano-Gómez et al. (2006)
Labor			Number of employees (annual financial reports)

[a] Dependent variable in Equation 2.
[b] Dependent variable in Equation 3.
[c] Dependent variable in Equation 4.
[d] Dependent variable in Equation 5.

Table 12.2 Descriptive statistics and correlations

Variables	Mean	SD	EX	IE	PM	FS	CI	IS	WG	CR	EI	ES	AGE	TEC	EFC
EX	.523	.496													
IE	15.812	10.353	.328												
PM	806.574	498.930	-.154	-.056											
FS	.112	.161	.261	.002	.017										
CI	11.030	8.759	.063	.017	.008	.070									
IS	8.064	6.146	-.012	.000	.073	.002	.037								
WG	21.836	5.470	.009	-.008	.000	-.001	.022	.000							
CR	.091	.042	.038	.032	-.002	.019	.009	.024	-.005						
EI	9.047	7.915	.054	.003	.005	.008	.016	.002	.014	.052					
ES	12872.385	8107.314	.018	.001	.000	.003	.003	-.000	.000	.041	.069				
AGE	8.437	4.691	.031	.056	.004	.000	.001	.001	.003	.008	.040	.075			
TEC	1.013	.184	.416	.018	.000	.368	.509	.088	.161	.186	.068	.012	.007		
EFC	1.005	.236	.189	.021	.007	.206	.286	.142	.302	.215	.194	.025	.046	.054	
EP	1.008	.312	.085	.009	.011	.091	.207	.091	.277	.246	.637	.217	.104	.017	.039
Value added	297.418	253.675													
Waste	89.752	114.272													
Capital	472.547	388.748													
Labor	88.371	58.190													

[a] Correlations greater than .065 or less than -.065 are statistically significant at $p < .05$.

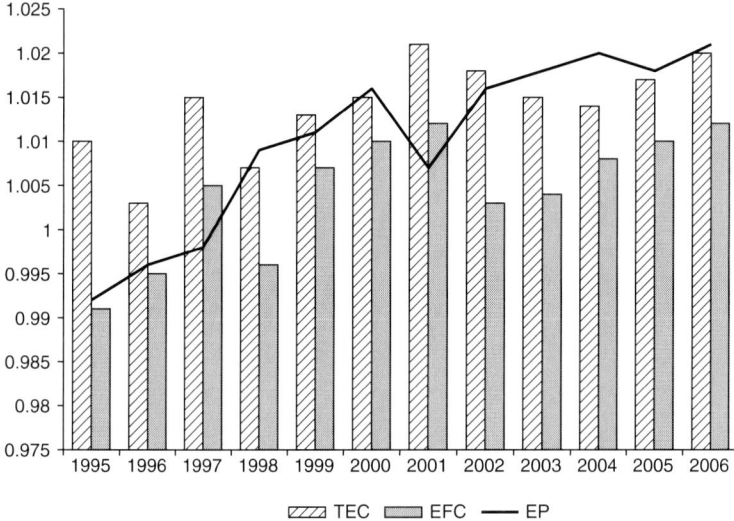

Figure 12.1 Productivity component estimations: Technological change *(TEC)*, efficiency change *(EFC)* and environmental productivity *(EP)*.

(*F*-test is used testing the fixed effects model) and introducing temporal dummy variables for the possible time effects. The vectors of instruments were constructed using the endogenous and exogenous variables lagged one period. To contrast the validity of the instrumental variables in each equation, the statistical test suggested by Sargan and Bhargava (1983) was applied. Table 12.3 shows the estimation results. Using the Sargan test, we are not able to reject the hypothesis that all instruments satisfy orthogonality conditions. By using *F*-statistic, the null hypothesis that firm fixed effects are equal cannot be rejected. The parameters have the expected signs and are significant in most cases.

The parameters estimated show a relatively low influence of productivity indicators on export intensity. Although the signs are positive, the efficiency change (coefficient = .081, $p < .10$) and environmental productivity (coefficient = .087, $p < .10$) coefficients are only marginally significant. Thus, we only found marginal support for Hypothesis 1, probably due to the recent application of environmental practices in the sector under analysis (this issue is analyzed later in this section). As expected, international experience in exporting activity (coefficient = .235, $p < .01$) and firm size (coefficient = .165, $p < .05$) have positive and significant parameters (Wagner, 2001). However the average market price has a negative impact (coefficient = −.104, $p < 0.05$) on competitive exports, showing the possible influence of the increasing international competitiveness in supplies to the European agro-food market over recent years (Pérez-Mesa, 2008).

Export intensity appears as a determinant of technological change (coefficient = .108, $p < .10$) and especially of efficiency change, with a

Table 12.3 Simultaneous equation GMM (Generalized Method of Moment) parameter estimates

Dependent variable	EX(1)†	EX(2)†	TEC	EFC	EP
Export intensity (EX)			.108* (1.805)	.168** (2.153)	.175** (2.251)
Technological change (TEC)	.049 (.438)	.036 (.295)			
Efficiency change (EFC)	.081* (1.693)	.077* (1.679)			
Environmental productivity (EP)	.087* (1.802)				
EP x Years of application (AGE)		.174** (2.279)			
International experience (IE)	.235*** (3.426)	.182** (2.385)			
Average market price (PM)	−.104** (−1.971)	−.113** (−1.984)			
Firm size (FS)	.165** (2.194)	.158** (2.147)	.204** (2.738)	.094** (1.969)	.107* (1.699)
Capital intensity (CI)			.297*** (3.052)	.210** (2.107)	
Imported inputs (IS)			.012 (0.064)	−.034 (−0.112)	
Wage rate (WG)			.153* (1.705)	.189*** (2.813)	.092* (1.704)
Concentration ratio (CR)			.206** (1.992)	.144** (1.981)	
Environmental investment intensity (EI)					.357*** (4.116)
Environmental spillovers (ES)					.135** (2.231)
Years of application (AGE)					.158** (1.983)
Constant	.178** (2.415)	.159** (2.326)	.247** (2.610)	.402*** (3.852)	.219** (2.136)
Adj R Square	.578	.683	.629	.672	.661
No. of observations	780	780	780	780	780
Sargan test (Prob. > χ^2)	.213	.198	.105	.146	.193
F-test (fixed effects)	68.744***	65.852***	74.691***	69.113***	8.618***

Note: t-tests are reported in parentheses. Using two-tailed test: ***p < .01; **p < .05; *p < .1.
†EX(1) refers to the model with EP as explanatory variable; EX(2) refers to the model with EP x AGE as explanatory variable."

positive coefficient (coefficient = .168, $p < .05$), meaning an efficiency improvement. These results suggest that exports have a positive effect on the three firm productivity components, and support the learning-by-exporting arguments. Technological and efficiency growth are also positively related to capital intensity and to the size of the firm, although this is not observed for imported inputs. The relationship between wages and EFC is particularly significant (coefficient = .189, $p < 0.01$). In line with others studies (Greenaway & Kneller, 2008), we also deduce that productivity growth is related to the concentration or collusion effect in the sector.

As regards environmental productivity (EP), the export intensity parameter is significant (coefficient= .175, p < 0.05), showing the positive impact of competitive exports on environmental performance (Hypothesis 2). The EP is directly related to the intensity of environmental investment (coefficient = .357, p < .01) and $AGE,$ experience in application of environmental practices (coefficient = .158, p < 0.05), indicating that there are efficiency gains in waste reduction (as shown in Table 12.2). In addition, we can observe a positive and significant spillover effect (coefficient = .135, p < 0.05) related to environmental knowledge within the firms of the sample.

One main problem in the empirical analyses is the possible heterogeneity of estimation results and the mis-specifications due to omitted variables. The robustness of the previous estimations has been checked in several ways. Regarding the relatively low significance of the effect of environmental productivity on export intensity (Equation 2), we have argued that this issue can be related to the recent application of environmental practices by some firms in the sample. The estimated effect of the variable $EP \times AGE$ (number of years applying environmental practices) on export intensity (column three, Table 12.3) is positive and significant (coefficient = .174; p < 0.05). This result shows the relevance of environmental experience in explaining the process of self-selection in firms' export activity (Hypothesis 1).

On the other hand, we use an alternative environmental performance indicator, which is the reduction in use of nitrates and pesticides per ton of marketed output (e.g., Van der Werf & Petit, 2002). This implies that environmental performance is not estimated as productivity component, but also from a more reduced sample, because these data are only available for 43 firms of the sample. The results (not included; available upon request) reveal that the signs of parameters are similar to those shown in Table 12.3, though the influence of export intensity on environmental performance is slightly less. Nevertheless, these results support the robustness of the previous results regarding environmental variables.

Additionally, other concerns are related to the control variables or firm specialization variables, particularly for the export-intensity model. Several empirical analyses show how skilled workers may explain the export heterogeneity among firms (Bernard et al., 2003; Wagner, 2007), and we use this as alternative explanatory variable (measured as percentage of skilled workers over total workers) in the export performance model. Some analyses also consider that age of the firm explains how younger firms may be more flexible, aggressive, and proactive in catering to world demand (Sterlacchini, 2001). Finally, we use system GMM (which combines the estimations of the model in first differences and in levels) as estimation method suggested by Blundell and Bond (1998).[4]

The estimations using alternative control variables in the export intensity model (not included; available upon request) show a positive and significant parameter for the skilled workers variable, though the differences in the estimated of firm age are not great with respect to international experience. On the whole, the results of the relationships between export intensity and the productivity components are not altered. Also, the system GMM estimates (not included; available upon request) reveal that there are few differences in the expected signs and significance of estimated parameters compared to the estimations in Table 12.3.

Discussion and Conclusion

The present study has focused on the application of an empirical framework in order to obtain insights into the relationship between export intensity and productivity components, including particularly an environmental productivity measure at firm level. We have argued that the different components of firm total productivity affect export intensity and that exporting increases productivity, including environmental productivity.

The results provide evidence of an increase in environmental productivity in our sample over recent years. This productivity component has a positive relationship with the export intensity for the firms in the sample, although this effect is weak when export intensity is evaluated as a dependent variable. This may be due to several factors associated with recent environmental programs and the failure in transmitting them effectively to the market, which is an issue linked to food products (Bellesi et al., 2005). Thus, the recent development of environmental practices in the sector, and the incidence of information asymmetry associated with environmental quality components of produce (Holleran, Bredahl, & Zeibat, 1999; Fu, 2005) may explain the slight relationship between firm environmental capabilities and export performance of agricultural firms (see Galdeano-Gómez, 2010). When the combined effect of environmental productivity and experience in the deployment of environmental practices is taken into account, the effect of environmental performance on export intensity is stronger. It may be argued that, in our empirical setting, the building of strategic resources associated with environmental capabilities (i.e., reputation) requires more time than in other settings. This is the case because the introduction of green technologies in the agricultural sector is relatively recent and such technologies involve highly complex processes (e.g., Pérez-Mesa, 2008).

On the other hand, export performance may be an important contributor to productivity growth in the analyzed firms. Specifically, there is evidence of its effect on cross-efficiency promotion and resource reallocation. Moreover, the productivity associated with environmental outputs improves with high export intensity.

Several factors may be linked to gains in efficiency: concentration in the sector, spillover effects due to the proximity of firms (Mazzanti &

Zoboli, 2006), product specialization of the export-oriented firms, and the development of environmental practices (Christmann & Taylor, 2001; Martín et al., 2008), as represented by the incorporation of more qualified staff (engineers, technicians, etc.).

As the main limitations of this research, we can consider the focus on a single sector and the fact that the findings obtained are related to the particularities of Spanish food firms. Therefore, researchers should interpret these results with care when extrapolating them to other organizational contexts with different levels of export activity and different regulatory, competitive, and technological conditions.

Despite these limitations, we can derive some implications from the analytical framework. Consequently, research projects related to these issues should bear in mind the bi-directional effects of variables in order to obtain suitable results of the relationships between exports and the firm's performance indicators. The combined use of different hypotheses in studies on international trade, such as resource endowment and productivity heterogeneity, can better explain the behavior of firms' exports and environmental performance. Results of this study are in part different from the big picture that emerges from the findings of previous empirical analysis about the relationship between productivity, environmental performance, and export at an aggregate level. If these findings support the self-selection arguments, our results show that firm export activity has a positive effect on their environmental performance, supporting environmental "learning by exporting" arguments. Exporters get access to knowledge about green practices and technologies at international markets. In turn, this knowledge supports an environmental performance improvement process. Over time, these improvements positively affect firm export intensity.

Beyond the specifics of the agricultural sector, more research is needed to determine the conditions that environmental capabilities have to meet to confer an international competitive advantage on firms. Complementary strategic capabilities such as innovation (Christmann, 2000); contingent external factors such as perceived uncertainty (Martín et al., 2008); institutional setting of the focal firm and target market (Madsen, 2009); and the potential to differentiate the product or the presence of information asymmetries should be analyzed as moderating variables to explain the connection between firm environmental productivity and international success.

Note: We acknowledge financial aid from Spanish MCINN and FEDER (projects ECO2008-02258 and ECO2008-03445) and from the Andalusian Government (project SEJ-5827).

Notes

1. Another commonly accepted hypothesis is the "pollution haven hypothesis" (Copeland & Taylor, 2003; Constantini & Crespi, 2008), which indicates that exporting firms may relocate to countries with less stringent environmental legislation, with a view to reducing production costs. Nevertheless, with few exceptions (e.g., Levinson & Taylor, 2008), there is scant empirical evidence of this assumption.

2. Packaging material represents the main intermediate inputs. Unlike other intermediate inputs that are acquired from the local or national industry, the packaging material comes from both local and multinational firms.
3. Some features of the sector under analysis imply the relevance of environmental spillovers: the recent application of environmental practices in the food industry and the agglomeration of firms, as recently pointed out by Galdeano-Gómez and Céspedes-Lorente (2008).
4. No consensus has been reached regarding which estimation method (GMM or system GMM) is most suitable (e.g., Bun & Windmeijer, 2010).

References

Agricultural Council of Andalusia (2007). *Statistical Yearbook*. Seville: Andalusian Government.
Anton, W. R. Q., Deltas, G., & Khanna, M. (2004). Incentives for environmental self-regulation and implications for toxic releases. *Journal of Environmental Economics and Management*, 48(1), 632–654.
Aragón, A. (1998). Strategic proactivity and firm approach to the natural environment. *Academy of Management Journal*, 41(5), 558–567.
Aragón, A., & Sharma (2003). A contingent resource-based view of proactive corporate environmental strategy. *Academy of Management Review*, 28(1), 71–88.
Arellano, M., & Bond, S. (1991). Some tests of specification for panel data: Monte Carlo evidence and an application to employment equations. *Review of Economics Studies*, 58, 277–297.
Ball, V. E., Lovell, C. A. K., Luu, H., & Nehring, R. (2004). Incorporating environmental impacts in the measurement of agricultural productivity growth. *Journal of Agricultural and Resource Economics*, 29(3), 436–460.
Barney, J. B. (1991). Firm resources and sustained competitive advantage. *Journal of Management*, 17, 99–120.
Beise, M., & Rennings, K. (2005). Lead markets and regulation: A framework for analizyng the international diffusion of environmental innovations. *Ecological Economics*, 52, 5–17.
Bellesi, F., Lehrer, D., & Tal, A. (2005). Comparative advantage: The impact of ISO 14001 environmental certification on exports. *Environmental Science and Technology*, 39, 1943–1953.
Bernard, A. B., Eaton, J., Jensen, B. J., & Kortum, S. (2003). Plants and productivity in international trade. *American Economic Review*, 93, 1268–1290.
Blundell, R., & Bond, S. (1998). Initial conditions and moment restrictions in dynamic panel data models. *Journal of Econometrics*, 87, 115–143.
Bun, M. J. G., & Windmaijer, F. (2010). The weak instrument problem of the system GMM estimator in dynamic panel data models. *Econometrics Journal*, 13(1), 95–126.
Caves, D. W., Christensen, L. R., & Diewert, W. E. (1982). The economic of index numbers and the measurement of input, output, and productivity. *Econometrica*, 50, 1393–1414.
Céspedes-Lorente, J., & Galdeano-Gómez, E. (2004). Environmental practices and the value added of horticultural firms. *Business Strategy and the Environment*, 13(3), 403–414.
Christmann, P. (2000). Effects of "best practices" of environmental management on cost advantage: the role of complementary assets. *Academy of Management Journal*, 4, 663–680.
Christmann, P., & Taylor, G. (2001). Globalization and the environment: Determinants of firm self-regulation in China. *Journal of International Business Studies*, 32(3), 439–458.
Chung, Y., Färe, R., & Grosskopf, S. (1997). Productivity and undesirable outputs: a directional function approach. *Journal of Environmental Management*, 51, 229–240.
Clerides, S. K., Lauch, S., & Tybout, J. R. (1998). Is learning by exporting important? Micro dynamic evidence from Columbia, Mexico and Morocco. *Quaterly Journal of Economics*, 113, 903–948.
Cole, M. A., & Elliot, R. J. R. (2003). Do environmental regulations influence trade patterns? Testing old and new trade theories. *The World Economy*, 26, 1163–1186.
Constantini, V., & Crespi, F. (2008). Environmental regulation and the export dynamics of energy technologies. *Ecological Economics*, 66, 447–460.
Copeland, B. R., & Taylor, M. S. (1994). North-south trade and the environment. *Quarterly Journal of Economics*, 109, 755–787.

Copeland, B. R., & Taylor, M. S. (2003). *Trade and the environment*. Princeton, NJ: Princeton University Press.

Egan, M. L., & Mody, A. (1992). Buyer-seller links in export development. *World Development, 20*, 321–334.

Ederington, W. J., & Minier, J. (2003). Is environmental policy a secondary trade barrier? An empirical analysis. *Canadian Journal of Economics, 36*(1), 137–154.

Färe, R., Grosskopf, S., Norris, M., & Zang, Z. (1994). Productivity growth, technical progress, and efficiency change in industrialized countries. *American Economic Review, 84*, 66–83.

Frankel, J., & Rose, A. (2002). Is trade good or bad for the environment? Sorting out the causality. NBER Working Paper No. 9021, NBER Research Associates. The National Bureau of Economic Research: http://www.nber.org/papers/w9201

Fu, X. (2005). Exports, technical progress and productivity growth in a transition economy: A nonparametric approach for China. *Applied Economics, 37*, 725–739.

Galdeano-Gómez, E. (2010). Exporting and environmental performance: A firm-level productivity analysis. *The World Economy, 33*(1), 60–88.

Galdeano-Gómez, E., & Céspedes-Lorente, J. (2008). Environmental spillover effects on firm productivity and efficiency. *Ecological Economics, 67*, 131–139.

Galdeano-Gómez, E., Céspedes-Lorente, J., & Rodríguez-Rodríguez, M. (2006). Productivity and environmental performance in marketing cooperatives: An analysis of the Spanish horticultural sector. *Journal of Agricultural Economics, 57*, 479–500.

Galor, O., & Moav, O. (2000). Ability-biased technological transition, wage inequality, and economic growth. *Quarterly Journal of Economics, 115*(2), 466–497.

Greenaway, D., & Kneller, R. (2008). Exporting, productivity and agglomeration. *European Economic Review, 52*, 919–939.

Grossman, G. M., & Helpman, E. (1991). Trade, knowledge spillovers, and growth. *European Economic Review, 35*(2–3), 517–526.

Harmse, C., & Abuka, C. A. (2005). The links between trade policy and total factor productivity in South Africa's manufacturing sector. *South African Journal of Economics, 73*(3), 389–405.

Hart, S. L. (1995). A natural-resource-based view of the firm. *Academy of Management Review, 20*(4), 986–1014.

Helpman, E., Melitz, M. J., & Yeaple, S. R. (2004). Exports versus FDI. *American Economic Review, 94*, 300–316.

Holleran, E., Bredahl, M. E., & Zeibat, L. (1999). Private incentives for adopting food safety and quality assurance. *Food Policy, 24*, 669–683.

International Study Group on Exports and Productivity (2008). Exports and productivity—comparable evidence for 14 countries. *Review of World Economics, 144*(4), 596–635.

Jaffe, A. B., Peterson, S. R., Portney, P. R., & Stavins, R. N. (1995). Environmental regulation and the competitiveness of U.S. manufacturing: What does the evidence tell us? *Journal of Economic Literature, 33*(1), 132–163.

Kaneko, S., & Managi, S. (2004). Environmental productivity in China. *Economics Bulletin, 17*(2), 1–10.

Levinson, A., & Taylor, M. S. (2008). Unmasking the pollution haven effect. *International Economic Review, 49*(1), 223–254.

Liddle, B. (2001). Free trade and the environment-development system. *Ecological Economics, 39*, 21–36.

Madsen, P. M. (2009). Does corporate investment drive a "race to the bottom" in environmental protection? A reexamination of the effect of environmental regulation on investment. *Academy of Management Journal, 52*, 1297–1318.

Managi, S., & Karemera, D. (2005). Trade and environmental damage in US agriculture. *World Review of Science, Technology and Sustainable Development, 2*(2), 168–190.

Managi, S., Opaluch, J. J., Jin, D., & Grigalunas, T. A. (2004). Environmental regulations and technological change in the offshore oil and gas industy. *Land Economics, 81*(2), 303–319.

Martín, I., Aragón, A., & Senise, M. E. (2008). Being green and export intensity of SMEs: The moderating influence of perceived uncertainty. *Ecological Economics, 68*(1–2), 56-67.

Martínez, E., Díaz, R., Navarro, M., & Ravelo, R. (1999). A study of the efficiency of Spanish port authorities using data envelopment analysis. *International Journal of Transport Economics*, 2, 237–253.

Mazzanti, A., & Zoboli, R. (2006). Economic instruments and induced innovation. The European policies on end of life vehicles. *Ecological Economics*, 58(4), 867–896.

Pérez-Mesa, J. C. (2008). Should Almería (Spain) have to be worried, thinking that their tomato export is currently affected by international competition? *Agricultural Economic Review*, 8(2), 42–54.

Porter, M. E., & van der Linde, C. (1995). Toward a new conception of the environment-competitiveness relationship. *Journal of Economic Perspectives*, 9(4), 97–118.

Ray, S. C., & Desli, E. (1997). Productivity growth, technical progress and efficiency change in industrialized countries: Comment. *American Economic Review*, 87(5), 1033–1039.

Reinhard, S., Lovell, C. A. K., & Thijssen, G. (1999). Econometric estimation of technical and environmental efficiency: An application to Dutch dairy farms. *American Journal of Agricultural Economics*, 81(1), 44–60.

Russo, M. V. & Fouts, P. (1997). A resource-based perspective on corporate environmental performance and profitability. *Academy of Management Journal*, 40(3), 534-559.

Salomon, R., & Jin, B. (2010). Do leading or lagging firms learn more from exporting? *Strategic Management Journal*, 31(10), 1088–1113.

Sargan, J., & Bhargava, A. (1983). Testing residuals from least squares regression for being generated by the Gaussian Random Walk. *Econometrica*, 51, 153–174.

Sharma, S., & Vredenburg, H. (1998). Proactive corporate environmental strategy and the development of competitively valuable organizational capabilities. *Strategic Management Journal*, 19, 729–753.

Shepard, R. W. (1970). *Theory of cost and production functions*. Princeton, NJ: Princeton University Press.

Shortle, J. S. & Abler, D. (2001). *Environmental Policies for Agricultural Pollution Control*. New York: CAB International Publishing.

Sterlacchini, A. (2001). The determinants of export performance: A firm-level study of Italian manufacturing, *Weltwirschaftliches Archiv*, 137, 450–472.

Tobey, J. A. (1990). The effects of domestic environmental policies on patterns of world trade: An empirical test. *Kyklos*, 43(2), 191–209.

Tyteca, D. (1997). Linear programming models for the measurement of environmental performance of firms – Concepts and empirical results. *Journal of Productivity Analysis*, 8(2), 183–197.

Tyteca, D., Carlens, J., Berkhout, F., Hertin, J., Wehrmeyer, W., & Wagner, M. (2002). Corporate environmental performance evaluation: Evidence from the MEPI report. *Business Strategy and the Environment*, 11, 1–13.

Van Beers, C., & Van den Bergh, J. C. J. M. (1997). An empirical multi-country analysis of the impact of environmental regulations on foreign trade flows. *Kyklos*, 50(1), 29–46.

Van der Werf, H., & Petit, H. (2002). Evaluation of the environmental impact of agricutlure and the farm level: Comparison and analysis of 12 indicators-based methods. *Agriculture, Ecosystems and Environment*, 93, 131–145.

Verwaal, E., & Donkers, B. (2002). Firm size and export intensity: Solving an empirical puzzle. *Journal of International Business Studies*, 3, 603–613.

Wagner, J. (2001). A note on the firm size-export relationship. *Small Business Economics*, 17, 229–237.

Wagner, J. (2007). Exports and productivity: A survey of the evidence from firm-level data. *The World Economy*, 30, 60–82.

CHAPTER THIRTEEN

Internationalization, Innovativeness, and Proactive Environmental Strategy among Small and Medium Enterprises

JAVIER AGUILERA-CARACUEL, M. ÁNGELES
ESCUDERO-TORRES, EULOGIO CORDÓN-POZO,
AND NURIA ESTHER HURTADO-TORRES

To date, technological innovations in some industries have resulted in significantly reduced emissions and increased shares of nonfossil fuels in energy production (Kivimaa, 2008). Nevertheless, several environmental challenges persist related to climate change, biodiversity, and chemicals dispersed through consumer products. Green innovation can be defined as hardware or software innovation that is related to green products or processes, including the innovation in technologies that are involved in energy saving, pollution prevention, waste recycling, green product designs, and corporate environmental management (Chen, Lai, & Wen, 2006; Shrivastava, 1995). This type of innovation involves the recognition by firms that environmental problems arise from the development, manufacture, distribution, and consumption of their products and services. Furthermore, green innovation requires the integration of environmental issues into the firm's strategic planning process (Banerjee, 2002).

It is widely recognized that firms that adopt a proactive environmental strategy (PES) need to incorporate valuable green innovation in their internal network (Russo & Fouts, 1997). A PES refers to the identification and analysis of the natural environmental aspects of a firm's products and services, and the establishment of comprehensive and preventive management programmes (Aragón-Correa, 1998). This strategy requires the complex coordination of skills and heterogeneous resources (Amit & Schoemaker, 1993) to reduce environmental impacts and increase firm competitiveness (Aragón-Correa, Hurtado-Torres, Sharma, & García-Morales, 2008).

Research based on samples from larger firms has shown that organizations of a larger size are more likely to undertake the most PES (Aragón-Correa, 1998; Russo & Fouts, 1997; Sharma, 2000). Scholars have consequently argued that because a PES requires accumulation of, and complex interaction among, skills and resources such as physical assets, technologies, and people (Russo & Fouts, 1997; Sharma, 2000; Shrivastava, 1995), the limited resources of small and medium enterprises (SMEs) might prevent them from adopting such practices (Russo & Fouts, 1997). Studies of SMEs have often highlighted their poor level of environmental commitment, describing them as mainly interested in controlling emissions of pollution to comply with environmental regulations (e.g., Schaper, 2002; Williamson & Lynch-Wood, 2001). In this context, SMEs' internationalization process may contribute to getting access to valuable and innovative knowledge abroad and consequently adopting an advanced PES.

The creation of new knowledge acquired by the firms' international activity in diverse foreign markets encourages them to develop organizational capabilities (Barkema & Vermeulen, 1998; Nelson & Winter, 1982). Firms can create experiences that allow them to explore and search for new knowledge through interacting with new cultures, demographics, regulations, and technologies (Levitt & March, 1988). One of the capabilities that can be generated through the internationalization process is innovation capability. Innovation capability is the ability of an organization to adopt or implement new ideas, processes, or products successfully (Hurley & Hult, 1998). It has been argued that implementation of new ideas can increase company productivity and efficiency, and lead to higher firm performance (e.g., Edosomwan, 1989). In addition, firms can benefit from increased productivity and adaptability, owing to process improvements (Sankar, 1991). Consequently, innovative firms are those firms that embrace innovation by constantly introducing change, such as new work structures, new work procedures, human resource management strategies, and creation of a work environment that will spur innovation (Terziovski, 2002). Hence, innovation capability is one of the most important dynamics that enables SMEs to achieve a high level of competitiveness, both in the national and international market.

Hurley, Hult, and Knight (2004) distinguished between "innovativeness," which is a cultural readiness and appreciation for innovation, and "innovation capability," which is the degree of innovation actually produced or adopted by organizations. Innovativeness, as a cultural precursor, providing the social capital to facilitate innovative behavior, is central to understanding how to create innovative organizations (Hurley & Hult, 1998). Underneath the innovativeness of the organization's culture resides a series of individual and group level properties that are characteristics of individual and group idea generation, learning, creativity, and change. Therefore, innovative ideas occur to individuals, not organizations, but learning is manifest in the organization only when ideas are shared, actions taken, and common meaning developed at the group and organization level (Hurley et al., 2004).

The purpose of this chapter is to analyze whether the internationalization process by itself may contribute to increasing the SMEs' innovativeness and to determine whether this innovativeness has a positive influence on the adoption of an advanced and proactive posture to environmental issues. In terms of data, our research employed a sample of 155 Spanish export firms from the food industry that had international presence in different regions. We answered two research questions. First, we studied whether international diversification and international learning orientation have an influence on innovativeness in internationalizing SMEs. Second, we studied whether SMEs' innovativeness was positively related to the generation of proactive environmental practices.

Internationalization, Innovativeness and PES

Internationalization entails the expansion of firms across national boundaries for the purpose of selling and producing products and services (Hitt, Hoskisson, & Ireland, 1994; Hitt, Ireland, & Hoskisson, 2007). Although the majority of empirical studies have focused on the relationship between internationalization and performance (e.g., Lu & Beamish, 2004), recent research has suggested other learning outcomes, recognizing the "knowledge-seeking" motive of international expansion (Hitt et al., 1997; Zahra, Ireland, & Hitt, 2000). Nachum and Zaheer (2005) explained that the motivation to expand internationally is not only financial performance, but also access to knowledge and resources.

The firm can acquire valuable international knowledge through its international process (Nelson & Winter, 1982). In relation to SMEs, it has been argued that knowledge about foreign markets gives these organizations the expertise to understand foreign competitors, develop effective business models, select viable modes of entry, and choose the appropriate time for foreign market entry (e.g., Zahra, Neubaum, & Naldi, 2007). Seeing internationalization as a learning process implies that the knowledge that is acquired by operating in international markets makes the development of certain organizational capabilities easier. Current studies have shown that export has a positive effect on a firm's innovation (e.g. Aw, Roberts, & Winston, 2007; Salomon & Jin, 2008). Barkema and Vermeulen (1998) show that firms may improve their innovation capability by simultaneously operating in several national markets. Indeed, firms may increase their opportunities to discover, for instance, new business or technological sources. Additionally, firms may also gain experience and knowledge that make product innovation easier (Craig & Douglas, 2000; Hitt, Hoskisson & Kim, 1997). In sum, operating in international markets may be an excellent way to improve firms' innovation capability (Kotabe, 1990). Finally countries' institutional profile may also condition the firms' knowledge acquisition in foreign markets (Kostova, Roth, & Dacin, 2008).

By operating in international markets, firms may obtain first-hand foreign market knowledge through direct relationships with competitors, clients, and suppliers of those markets (Birkinshaw, 2000; Zahra et al., 2000).

The sources of international knowledge vary greatly. Indeed, these have been exhaustively researched in the literature concerning internationalization, specifically with regard to export behavior. Yeoh (2005) carried out a detailed assessment of the literature on the antecedents and performance of alternative sources of foreign information. In our study we considered two groups of information sources. On the one hand, we analyzed internal information that was obtained through the diversity of international markets. Indeed, firms that operate in different and diverse markets may integrate valuable resources and capabilities within their internal network (Hitt, Hoskisson, & Kim, 1997). On the other hand, we pay special attention to the external information that is obtained through the international distribution channel. Many manufacturers, in particular SMEs, find it impractical to integrate vertically into international distribution (i.e. with foreign subsidiaries and overseas sales offices). Consequently, nonintegrated modes, such as independent, foreign-based distributors are often relied upon to enter foreign markets (Bello, Chelariu, & Zhang, 2003). Such distributors not only offer a relatively easy, low-cost way to distribute globally, but also provide manufacturers with key contacts with foreign buyers, important local-market knowledge, and sophisticated marketing services. For these reasons, in this chapter we consider these two sources of international knowledge and their influence on the firms' innovativeness.

Hitt, Hoskisson and Ireland (1994, p. 298) define international diversification as *"expansion across* country borders into geographic locations (e.g., markets) that are new to the firm," while Li and Qian (2005, p. 7) define international diversification as "expansion across borders of global regions and countries into different geographic locations or markets." Tihanyi, Ellstrand, Daily and Dalton (2000) refer to it as diversification of business activities across national borders. Firms that adopt an internationalization strategy pursue new opportunities to leverage core competences across a broader range of markets (Zahra et al., 2000). The firm's involvement in multiple markets contributes to giving it access to relationships that enhance learning and innovations (Kotabe, 1990; Miller & Chen, 1996). Consequently, companies with a diversity of target markets and products have better opportunities for learning, since they are exposed to different customer needs, rivals, suppliers, and partners (Barkema & Vermeulen, 1998). Therefore, international diversification may greatly contribute to using the selective advantages of multiple countries (Hitt et al., 1997), and innovation can help firms overcome local disadvantages and achieve a competitive advantage in international markets (Porter, 1990). In sum, firms may learn and innovate through international diversification.

Many firms, especially SMEs, have not diversified into international markets. Studies indicate that SMEs' international diversification may be restricted by factors such as limited financial and managerial resources, lack of knowledge of international opportunities, perceptions of risk, and lack of managerial experience (e.g., Javalgi, White, & Lee, 2000). Zimmerman, Barsky and Brouthers (2009) suggest that social networks are an important factor that can influence the international diversification decision. They find that both the strength of the ties to international firms and the size of international networks influence the SME's decision to diversify internationally.

We state that SMEs can increase their level of innovativeness by participating in diverse regions and interacting with different agents. In fact, exposure to and greater involvement with foreign customers and businesses in multiple and diverse markets may promote the creation of openness to new ideas as an aspect of a firm's culture. Thus, SMEs can benefit from using and applying the selective advantages of multiple regions in order to reinforce their innovativeness. Consequently, we propose the following hypothesis.

Hypothesis 1: *The SMEs' international diversification is positively related to their innovativeness (H1).*

Manufacturers, especially those of smaller size, typically expand into international markets through independent distributors and agents. In an increasingly competitive global economy, classical marketing tools such as price and product quality are susceptible to imitation by rivals, which supports the notion that a more enduring source of advantage may stem from mutually beneficial and trust-based relationships with local distributors (Zhang, Cavusgil, & Roath, 2003). Such distributors not only offer a relatively easy, low-cost way to distribute globally, but also provide manufacturers with key contacts with foreign buyers, important local-market knowledge, and sophisticated marketing services. Information coming from the distributor keeps the manufacturer informed about changes that occur in the specific foreign market, ensuring faster and better adaptation (Bello et al., 2003).

A strong relationship with distributors enhances international venture performance through reduced transaction costs (Slater & Narver, 1995). At the same time, mutual benefits may accrue to the business partners. These benefits may take the form of rich market and process information exchange, which could boost the partners' ability to respond quickly to the operating environment. In particular, since SMEs usually lack human, physical, and/or financial resources (Martínez-Costa & Jiménez-Jiménez, 2009), their negotiating power is less than that of big firms (Lee, 2004), they may rely more heavily on an external knowledge source as an input to innovation than do large firms.

Within the context of the international supply chain, learning orientation can be integrated between the firms to represent one of the most essential ways to develop strategic capability and competitive advantage (Nonaka, 1994). Learning orientation facilitates the generation of resources and skills essential for firm performance (Calantone, Cavusgil, Zhao, 2002). Learning orientation is an attempt to instil a high value on learning activities, which include seeking and sharing market knowledge. In fact, it increases the potential to confront successfully changing circumstances because the knowledge-sharing routines that are practiced between the organizations represent an effort to collaborate in good faith. A result of cooperative behavior between the partners is information exchange. Knowledge-sharing routines through learning orientation also increase access to a greater number of information sources, which offer alternative interpretations of market information (Slater & Narver, 1995). Consequently, the coordination, manifested through a learning orientation, provides the advantage of increased efficiency in information dissemination.

Calantone et al. (2002) demonstrated the direct impact of learning orientation on a firm's innovativeness. Indeed, they argued that learning orientation has a direct impact on firm innovativeness in three ways: (1) since learning occurs through organizational observation and interaction with the environment, it is more likely that firms will be committed to innovation; (2) since learning organizations are linked to their environment, they have the knowledge and ability to understand and anticipate customer needs and emerging markets; (3) since organizations closely monitor the competitors' actions in the market, they can understand the strengths and weaknesses of rivals, and learn not only from their successes but also from their failures. All of this contributes to a high innovation capability. Hurley and Hult (1998) found evidence to suggest that higher levels of innovativeness are associated with cultures emphasizing learning and development. Consequently, we propose the following hypothesis:

Hypothesis 2: The SMEs' international learning orientation is positively related to their innovativeness (H2).

Firms that adopt a proactive environmental strategy (PES) need to incorporate valuable green innovations in their internal network. This strategy implies that firms need to redesign their production and service delivery processes. Such a redesign would likely involve the acquisition and installation of new technologies (Russo & Fouts, 1997). Management, R&D, production, and marketing all must be involved and committed if a firm is to implement a policy of using clean technologies (Craig & Dibrell, 2006; Hart, 1995). Moreover, use of clean technologies also adds complexity to production or delivery processes and requires increased skills from workers at all levels of the firm (Groenwegen & Vergragt, 1991).

Although much of the past research has focused on the impact of large companies on the environment, it has been suggested that that the estimated collective impact of SMEs on the environment is substantial

(Hillary, 2000) and could outweigh the combined environmental impact of large companies (McKeiver & Gadenne, 2005). Research related to the SMEs' environmental management is controversial. On the one hand, descriptive studies of SMEs have often highlighted their poor level of environmental commitment, describing them as mainly interested in controlling emissions of pollution to comply with environmental regulations (Schaper, 2002; Williamson & Lynch-Wood, 2001). Other studies argue that SMEs' poor level of environmental development is due to the lack of coherence in their organizational structure (Alberti, Caini, Calabrese, & Rossi, 2000), the lesser influence of the managers' poor environmental training (Azzone, Bertele, & Noci, 1997), the employees' limited involvement and training in this area (Azzone & Noci, 1998), the lower capacity to give rise to highly radical innovations (Leonard, 1985), the lack of centralization of the information concerning the research effort in a unique department (Sánchez, 1997), and finally, the lack of capacity in relation to external pressures, essential for success of the environmental approach (Noci & Verganti, 1999). On the other hand, a few studies in several countries contradict this assumption and have shown that SMEs may successfully implement environmental strategies consistent with the advanced environmental practices of big firms (e.g., Bianchi & Noci, 1998; Carlson-Skalak, Leschke, Sondeen, & Gelardi, 2000) including innovations that prevent pollution at the source rather than pollution control at the end of the pipeline. It has been shown that SMEs' potential to adopt proactive environmental practices is associated with specific organizational capabilities based on their unique strategic characteristics of shorter lines of communication and closer interaction, the presence of a founder's vision, flexibility in managing external relationships, and an entrepreneurial orientation. These capabilities are shared vision, stakeholder management, and strategic proactivity (Aragón-Correa et al., 2008). However, little attention has been paid to the influence that innovativeness of internationalizing SMEs may have on the adoption of a PES.

Shrivastava (1995) suggests that the implementation of a PES requires significant employee involvement, cross-disciplinary coordination and integration, and a forward-thinking managerial style. Bansal and Roth (2000) show that firms that employ an environmental policy that is positive toward the natural environment are more likely to emphasize firm innovation. Christmann (2000) finds that firms that use pollution prevention technologies without possessing capabilities for process innovation and implementation might not be able to generate cost savings from adopting proactive environmental practices. In fact, changing a well-running production process without having the capability for process innovation and implementation might make the process less efficient and more risky than it was previously. Considering the innovative nature of advanced and proactive environmental practices, we suggest that SMEs' innovativeness is required in order to be able to generate a PES. In fact, those SMEs with a high level of innovativeness are very open-minded to

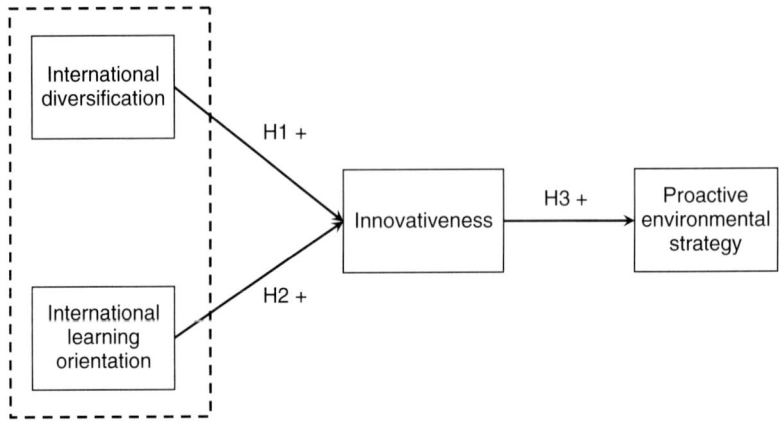

Figure 13.1 Proposed model of the effect of international diversification and innovation environmental strategy.

new ideas, and consequently are more willing to take advantage of all the positive benefits that green innovation can provide. They can not only minimize production waste and increase productivity, but also improve the overall productivity (Porter & Van der Linde, 1995; Shrivastava, 1995); increase corporate reputation transparency (Berry & Rondinelli, 1998; Christmann, 2004); and increase legitimacy (Bansal, 2005; Kostova et al., 2008). Stated differently, integrating environmental considerations into company R&D is a potential way to generate valuable environmental innovations, which are sources of competitive advantage. Consequently, we propose the following hypothesis.

Hypothesis 3: The SMEs' innovativeness is positively related to the adoption of a PES (H3).

The linkages proposed between the different constructs are illustrated in Figure 13.1.

Methods

We focused on export SMEs located in Spain and belonging to the food industry. Focusing on one specific sector and country is often suggested in the resource-based view literature to remove the possible disturbing influence exerted by specific peculiarities related to the context (Barney, 2001). We chose the food industry for our analysis because of its large contribution to gross domestic product (GDP) all over the world. In addition, it is widely influenced by the firms' internationalization process. According to the World Bank in 2006, the export percentage of products from the food industry is nearly 9 percent of total exports in the world, with a total value of $850 billion in 2005. Finally, the food industry has a strong impact on the natural environment (Sánchez-Chóliz, Duarte, & Mainar, 2008). This type of industry generates a large amount of industrial waste water, solid wastes, gas effluents, and transport emissions, and it consumes quite a lot of energy.

Our sample was chosen from the Dun & Bradstreet (D&B) at the end of 2004. This database yielded 1556 export firms, mostly SMEs. Using a random sampling method, our sample consisted of 155 Spanish export firms (9.99 percent of the target population). The sampled firms clearly reflect the situation of the Spanish firms from that sector well. The selected firms had an average size of 54.66 employees and mostly exported to the European Union (EU), but also to Latin America (11.70 percent), the US and Canada (6.92 percent), Eastern Europe (4.30 percent), Africa (4.60 percent), and Asia, Australia, and New Zealand (4.00 percent). We did not find significant differences between the descriptive characteristics of the firms finally included in the study (e.g., location, activities, and size) and the original population.

Since data on SMEs' environmental strategies and environmental performance are not available from published sources, we used a questionnaire to evaluate the PES. We also measured international diversification, international learning orientation, and innovativeness through the same questionnaire. The questionnaire was constructed using validated scales obtained from a review of the literature, adapting them to the activity of the sector. The CEO of each of the firms responded to the questionnaire during a personal interview carried out by the survey company TNS in December 2004 and January 2005. CEOs are the people who may best capture a firm's strategic and environmental approach (e.g., Chandler & Hanks, 1993; Christmann, 2000; Lyon, Lumpkin, & Dess, 2000). The use of questionnaires and personal interviews have great potential and have been used previously by other researchers (Martín-Tapia, Aragón-Correa, & Rueda-Manzanares, 2010). Next, we explain the variables used in the analysis.

Proactive Environmental Strategy (PES). Empirical studies have usually measured the diverse components of PES via managerial perceptions of their practices (e.g., Aragón-Correa, 1998; Christmann, 2000; Sharma & Vredenburg, 1998). We adopted the 14 items used by Aragón-Correa (1998) for a multi-activity sample to measure PES, adding two items related to the food industry. Such items include the ecological measure of products, based on their use of ecological ingredients, and the ease of returning and recycling bottles and containers. Items that reflect these dimensions are good indicators of the firm's degree of environmental proactivity (Buysse & Verbeke, 2003). We finally constructed a 16-item scale to evaluate firm environmental proactivity (see Appendix). Using a seven-point Likert scale, interviewees were asked to assess their firm's degree of development in relation to the environmental activities mentioned, as well as to compare their activities with those of their competitors.

International Diversification. Certain papers have studied international diversification at the country level, which assesses the firm's expansion into individual foreign countries. Other papers consider regional diversification as a measure of international diversification, which refers to the firm's expansion into different global regions or areas (Hitt et al., 1997). It is highly relevant to point out that low levels of country diversification can be risky or costly if a firm spreads its limited markets across regions that are different in terms of psychic distance, competition intensity, demand patterns, and consumer cultures (Li & Qian, 2005). Therefore, considering the aim of this chapter, we analyzed the regional international diversification. This type of diversification has been measured through the number of different regions where SMEs sell their products. According to the World Bank (2001), the regions that have been included in the analysis are: Northern Europe (Sweden, Norway, Denmark, Finland, and the Netherlands); Central Europe (Germany, Austria, Belgium, France, United Kingdom, and Ireland); Southern Europe (Italy, Portugal, and Greece); Eastern Europe (Russia, Poland, and Czech Republic); North America (USA and Canada); Latin America, Asia (excluding Japan); Japan and Oceania; and finally, Africa (excluding South Africa).

International Learning Orientation. Measures for the international learning orientation scale were adapted from Roath and Sinkovics (2006) and Hult, Tomas and Ferrell (1997), who build on conceptual work from Lyles and Schwenk (1994). We constructed a seven-point Likert scale with four items and asked managers to express their level of agreement or disagreement with various questions (see Appendix).

Innovativeness. In relation to this construct, there are some differences among the assumptions, procedures, and objectives of previous measures. Because in this chapter we define innovativeness as a cultural readiness and appreciation for innovation that allows firms to be open-minded to new ideas (Hurley & Hult, 1998), our scale was adapted from Hurley and Hult (1998). We used a seven-point Likert scale with five items and asked managers to express their level of agreement or disagreement with a set of questions (see Appendix).

Results

Before estimating the model, we conducted a confirmatory factor analysis to verify the psychometric properties of the different scales of measurement used to ensure validity and reliability. To do this, we used the Lisrel 8.54 program and confirmatory factor analysis on each construct. Following the recommendations of the literature (Hair, Black, Babin, & Anderson, 2009; Sharma, 1996), we found that indicators were reliable, resulting in standardized factor loadings above 0.50, and were significant at 5 percent (t-value > 1.96). However, the individual reliability of one of the items on the scale of innovativeness did not reach the minimum of 0.50 recommended in the literature, so it was dropped from the final measurement instrument used (the final scale consisted of four innovativeness items). Table 13.1 reports the means and standard deviations for all of the constructs, as well as the inter-factor correlations matrix for the study variables. The Appendix shows information about the composition of the final measurement scales and indicators of their validity and reliability.

Consistent with the two-step approach advocated by Anderson and Gerbing (1988), we created a measurement model before examining structural model relationships. Table 13.2 shows the standardized loadings that allowed us to assess the measurement model. The lowest loading obtained is 0.54, and all parameter estimates are significant at 5 percent (t-value > 1.96).

Table 13.1 Descriptive statistics and correlations for the constructs

Construct	Mean	SD	1	2	3	4
1. International diversification	2.53	1.92	1.00			
2. International learning orientation	4.01	1.53	0.192*	1.00		
3. Innovativeness	5.6	1.16	0.207**	0.259**	1.00	
4. Proactive environmental strategy	3.77	1.39	0.206*	0.163	0.336**	1.00

* $p < 0.05$; ** $p < 0.01$ (two-tailed).

Table 13.2 Standardized factor loadings, Average Variance Extracted, and reliability estimates

Construct	Items	Standardized loadings	t-value	CR	AVE
International learning orientation	X2	0.80	–	0.87	0.64
	X3	0.54	6.29		
	X4	0.94	12.97		
	X5	0.86	12.89		
Innovativeness	Y1	0.85	–	0.91	0.72
	Y2	0.90	15.98		
	Y3	0.85	12.24		
	Y4	0.78	11.76		
Proactive environmental strategy	Y5	0.86	–	0.95	0.55
	Y6	0.67	5.54		
	Y7	0.57	6.50		
	Y8	0.80	6.38		
	Y9	0.76	5.12		
	Y10	0.86	5.72		
	Y11	0.77	5.46		
	Y12	0.77	5.68		
	Y13	0.71	5.09		
	Y14	0.87	5.73		
	Y15	0.69	5.80		
	Y16	0.56	4.77		
	Y17	0.57	4.55		
	Y18	0.90	7.02		
	Y19	0.54	4.44		
	Y20	0.61	4.51		

CR = construct reliability; AVE = Average Variance Extracted.

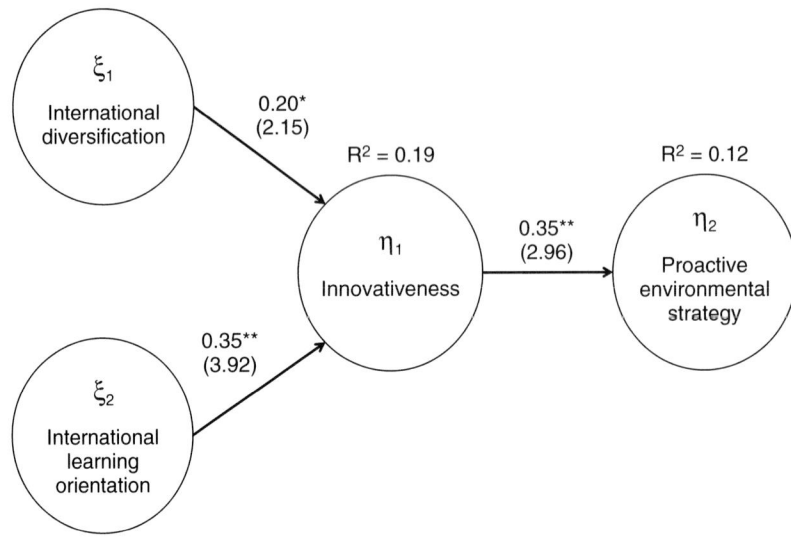

Figure 13.2 Structural model estimation (standardized solution)

The Average Variance Extracted (AVE) estimate ranged from 0.55 for Proactive Environmental Strategy, to 0.72 for Innovativeness (all exceed the 50 percent rule of thumb suggested by Hair et al., 2009). Construct reliabilities (CR) range from 0.87 to 0.95 and, again, these exceed 0.70 recommended by the literature. Figure 13.2 shows the structural model estimation (including the standardized solution).

The information in Table 13.3 shows the statistical overall fit from testing the model. With the exception of the chi square, all the other indicators of the goodness of fit are indicative of good model fit. Since all the estimated parameters in the structural model are significant at 5 percent (t-value > 1.96), the three hypotheses have been verified by our data.

Discussion and Conclusions

It has been traditionally suggested that small and medium enterprises have more trouble in implementing a proactive environmental strategy than larger firms do (Russo & Fouts, 1997). Scholars have argued that the generation of a PES requires accumulation of, and interaction between, resources such as physical assets, technologies, and people (Shrivastava, 1995; Russo & Fouts, 1997; Sharma, 2000). Studies suggest that, compared with multinational enterprises, SMEs face particular problems in the formulation of their innovation strategies because of the deficiencies arising from their limited resources and range of technological competencies, influence of their owners/managers in decision-making, dependence on small numbers of customers and suppliers, and focus on the efficiency

Table 13.3 Structural model results (standardized structural coefficients)

Effect of	On	Standardized solution	t-value	Related hypothesis
International diversification	Innovativeness	0.20	2.15*	H1
International learning orientation	Innovativeness	0.35	3.92**	H2
Innovativeness	Proactive Environmental Strategy	0.35	2.96**	H3

* $p < 0.05$; ** $p < 0.01$ (two-tailed).

Goodness of fit statistics
$\chi^2_{265\,df} = 669.88$ ($p = 0.0$)
Root Mean Square Error of Approximation (RMSEA) = 0.06
Root Mean Square Residual (RMR) = 0.08
Standardized RMR = 0.08
Goodness of Fit Index (GFI) = 0.95
Adjusted Goodness of Fit Index (AGFI) = 0.94
Normed Fit Index (NFI) = 0.93
Non-Normed Fit Index (NNFI) = 0.98
Parsimony Normed Fit Index (PNFI) = 0.83
Comparative Fit Index (CFI) = 0.98
Incremental Fit Index (IFI) = 0.98
Normed Chi Square = 2.52

of current operations (Chaston, Badger, & Sadler-Smith, 2001). However, SMEs can also be very interested in being environmentally proactive since they can increase their business coherence, reputation (Christmann, 2004), and legitimacy (Bansal, 2005) by doing so.

In this chapter we found that internationalizing SMEs can reinforce their level of innovativeness through international expansion. Operating in diverse regions (international diversification) and having an international learning orientation for the supply chain are two valuable sources of international knowledge that have a very positive influence on the SME's innovativeness. In addition, high levels of innovativeness can be translated into adoption of a more proactive and advanced environmental approach.

First, we showed that export firms can reinforce their innovativeness through their international diversification. Export firms that have operations in different regions can strengthen their innovation capabilities through their interaction with different agents, such as governments, public regulators, or competitors. Therefore, these firms can acquire valuable knowledge through their international expansion and, at the same time, can increase their social networks abroad (Zimmerman et al., 2009).

Second, we found that an international learning orientation has a relevant influence on the degree of firms' innovativeness. A strong channel relationship enhances international venture performance, and mutual benefits may accrue to the business partners. These benefits may take the form of rich market and process information exchange, which could boost the partners' ability to respond quickly to the operating environment. In the case of SMEs, this external knowledge source is greater than in larger firms. Since learning orientation allows firms to cooperate with different partners in the supply chain and exchange valuable information (Slater & Narver, 1995), a higher level of SMEs' international learning orientation will allow them to increase their level of innovativeness.

Third, we observe that those internationalizing SMEs with a high level of innovativeness are effectively interested in paying special attention to the development of a proactive environmental strategy. Although the generation of proactive environmental practices may be initially risky and costly (Porter & van der Linde, 1995; Shrivastava, 1995), there is a great variety of advantages that can be reached more easily through the firms' innovativeness. These advantages are related to a reduction in operation costs (Berry & Rondinelli, 1998; Christmann, 2000; Shrivastava, 1995) and an increase in their level of legitimacy and reputation (Bansal, 2005; Christmann, 2004).

This chapter encourages us to think about how the process of internationalizing SMEs can contribute to increasing their innovativeness and, at the same time, create proactive environmental practices. We indicate that a strong factor that discriminates between the more and less successful internationalizing firms that want to increase their level of innovativeness is the former group's ability to learn by actively seeking knowledge about different markets with different potential customers, competitors, and issues of operations management in distant and unfamiliar environments. Having strong relationships with international distributors is also required to reinforce their innovativeness level. Finally, SMEs' innovativeness, directly derived from their internationalization process, is an essential determinant in adopting a PES in the different locations where they operate.

From a managerial viewpoint, this research encourages managers about the increasing importance that the internationalization process of firms is having nowadays. Operating in foreign markets can contribute to acquiring valuable innovative knowledge that can be assimilated and integrated within firms' internal organizational culture. An international learning orientation can provide these firms with the possibility of increasing their level of innovativeness through cooperation with distributors and with other participants in the supply chain. Hence, managers should make great efforts to internationalize the firm's management activities, share valuable knowledge with the other agents of the supply chain, and take advantage of the acquisition of environmental knowledge in the different areas where the firm operates.

From a government perspective, our results imply that regulators and public organizations should design special programs and incentives to help internationalizing SMEs to expand their innovation practices abroad without barriers, and should make efforts to create of proactive and innovative environmental practices, which are sources of competitive advantage. Governments are required to find effective policies to encourage firms to internationalize their activities and develop a set of "best environmental practices" worldwide.

In this chapter there is a set of limitations that should be addressed by future research. First, because of insufficient information or lack of public information about the SMEs' environmental management, international learning orientation, and innovativeness, we used questionnaires. The main disadvantage of this method is that responses could be biased because of misrepresentation of environmental and business reality by respondents. Second, we used cross-sectional data because we did not include observations in different years. Future analysis would benefit from using longitudinal data. Finally, we caution against generalizing our results, given the home country and industry of our sample.

It would be highly relevant for future work to consider other internationalizing firms (SMEs, multinational enterprises) from different industries, to determine the effect that industry type may have on the established relationships. Furthermore, the inclusion of developing countries in the analysis can provide tools about how the countries' institutional profile may influence the SMEs' internationalization process (Chiao, Yang, & Chwo-Ming, 2006) and the adoption of proactive environmental strategies in foreign markets. Finally, the inclusion of objective data related to the firms' environmental performance (releases, recycling, and use of specific environmental management programs) would overcome the managers' bias derived from the use of the questionnaire.

Note: This research was partially funded by the Foundation Banco Bilbao Vizcaya Argentaria through the project: "Strategies in internationalizing European firms and the natural environment: human resources, production and management analysis (SEJ2007-67833)" and the project "Estrategias medioambientales y empresas internacionales: La influencia de la innovación" ["Environmental strategies and international firms: The influence of innovation"] (ECO2010-20483). We want to thank members of group "Investigación, Sostenibilidad y Desarrollo Empresarial" (ISDE) of the University of Granada (Spain) for their insightful comments.

Appendix: Measurement of Constructs

Proactive Environmental Strategy

(Cronbach's Alpha = 0.90; Composite Reliability = 0.93; Average Variance Extracted = 0.82)

Compared with your competitors and using a 1 to 7 scale, please specify the degree of development in your firm of the following activities related to the natural environment:

1. Natural environmental aspects in administrative work (toner recycling, etc.).
2. Periodic natural environmental audits.

3. Recycling of residues and waste produced by the organization.
4. Purchasing manual with ecological guidelines.
5. Natural environmental seminars for executives.
6. Natural environmental training for the firm's employees.
7. Total quality program including natural environmental aspects.
8. Prevention systems to cover possible environmental accidents and emergencies caused by the organization.
9. Natural environmental management manual for internal use.
10. Sponsorship of natural environmental events.
11. Use of natural environmental arguments in marketing.
12. Natural environmental information and training programs for our distributors and customers.
13. Filters and controls for emissions and discharges.
14. Systematic control of energy consumption so as to reduce the organization's demand.
15. Recycling of the water used by the organization with the purpose of reusing it in other processes and/or before throwing it down the drain.
16. Use of ecological ingredients in the manufacture of our products.

International Learning Orientation

(Cronbach's Alpha = 0.85; Composite Reliability = 0.87; Average Variance Extracted = 0.64)

Compared with your competitors and using a 1 to 7 scale, please specify the degree of agreement with the next statements:

1. We strongly encourage our employees to share fresh ideas with distributors.
2. Our company works with distributors to continually improve capabilities.
3. We encourage our distributors to participate actively in joint marketing activities.
4. Joint training programmes with our distributors are designed to improve mutual learning.

Innovativeness

(Cronbach's Alpha = 0.79; Composite Reliability = 0.91; Average Variance Extracted = 0.73)

Compared with your competitors and using a 1 to 7 scale, please specify the degree of agreement with the next statements:

1. Technical innovation, based on research results, is readily accepted *(Item dropped)*
2. Management actively seeks innovative ideas.

3. Innovation is readily accepted in program/project management.
4. Innovation is perceived as too risky and is resisted.
5. People are penalized for new ideas that do not work.

References

Alberti, M., Caini, L., Calabrese, A. & Rossi, D. (2000). Evaluation of the costs and benefits of an environmental management system. *International Journal of Production Research, 38*(17), 4455–4466.

Amit, R., & Schoemaker, P. J. H. (1993). Strategic assets and organizational rent. *Strategic Management Journal, 14*(1), 33–46.

Anderson, J. C., & Gerbing, D. W. (1988). Structural equation modeling in practice: A review and recommended two-step approach. *Psychological Bulletin, 103*(3), 11–23.

Aragón-Correa, J. A. (1998). Strategic proactivity and firm approach to the natural environment. *Academy of Management Journal, 41*(5), 556–567.

Aragón-Correa, J. A., Hurtado-Torres, N. E., Sharma, S., & García-Morales, V. J. (2008). Environmental strategy and performance in small firms: A resource-based perspective. *Journal of Environmental Management, 86*(1), 88–103.

Aw, B. Y., Roberts, M. J., & Winston, T. (2007). The complementary role of exports and R&D investments as sources of productivity growth. *The World Economy, 30*(1), 83–104.

Azzone, G., Bertele U., & Noci, G. (1997). At last we are creating environmental strategies which work. *Long Range Planning, 30*, 562–571.

Azzone, G., & Giuliano N. (1998). Seeing ecology and "green" innovations as a source of change. *Journal of Organizational Change Management, 11*(2), 94–111.

Azzone, G., & Noci, G. (1998). Seeing ecology and "green" innovations as a source of change. *Journal of Organizational Change Management, 11*(2), 94–111.

Banerjee, S. B. (2002). Corporate environmentalism: The construct and its measurement. *Journal of Business Research, 55*(2), 177–191.

Bansal, P. (2005). Evolving sustainably: A longitudinal study of corporate sustainable development. *Strategic Management Journal, 26*(3), 197–218.

Bansal, P., & Roth, K. (2000). Why companies go green: A model of ecological responsiveness. *Academy of Management Journal, 43*(4), 717–736.

Barkema, H. G. & Vermeulen, F. (1998). International expansion through start-up or acquisition: A learning perspective. *Academy of Management Journal, 41*(1), 7–26.

Barney, J. B. (2001). Is the resource-based theory a useful perspective for strategic management research? Yes. *Academy of Management Review, 26*(1), 41–56.

Bello, D. C., Chelariu, C., & Zhang, L. (2003). The antecedents and performance consequences of relationalism in export distribution channels. *Journal of Business Research, 56*(1), 1– 16.

Berry, M. A., & Rondinelli, D. A. (1998). Proactive corporate environmental management: A new industrial revolution. *The Academy of Management Executive 12*(2), 38–51.

Bianchi, R., & Noci, G. (1998). Greening SMEs' competitiveness. *Small Business Economics, 11*(3), 269–281.

Birkinshaw, J. (2000). *Entrepreneurship in the global firm* (2nd ed.). Thousand Oaks, CA: Sage Publications.

Buysse, K. & Verbeke, A. (2003). Proactive environmental strategies: A stakeholder management perspective. *Strategic Management Journal, 24*(5), 453–470.

Calantone, R. J., Cavusgil, S. T., & Zhao, Y. (2002). Learning orientation, firm innovation capability, and firm performance. *Industrial Marketing Management, 31*(6), 515–524.

Carlson-Skalak, S., Leschke, J., Sondeen, M., & Gelardi, P. (2000). E-media's global zero: Design for environment in small firm. *Interfaces, 30*(3), 66–82.

Chandler, G. N. & Hanks, S. H. (1993). Measuring the performance of emerging businesses— a validation study. *Journal of Business Venturing, 8*(5), 391–408.

Chaston, I., Badger, B., & Sadler-Smith, E. (2001). Organizational learning: An empirical assessment of process in small UK manufacturing firms. *Journal of Small Business Management, 39*(2), 139–152.

Chen, Y. S., Lai, S. B., & Wen, C. T. (2006). The influence of green innovation performance on corporate advantage in Taiwan. *Journal of Business Ethics, 67*(4), 331–339.

Chiao, Y. C., Yang, K. P., & Chwo-Ming, J. Y. (2006). Performance, internationalization, and firm-specific advantages of SMEs in a newly-industrialized economy. *Small Business Economics, 26,* 475–492.

Christmann, P. (2000). Effects of best practices of environmental management on cost advantage: The role of complementary assets. *Academy of Management Journal, 43*(4), 663–680.

Christmann, P. (2004). Multinational companies and the natural environment: Determinants of global environmental policy standardization. *Academy of Management Journal, 47*(5), 747–760.

Craig, J., & Dibrell, C. (2006). The natural environment, innovation and firm performance. A comparative study. *Family Business Review, 19*(4), 275–289.

Craig, S. C., & Douglas, S. P. (2000). *International marketing research* (2nd ed.). Hoboken, NJ: Wiley-Interscience.

Edosomwan, J.A. (1989). *Integrating innovation and technology management.* New York: Wiley-Interscience.

Groenwegen, P., & Vergragt, P. (1991). Environmental issues as threats and opportunities for technological innovation. *Technology Analysis and Strategic Management, 3*(1), 43–55.

Hair, J. F., Black, W. C., Babin, B.J., & Anderson, R. E. (2009) *Multivariate data analysis* (7th ed.). Trenton, NJ: Pearson Prentice Hall.

Hart, S. L. (1995). A natural resource-based view of the firm. *Academy of Management Review, 20*(4), 986–1014.

Hillary, R. (2000). *Small and medium-sized enterprises and the environment: Business imperatives.* Sheffield, UK: Greenleaf Publishing.

Hitt, M. A., Hoskisson, R. E., & Ireland, R. D. (1994). A mid-range theory of the interactive effects of international and product diversification on innovation and performance. *Journal of Management, 20*(2), 297–326.

Hitt, M.A., Hoskisson, R .E., & Kim, H. (1997). International diversification: Effects on innovation and firm performance in product-diversified firms. *Academy of Management Journal, 40*(4), 767–799.

Hitt, M. A., Ireland, R. D., & Hoskisson, R. E. (2007). *Strategic management: Competitiveness and globalization* (7th ed.). Mason, OH: South-Western.

Hoffman, A. J. (1999). Institutional evolution and change: Environmentalism and the US chemical industry. *Academy of Management Journal, 42*(4), 351–371.

Hult, G., Tomas, M., & Ferrell, O. C. (1997). A global learning organization structure and market information processing. *Journal of Business Research, 40*(2), 155–166.

Hurley, R. F., & Hult, G. T. M. (1998). Innovation, market orientation, and organizational learning: An integration and empirical examination. *Journal of Marketing, 62*(3), 42–54.

Hurley, R. F., Hult, G. T. M., & Knight, G. A. (2004). Innovativeness and capacity to innovate in a complexity of firm-level relationships: A response to Woodside. *Industrial Marketing Management, 34*(3), 281–283.

Javalgi, R. G., White, D. S., & Lee, O. (2000). Firm characteristics influencing export propensity: An empirical investigation by industry type. *Journal of Business Research, 47*(3), 217–228.

Kivimaa, P. (2008). Integrating environment for innovation: Experiences from product development in chapter and packaging. *Organization and Environment, 21*(1), 56–75.

Kostova, T., Roth, K., & Dacin, M. T. (2008). Institutional theory in the study of multinational corporations: A critique and new directions. *Academy of Management Review, 33*(4), 994–1006.

Kotabe, M. (1990). The relationship between offshore sourcing and innovativeness of US multinational firms: An empirical investigation. *Journal of International Business Studies, 21*(4), 623–639.

Lee, C. Y. (2004). Perception and development of total quality management in small manufacturers: An exploratory study in China. *Journal of Small Business Management, 42*(1), 102–115.

Leonard, H. J. (1984). *Are environmental regulations driving US industry overseas?* Washington, D.C.: Conservation Foundation.

Levitt, B., & March, J. G. (1988). Organizational learning. *Annual Review of Sociology, 14*(3), 319–340.

Li, L., & Qian, G. (2005). Dimensions of international diversification: Their joint effects on firm performance. *Journal of Global Marketing, 18*(3–4), 7–35.

Lu, J. W., & Beamish, P. W. (2004). International diversification and firm performance: The S-curve hypothesis. *Academy of Management Journal, 47*(4), 598–609.

Lyles, M. A. (1994). The impact of organizational learning on joint venture formations. *International Business Review, 3*(4), 459–476.

Lyles, M. A., & Schwenk, C. R. (1994). The impact of organizational learning on joint venture formations. *International Business Review, 3*(4), 459–476.

Lyon, D. W., Lumpkin, G. T., & Dess, G. D. (2000). Enhancing entrepreneurial orientation research: Operationalizing and measuring a key strategic decision making process. *Journal of Management, 26*(5), 1055–1085.

Martín-Tapia, I., Aragón-Correa, J. A., & Rueda-Manzanares, A. (2010). Environmental strategy and exports in medium, small and micro-enterprises. *Journal of World Business, 45*(3), 266-275.

Martínez-Costa, M., & Jiménez-Jiménez, D. (2009). The effectiveness of TQM: The key role of organizational learning in small-businesses. *International Small Business Journal, 27*(1), 98–125.

McKeiver, C., & Gadenne, D. (2005). Environmental management systems in small and medium businesses. *International Small Business Journal, 23*(5), 513–537.

Miller, D., & Chen, M. J. (1996). The simplicity of competitive repertoires: An empirical analysis. *Strategic Management Journal, 17*(6), 419–439.

Nachum, L., & Zaheer, S. (2005). The persistence of distance? The impact of technology on MNE motivations for foreign investment. *Strategic Management Journal, 26*(8), 747–767.

Nelson, R. R., & Winter, S. G. (1982). The Schumpeterian trade off revisited. *The American Economic Review, 72*(1), 114–133.

Noci, G., & Verganti, R. (1999). Managing green product innovation in small firms. *R&D Management, 29*(1), 3–15.

Nonaka, I. (1994). A dynamic theory of organizational knowledge creation. *Organization Science, 5*(1), 14–29.

Porter, M. E. (1990). The competitive advantage of nations. *Harvard Business Review, 68*(2), 73–94.

Porter, M.E., & Van der Linde, C. (1995). Green and competitive: Ending the stalemate. *Harvard Business Review, 73*(5), 120–137.

Roath, A. S., & Sinkovics, R. R. (2006). Utilizing relational governance in export relationships: Leveraging learning and improving flexibility and satisfaction. *Advances in International Marketing, 16*, 157–185.

Russo, M. V., & Fouts, P. A. (1997). A resource-based perspective on corporate environmental performance and profitability. *Academy of Management Journal, 40*(3), 534–559.

Salomon, R., & Jin, B. (2008). Does knowledge spill to leaders or laggards? Exploring industry heterogeneity in learning by exporting. *Journal of International Business Studies, 39*(1), 356–373.

Sánchez, C. M. (1997). Environmental regulation and firm-level innovation. The moderating effects of organizational and individual level variables. *Business and Society, 36*(2), 140–168.

Sánchez-Chóliz, J., Duarte, R., & Mainar, A. (2008). Environmental impact of household activity in Spain. *Ecological Economics, 62*(2), 308–318.

Sankar, Y. (1991). Implementing information technology: A managerial audit for planning change. *Journal of Systems Management, 42*(11), 32–37.

Schaper, M. (2002). Small firms and environmental management: Predictors of green purchasing in western Australian pharmacies. *International Small Business Journal, 20*(3), 235–251.

Sharma, S. (1996). *Applied multivariate techniques*. New York: John Wiley and Sons.

Sharma, S. (2000). Managerial interpretations and organizational context as predictors of corporate choice of environmental strategy. *Academy of Management Journal, 43*(4), 681–697.

Sharma, S., & Vredenburg, H. (1998). Proactive corporate environmental strategy and the development of competitively valuable organizational capabilities. *Strategic Management Journal, 19*(8), 729–753.

Shrivastava, P. (1995). Environmental technologies and competitive advantage. *Strategic Management Journal, 16* (summer special issue), 183–200.

Slater, S., & Narver, J. (1995). Market orientation and the learning organization. *Journal of Marketing, 59*(3), 63–74.

Terziovski, M. (2002). Achieving performance excellence through an integrated strategy of radical innovation and continuous improvement. *Measuring Business Excellence, 6*(2), 5–14.

Tihanyi, L., Ellstrand, A., Daily, C. M., & Dalton, D. R. (2000). Composition of the top management team and firm international diversification. *Journal of Management, 26*(6), 1157–1177.

Williamson, D., & Lynch-Wood, G. (2001). A new paradigm for SME environmental practice. *TQM Magazine, 13*(6), 424–32.

World Bank (2001). *World development report*. New York: Oxford University Press.

World Bank (2006). *World development indicators*. Washington, DC: World Bank Publications.

Yeoh, P. L. (2005). A conceptual framework of antecedents of information search in exporting: Importance of ability and motivation. *International Marketing Review, 22*(2), 165–199.

Zahra, S. A., Ireland, R. D., & Hitt, M. A. (2000). International expansion by new venture firms: International diversity, mode of market entry, technological learning, and performance. *Academy of Management Journal, 43*(5), 925–950.

Zahra, S. A., Neubaum, D. O., & Naldi, L. (2007). The effects of ownership and governance on SMEs' international knowledge-based resources. *Small Business Economics, 29*(3), 309–327.

Zhang, C., Cavusgil, S. T., & Roath, A. S. (2003). Manufacturer governance of foreign distributor relationships: Do relational norms enhance competitiveness in the export market? *Journal of International Business Studies, 34*(6), 550–566.

Zimmerman, M. A., Barsky, D., & Brouthers, K. D. (2009). Networks, SMEs, and international diversification. *Multinational Business Review, 17*(4), 143–163.

PART IV

New Forms of Cross-Sector Cooperation Matter

The forms of cross-sector cooperation that must flourish for a green economy to further emerge are many. They include cooperation within a firm amongst its functional units, cooperation across the value chain of firms, cooperation within industries, cooperation among industries, and collaborations among stakeholders across multiple sectors of society outside of business, including nongovernmental organizations, government bodies, global and cross-country institutions, research institutes, universities, and the media. The challenges to such cooperative endeavors are many. They include information-sharing, trust, learning, and joint problem-solving. What are the benefits of cooperation and when do they outweigh the costs? Conflict is an inherent danger. On the one hand, discrepant viewpoints and difference may hinder change. On the other hand, they may stimulate creativity and innovation.

Power is a force that has to be reckoned with. The sources of power are many. They include power that is based on knowledge, position in a hierarchy, financial clout and control of resources, organizational know-how, the capacity to communicate, and network understanding. Those with power are unlikely to want to give it up. Yet they may have to cede some power in the face of the aggressive opposition and demands. Thus, for better and for worse, politics plays an important role in cross-sector relations. Bargaining, negotiating, and building coalitions are the skills upon which leaders have to rely. All these elements can be found in the chapters that follow. These chapters have a distinct practical orientation. The final chapter derives lessons from researcher and practitioner perspectives.

CHAPTER FOURTEEN
Why Do Private Firms Invest in Public Goods?

GURNEETA VASUDEVA AND HILDY TEEGEN

Recent years have witnessed a growing debate surrounding the idea that firms that act in public interest will profit from doing so. Although the proponents of this idea have argued that public and private interests are inextricably linked for firms seeking sustainable competitive advantage (e.g., Kanter, 2009), the skeptics offer a different perspective. They point out that "in circumstances in which profits and social welfare are in direct opposition [as has been argued in the case of the oil and gas industry, for instance], managers are unlikely to act voluntarily in the public interest and against shareholder interests" (Karnani, 2010). Based on these observations, we develop a framework for why the private sector invests in the creation of public goods, how it organizes such activities, and how it benefits from these investments.

Our study was motivated by an important multinational technology venture involving the world's largest oil and gas firms: the CO_2 Capture Project (CCP), which has been in existence since 2000. In 2000, investing oil and gas firms, with support from national governments, funded a multimillion-dollar long-term project called the CCP. The CCP has the following goal: "To help develop next generation technologies that will reduce the costs of carbon capture and storage (CCS) and hence, help to make CCS a practical and cost effective option for reducing or eliminating CO_2 emissions to the atmosphere resulting from use of fossil fuels." (Recent activities and updates of the venture can be accessed through their website: http://www.co2captureproject.org/). CCS technologies contribute to not only public goods by creating a cleaner environment by decreasing the attendant emissions from the use of certain nonrenewable energy sources in power production, but are also aligned with the investing firms' core oil and gas business and thus have an impact on private benefits that are particular to individual firms, common benefits that are shared by the collaborating firms, and public goods that are shared by global publics.

Interviews with executives from the investing oil and gas firms, as well as archival data, helped identify the key reasons and organizational strategies for investing in a venture resulting in private, common and public goods types of benefits. Our interviews revealed that although firms realize that they stand to benefit when they combine public and private interests, developing capabilities and organizational structures that allow firms to achieve these dual objectives is not an easy task (e.g., Marcus & Geffen, 1998). Firms' propensity to successfully combine public interests with profit objectives depends on their ability to organize collaborative relationships with industry participants, governments, and research institutions, and on the extent to which positive spillovers from the public and common benefits accruing to participating members of the industry can be leveraged for private benefits.

When it comes to creating public goods like clean air, firms have little incentive to go it alone as significant costs are associated with such an endeavor and the benefits from environmental improvement are shared broadly. Collaboration through pooled equity investments is a core element of the CCP venture. It is financially supported by public agencies like national governments, and actively engages with societal stakeholders such as environmental and technical communities, through joint conferences and publications. Such engagement allows for building broad-based public support and understanding for the CO_2 mitigation initiative, and it provides economic and technical insights for regulators and policymakers. Further, ventures like the CCP are established under the firms' alternative energy portfolio that includes businesses and future growth options outside their core capabilities in oil and gas. Our framework suggests an interplay among three types of benefits— private, common and public— that accrue from such collaboration. The expectation of leveraging the positive externalities of the public and common benefits for increasing the net private benefits serves as an important driver for investing firms.

Private Investments in Public Goods:
The Case of the CO_2 Capture Project

The CO_2 Capture Project is a collaborative venture of leading oil and gas firms, largely from industrialized countries. The project was established at the initiative of BP in 2000 to develop technologies to reduce the cost of capturing CO_2—a byproduct of fossil fuel combustion contributing to global warming—and to safely store it underground. The project grew out of BP's participation in the International Energy Agency's Greenhouse Gas Program and the U.S. Department of Energy's initiative for CO_2 capture and sequestration through a partnership program. At this point in time, BP's participation in these clean technology efforts was seen as an effort to reposition itself as firm committed to a low-carbon future, consistent with its "beyond petroleum" branding campaign.

At the beginning of 2000, the CCP started as a 3-year development program with the goal of bringing candidate technologies to pilot plant or demonstration stage. The program rapidly grew to a $50 million long-term project with 70 percent funding from participant firms and the remaining financial support from governments. In addition to BP, headquartered in the United Kingdom, EnCana from Canada, Statoil from Norway, and Repsol from Spain were the earliest firm participants in the project. The project is currently in the third of its three phases. The focus of Phase 3 (2009–2013) is to prepare the ground for widespread deployment of CO_2 capture and storage. The Phase 3 industry participants in the CCP include competing oil majors: BP, Shell, Chevron, ENI, ConocoPhillips, Suncor, and Petrobras. Government contributions come from the U.S. Department of Energy, the European Union, and Norway.

Although CCP activities are carried out through the close cooperation and shared decision-making of technical teams in four areas (capture, storage, monitoring, and verification, policy and incentives, and communications), from a collaborative perspective, the creation of government policies and incentives are most crucial, since no firm can do it alone. The teams are composed of technologists and global experts from CCP member firms and external organizations who investigate advances, monitor development of technologies and policies, look for ways to integrate best technology advances from the program, and present results at technology forums and industry and academic conferences. The program is led by and operates through an executive board composed of representatives from each full member organization. Individual firms' interests are thus accounted for through shared decision-making, stemming from joint ownership and governance of the venture. This executive board selects from the many opportunities for technology improvements and funds those developments. An advisory board, composed of experts from academia, consulting organizations, and other independent bodies, also reviews and recommends changes to the program and potential new areas for exploration on a regular basis.

According to a recent report published by Pricewaterhouse Coopers, carbon capture and storage technologies are estimated to result in the largest reduction in atmospheric CO_2, though remaining cost and technological obstacles are nontrivial. These technologies will be applicable to a large fraction of CO_2 sources around the world, such as power plants and other large-scale industrial processes. Implementing these new technologies will reduce the impact of continued fossil energy use while cleaner energy sources are being developed. In addition to financial equity, participating firms also provide governance oversight and technical assistance in screening candidate CO_2 capture technologies for focused development and scale-up. In return for the investments they make, participating firms enjoy preferential royalty-free access to the technology developed by the venture.

Managerial support to the venture includes keeping the venture's executive board (which draws members from all investing organizations) abreast of national and global level environmental technologies, policies, regulations, and incentives that may affect the costs and benefits of the carbon capture technology. In addition, a team of executives from investing firms is responsible for communicating the benefits of the technology and creating public acceptance by working with technical and nontechnical communities including nongovernmental organizations and universities. This public advocacy role is crucial, as carbon capture is controversial for some opponents, who deem any use of carbon-based energy stocks as inherently damaging; also, the ultimate technical and economic feasibility of the carbon capture and sequestration approach is not fully proven.

What motivates the establishing of a venture for creating public goods? What type of organizational strategy is likely to arise? We argue that it depends on the types of benefits, their relative importance, and their interplay in such ventures.

The CO_2 Capture Project Interviews

To ground our understanding of firm investments in an environmental technology development venture that aims at creating a public good, we undertook interviews with executives who had been the protagonists of this venture and were therefore actively engaged in managing the CCP on behalf of their respective firms. Our qualitative approach allowed us to gather insights from a real project setting and gave us the opportunity to analyze a novel and apparently successful instance of collaboration among competing firms (Browning, Beyer & Shetler, 1995; Gephart, 2004).

We conducted two rounds of interviews in 2001 and 2003 with executives from investing firms responsible for managing the venture, yielding a total of 16 interviews. Interviews with executives from BP and Chevron were face-to-face, and with other executives, the interviews were conducted via phone or e-mail. Our research greatly benefited from interactions with the BP executive, who had played a lead coordinating role with other firm and government investors since the inception of the project.

We followed an exploratory approach (e.g. Browning, et al., 1995), wherein respondents were encouraged to discuss factors that they thought were important for the creation of the venture. Our understanding of the phenomenon under study emerged from these discussions. To further refine and validate our framework, we compared data obtained from interviews with the data available from archival reports and our own observations at the CCP meetings. We also employed time triangulation, so that the information about the early history of the venture was compared to more recent publicly available information.

The framework that emerged from the information provided by the respondents, archival data obtained from the CCP project website, and our own observations yield answers to three issues of interest—the motivation to invest in environmentally clean technologies, the organization of such a venture, and benefit types.

Why Private Investments in Public Goods? Our observations and interpretations of the interviews and secondary reports suggest that the incentives for investment in environmentally friendly technologies stem mainly from two sets of reasons.

First, in recent years, combustion of fossil fuels such as oil and gas that contribute to concentrations of CO_2 and other harmful gases in the atmosphere has become the leading anthropogenic cause of global warming and other forms of environmental damage. Viewed from a public goods perspective, the natural environment is a classic example of public goods where the condition of joint supply is often compromised because excessive exploitation of natural resources leading to their degradation or depletion will prevent others from accessing the same quality or quantity of natural resources. This situation then imposes a greater burden for environmental protection on those firms that are either resource-intensive in their operations, thereby bearing responsibility for resource depletion, or on those that are the largest contributors to environmental degradation. This has led to a greater push by the public—both consumers and the government—for the development of cleaner alternatives, such as renewable energy sources by energy producers.

Based on these consumer and governmental pressures, any growth in business that results from increased fossil fuel consumption, and that contributing to increases in environmentally harmful CO_2 concentrations, is likely to be impeded by policies and regulations that aim at environmental protection. Thus, increasingly, oil and gas firms have come to realize that producing cleaner fossil fuels and alternative technologies are important for their long-term viability. Consequently, CCS technologies that make fossil fuel combustion more socially and environmentally acceptable via greater environmental protection, technologies for efficient utilization of resources, or cost-effective energy processes are regarded as critical to the long-term sustenance and profitability of oil and gas producers (Sethi, 1995; Sharma, Pablo, & Vredenburg, 1999). Thus, investments in environmental technologies offer a source of competitive advantage along several points in a firm's value chain and enhance its ability to cope with changes in its external conditions (Shrivastava, 1995). Firms can build sustainable competitive advantage through many mechanisms, including positive consumer reception stemming from perceived proenvironmental initiatives and from investments that provide cost savings in new regulatory environments that favor cleaner processes.

The second reason for investment is as follows. The oil and gas industry has been under pressure from the public, from governments, and from their own shareholders to consider the adverse economic effects emanating from the depletion of natural resources. Hence, there is growing interest in intervening in a proactive manner. Pollution is increasingly viewed as a sign of economic waste associated with unproductive use of resources. Inherent in the concept of an economic cost of waste is the strong rationale for efficient management of the natural resource portfolio, because many of the goods and services supplied by the ecosystem are not replaceable at reasonable prices—and some are entirely irreplaceable. Incentives for environmental conservation are further strengthened by the realization that ecosystem goods and services must be treated as assets with a market value rather than as "free" goods. Markets are developing for these goods and services that ecosystems provide, through creation of new property rights, trading CO_2 emission credits, water pricing schemes, and other economic incentive systems that include the valuation of more complex service functions provided by the natural environment.

Coastal wetlands serving as estuaries central to food production and as natural storm damage mitigation devices exemplify this heightened valuation of natural resources impacted directly and indirectly by carbon-emitting technologies. Although relevant regulatory systems are still in their infancy, and in many settings have yet to materialize to foster economic valuation of these benefits, there is a call (even by firms in traditionally polluting industries) for increased future certainty associated with the marketing of these benefits. As markets emerge, competitive advantage will go to businesses that can most quickly and effectively reduce environmental costs and liabilities in their products.

In sum, the oil and gas industry recognizes that environmentally friendly energy technologies offer a tremendous market opportunity as the cost competitiveness of these products and technologies increases. Large well-established firms often invest in new technological ventures for strategic renewal (Wadhwa & Kotha, 2006; Dushnitsky & Lenox, 2005a; 2005b; Stopford & Baden-Fuller, 1994; Kanter, Richardson, North, & Morgan, 1991). Of course, not all firms invest in such ventures, because other than profitability and growth orientation, firms must also be willing to take risks, demonstrate organizational flexibility and recognize new opportunities (Zahra, 1993; Barringer & Bluedorn, 1999; Phan & Foo, 2004). Vision statements in the annual reports of participating firms clearly indicate a strong corporate commitment to environmental protection and clean energy technology development through innovation.

How Does the Private Sector Organize Investments in Public Goods? When it comes to reducing CO_2 emissions in the atmosphere, one firm's efforts to reduce these emissions will be offset by continuing increases in emissions from other firms' operations, unless they all engage in similar CO_2 reduction efforts. Thus, although individual firms may possess sufficient technological and financial capabilities, they still must engage collectively. Therefore, firms will invest collaboratively in environmental technologies such as carbon capture, because efforts at the margin will not yield tangible gains from a public good standpoint. Without sufficient critical mass, the impact of a single firm will be deemed trivial and thus, not valued by the public.

A cooperative strategy among firms helps create larger public acceptance for the technology. Engaging with competitors as well as with others, including universities and nongovernmental organizations (NGOs), lends transparency to the innovation process and helps build on the knowledge of various stakeholders in the initial stages of technology development (Freeman, 1984). NGOs and governments steward broader public considerations and thus, the scale and scope of such efforts are of particular concern to them.

Cooperation between investing firms is generally desirable because it allows for access to valuable complementary assets (Gans, Hsu, & Stern, 2002) and the joint problem-solving and cost-sharing (McEvily & Marcus, 2005) critical for the commercialization of the technology. Other things being equal, joint investments through pooled equity to develop a new technology are preferred for investments in environmentally beneficial technologies, because participating firms operate in common geographies, face common problems, and deal with a fairly common set of environmental regulations (Monge, Fulk, Kalman, Flanagin, Parnasse, & Rumsey, 1998). To illustrate this point, it was not until after BP completed its offshore oil and gas exploration activities in northern Brazil that Petrobras, the state-owned Brazilian energy company joined the venture.

Despite the attractiveness of a collaborative strategy, as Monge et al. (1998) note, a key dilemma exists in jointly developing common technologies, because the impossibility of exclusion of the innovation's benefits may also extend benefits to free riders (Olson, 1965). Because of this joint environmental benefit, equity sharing provides defined and prescribed access to valuable benefits stemming from the innovation (e.g., royalty-free technology access) that otherwise would not be available. It is perhaps for this reason that prior experience working together and social networks of managers that induce trust have played an important role in determining which firms participate in the venture (e.g., Gulati & Gargiulo, 1999; Venkataraman, 2004; Ireland, Hitt, Camp, & Sexton, 2001). Our interviews also revealed close interaction between investing firms whose executives hold board positions, and firms that have dedicated liaison personnel involved with the CCP. Thus, in addition to equity sharing, informal normative social capital ties between collaborating parties helps to bind partners (Ring &Van de Ven, 1994), and constrain their opportunistic attempts to gain advantage without contributing valuably to the common good among the collaborators.

Rationale for Corporate Venture Creation. Our interviews suggest that investing firms prefer to establish a separate venture when the technology does not contribute to the firms' core business operations. In the case of the CCP, for example, developing a technology for carbon capture and storage, though important for the long-term sustainability of the firms, is not perceived as core to their oil and gas business. These findings appear consistent with previous studies, which suggest that by establishing a separate venture, firms can evaluate the merits of the technology and make decisions about integrating it into their existing business model subsequently (Chesbrough, 2002). By setting up a separate venture, firms are able to learn from the venture's successes and failures, and integrate the new technology with the core business, contingently reducing the risks to shareholders. Literature in corporate entrepreneurship similarly reveals that such ventures provide a unique opportunity to learn, especially when there is a close working relationship between the investing firms' scientific and managerial staff and the venture's operators (Chesbrough, 2002; Dushnitsky & Lenox, 2005a). The collaboration mode, then, provides partnering firms with an option to adopt carbon capture technologies but does not obligate them to do so. This strategic flexibility is particularly valuable in a nascent regulatory environment and a dynamic technology environment.

Investing firms did leverage their existing core competencies such as marketing and public relations to develop acceptance for technology among its prospective users and the public. Therefore, investing firms provided substantial value-added services (Block & MacMillan, 1993) and facilitated the new venture's access to complementary assets (Alvarez & Barney, 2001). Through cost-sharing and codevelopment of a favorable public perception and regulatory landscape, corporate ventures provide certain advantages over independent venture capital investments (Maula & Murray, 2001).

How do Private Investors Benefit? As is the case with any strategic decision, a firm will decide to participate in a collaborative venture for creating public goods when the private benefits of the action are sufficiently large to compensate for the attendant costs of doing so. The potential for private benefits confer the greatest incentive to the investing firm because no other firm enjoys these identical advantages, making them potentially very valuable. Private benefits here also include the extent of opportunities that individual firms have to apply innovations to contexts beyond the scope of the venture, in ways not available to other investing firms, owing to the firm's unique resources or other legacies. For instance, where the collaboration develops an innovation that favors a particular grade of coal or petroleum resource for subsequent carbon capture, that firm with proven reserves of said grade will differentially benefit from the innovation. Clearly, despite the learning, cost, and reputational advantages stemming from collaboration generally, not all collaborators will be equally positioned to take full regulatory and commercial advantage of the jointly developed innovation. Thus there exist important private benefits to investing firms. Investing governments can likewise uniquely gain vis-à-vis other national collaborators. For example, should a national government already have established a regulatory environment or possess national natural resource endowments favored by the innovative technology of the collaboration, it will differentially benefit ("privately" benefit) from the collaboration.

For investing firms, the common benefit is realized in the form of preferential royalty-free access to the technology and the opportunity for setting technological standards for the entire global industry (Khanna, Gulati, & Nohria, 1998). Common benefits confer additional advantages to the investing firm, but such advantages here are shared by all in the venture, and thus provide less relative advantage to a single firm than do private benefits. In addition, investing firms may also derive some residual public benefit as "citizens of the world" for investing in an environmentally clean technology and thereby contributing to societal welfare.

Societal benefits in the form of public goods made available through technology likely constitute the weakest incentives to the investing firm. Nonetheless, an investing firm is not forced to choose among these various benefits; rather the net advantage to the firm is the cumulative benefit that such participation provides. Table 14.1 highlights these specific benefits and advantages for investing firms and for governments. Table 14.2 highlights the rationale for collaborative ventures and the hierarchy of values to the investing partner of the types of benefits produced through such ventures.

We suggest an inherent interplay between and among the benefit types indicated in Table 14.1. The various benefit types can serve to enhance one another or detract from one another. Benefits from investments in technology ventures that help create public goods such as a clean environment often include the creation of social capital in the form of goodwill and reputation gains, for example, which in turn can be leveraged to gain support of key stakeholders or constituents, who influence public policy

Table 14.1 Specific benefits to investing firms and governments, and to the global public

	Private benefits	*Common/collective benefits*	*Public benefits*
Benefits to firms	- Learning as a function of firm specific resources and markets served - Reputation/leadership within the venture - Reputation with public - Diversification of geographic and product markets - Potential ability to monetize/leverage extant resources more effectively	-Preferential access to technology -Standard-setting -Creating entry barriers/technological lock-out for nonmembers - Resource-sharing and risk-sharing - Broader competitive insights	- Economically and socially productive future as a result of improved and sustainable business and natural environment - Efficient utilization of global resource base - Temporal extension of global resource base - Enhanced technological base - Global sustainable development
Benefits to governments	- Corporate intelligence and influence - Geopolitical leadership - Domestic political benefits - Fiscal benefits from technologies that favor particular natural resource endowments	- Seat at the table/voice in trajectory selection - Technology insights - Regulatory insights	

Table 14.2 Hierarchy and interplay of benefit types from investing in a collaborative technology development venture

Key: Up arrow = competitively more valuable benefit type (to investing party) ↑ Down arrow = competitively less valuable benefit type (to investing party) ↓	Benefit type, benefit interplay[1]	Benefit conditions	Private sector beneficiaries	Public sector beneficiaries
	Private Benefits ↕	Excludable[2] Rivalrous[3]	Individual investing firm	Individual investing government
	Common Benefits ↕	Excludable[4] Rivalrous/ nonrivalrous	All investing firms	All investing governments
	Public Benefits ↕	Nonexcludable Nonrivalrous	All noninvesting firms throughout the world	All noninvesting publics throughout the world

[1] Interplay denoted by upward and downward arrows refers to a potential whereby a given benefit type may positively or negatively impact another benefit type.
[2] Benefits excludable from firms both within and outside consortium.
[3] Benefits enjoyed by one member firm that may come at expense of another member firm.
[4] Benefits excludable from firms outside consortium.

(Hilman & Hitt, 1999). Firms may also receive compensation from governments for contributing to societal welfare (Arrow, 1962). For example, firms can leverage their investments to extract subsidies, buy-downs, and preferential loans from the government to underwrite technological risks, as well as influence government regulations in favor of their own technology standards (Sharma et al., 1999), and can thus increase their net private benefits. Through such private opportunities and benefits firms can compensate shareholders for their investments in creating public goods.

Similarly, firms may expect to maximize their net private benefits from such ventures by combining the new technology, which constitutes a common benefit, with their distinctive resources in a novel way (Kogut & Zander, 1992), and deploying the technology in markets that do not overlap with their competitors. Thus, the interplay of benefit types in such a setting provides for an important refinement of our current understanding of how firms justify collaborative investments for creating public goods.

Conclusions

This chapter contributes to an understanding of three questions that lie at the heart of the debate concerning private investments in public goods: Why do private actors such as oil and gas firms care about public goods such as the natural environment? How do they organize their investments in the creation of such public goods? How do they derive private benefits from such investments? Some firms are able to discover and exploit entrepreneurial and innovative opportunities through collaboration framed in public goods creation parlance, while others cannot (e.g., Venkatarman, 1997).

Our argument points to the emergence of new models of innovation that combine features of the predominant models: private-investment (Arrow, 1962), collective gains (Hargrave & Van de Ven, 2006) and public goods (e.g., Olson, 1965; Hardin, 1982; Sandler, 1992; Marwell & Oliver, 1993). Innovation benefits are neither a purely private good for the firms who retain the rights to consume them or sell them, nor are they purely a common good that is shared by industry participants, nor are they a public good that relies on public subsidy for the knowledge developed, but instead a combination of all three. More recently, scholars have investigated incentives for such hybrid private-collective innovation models (Von Hippel & Von Krogh, 2003; Gächter, Von Krogh, & Haefliger, 2010), wherein private innovators fund public good innovations voluntarily and privately. Our understanding is based on the premise that oil and gas firms are concerned about environmental protection because it contributes to their own long term sustainability. The shadow of a carbon-constrained future exerts a strong pressure on these firms to engage proactively in environmental protection by developing carbon capture and storage technologies. Such an approach also benefits investing firms by conferring differentiation-based competitive advantages upon them, and greater public acceptance and legitimacy in terms of their operations, especially in the eyes of nongovernmental organizations and advocacy groups, which have emerged as increasingly important stakeholders.

Though most firms participating in the CCP possess the financial resources and technological capability to develop carbon capture technology independently, they prefer to pool equity in a collaborative venture. The rationale for a collaborative approach for new venture creation is derived in large measure from the realization that no one firm—however large and resource-rich—can create a clean environment independently. Such a collaborative approach is especially applicable to settings where efforts at the margin are not likely to be productive. In such a situation, firms benefit most when their competitors also invest in the venture and, by extension, invest in a shared vision of a future technological state. Such an approach becomes vital, given the global impact of the products they sell. Pooling equity is also justified based on overlapping markets and geographies, such that participating firms are faced with similar environmental regulations and policies in the countries in which they operate. In addition, the large scale of operations of participating firms helps disseminate the carbon capture technology widely, making it a strong candidate for an emergent industry standard. Firms also expect knowledge-sharing and pooling of risks and resources, but what emerged prominently from our interviews is the importance of social networks that serve as the glue for bringing together competing firms in ways that promote common benefits to the participants.

The rationale for setting up an independent venture stems from at least three considerations. First, the carbon capture technology is not viewed as being core to the operations of energy firms, and so firms do not have strong incentives to own the technology so long as they can license it from the venture on a preferential royalty-free basis. Second, rather than focus on the development and protection of intellectual property in a noncore area, energy firms prefer to deploy their resources in creating a buy-in for the technology. Third, given the uncertainty involved with an untested technology, a new-venture approach allows for learning and subsequent integration, and at the same time reduces the need for the creation of a new organizational subunit whose failure could have a negative impact on the firm. Thus, flexible options for contingent adoption of the innovation are most valuable, given the regulatory and technological immaturity in this domain.

The collaborative pooled equity setting for technology development suggests that all three benefit types—private benefits that accrue to individual firms, common benefits that all investing firms enjoy, and public goods or gains from societal welfare—play an important role in incentivizing firm participation. Further, these benefit types provide relatively distinct values and are interdependent, simultaneously exerting positive and negative effects on one another. Thus, firms that maximize their net cumulative advantage are likely to benefit most from such a collaborative venture.

References

Alvarez, S. A., & Barney, J. B. (2001). How entrepreneurial firms can benefit from alliances with large partners. *Academy of Management Executive*, 15(1), 139–148.

Arrow, K. (1962). Economic welfare and the allocation of resources for innovations. In R. Nelson (Ed.), *The rate and direction of inventive activity: Economic and social factors* (pp. 609–625). Princeton, NJ: Princeton University Press.

Barringer, B. R., & Bluedorn, A. C. (1999). The relationship between corporate entrepreneurship and strategic management. *Strategic Management Journal*, 20, 421–444.

Block, Z., & MacMillan, I. (1993). *Corporate venturing: Creating new business within the firm*. Boston: Harvard Business School Press.

Browning, L. D., Beyer, J. M., & Shetler, J. C. (1995). Building cooperation in a competitive industry: Sematech and the semiconductor industry. *Academy of Management Journal*, 38(1), 113–151.

Chesbrough, H. W. (2002). Making sense of corporate venture capital. *Harvard Business Review*, 80, 90–99.

Dushnitsky, G., & Lenox, M. J. (2005a). When do firms undertake R&D by investing in new ventures? *Strategic Management Journal*, 26, 947–965.

Dushnitsky, G., & Lenox, M. J. (2005b). When do incumbents learn from entrepreneurial ventures? Corporate venturing capital and investing firm innovation rates. *Research Policy*, 34, 615–639.

Freeman, R. E. (1984). *Strategic management: A stakeholder approach*. Boston: Pitman.

Gächter, S., von Krogh, G., & Haefliger, S. (2010). Initiating private-collective innovation: The fragility of knowledge sharing. *Research Policy*, 39(7), 893–906.

Gans, J., Hsu, D., & Stern, S. (2002). When does start-up innovation spur the gale of creative destruction? *Rand Journal of Economics*, 33, 571–586.

Gephart Jr., R. P. (2004). Qualitative research and the academy of management journal. *Academy of Management Journal,* 454–462.

Gulati, R., & Gargiulo, G. (1999). Where do interorganizational networks come from? *American Journal of Sociology, 104,* 1439–1493.

Hardin, R. (1982). *Collective action.* Baltimore, MD: Johns Hopkins University Press.

Hargrave, T. J., & Van De Ven, A. H. (2006). A collective action model of institutional innovation. *Academy of Management Review, 31*(4), 864–888.

Hillman, A. J., & Hitt, M. A. (1999). Corporate political strategy formulation: A model of approach, participation, and strategic decisions. *Academy of Management Review, 24,* 825–842.

Ireland, R. D., Hitt, M. A., Camp, S. M. & Sexton, D.L (2001). Integrating entrepreneurship and strategic management actions to create firm wealth. *The Academy of Management Journal, 15,* 49–63.

Kanter, R. M., Richardson, L., North, J., & Morgan, E. (1991). Engines of progress: Designing and running entrepreneurial vehicles in established companies; the new venture process at Eastman Kodak, 1983-1989. *Journal of Business Venturing, 6,* 63–82.

Kanter, R. M. (2009). *Supercorp: How vanguard companies create innovation, profits, growth, and social good.* New York: Crown Business.

Karnani, A. (2010, August 23). The case against corporate social responsibility. *Wall Street Journal—Eastern Edition,* pp. R1–R4.

Khanna, T., Gulati, R., & Nohria, N. (1998). The dynamics of learning alliances: Competition, cooperation and relative scope. *Strategic Management Journal, 19,*193–210.

Kogut, B., & Zander, U. (1992). Knowledge of the firm, combinative capabilities and the replication of technology. *Organization Science, 3,* 383–397.

Marcus, A., & Geffen, D. (1998). The dialectics of competency acquisition: Pollution prevention in electric generation. *Strategic Management Journal, 19*(12), 1145.

Marwell, G., & Oliver, P. (1993). *The critical mass in collective action: A micro social theory.* New York: Cambridge University Press.

Maula, M., & Murray, G. (2001). Corporate venture capital and the creation of us public companies: the impact of sources of venture capital on the performance of portfolio companies. In M. A. Hitt, R. Amit, C. Lucier, & B. Shelton (Eds.), *Strategy in the entrepreneurial millennium.* Hoboken, NJ: Wiley.

McEvily, B., & Marcus, A. (2005). Embedded ties and the acquisition of competitive capabilities. *Strategic Management Journal, 26*(11), 1033–1055.

Monge, P. R, Fulk, J., Kalman, M. E., Flanagin, A. J., Parnassa, C., & Rumsey, S. (1998). Production of collective action in alliance based interorganizational communication and information systems. *Organization Science, 9*(3), 411–433.

Olson, M. (1965). The logic of collective action: Public goods and the theory of groups. Cambridge, MA: Harvard University Press.

Phan, P. H., & Foo, M. D. (2004). Technological entrepreneurship in emerging regions. *Journal of Business Venturing, 19* (1), 1–5.

Ring, P. S., & Van de Ven, A. H. (1994). Developmental processes of cooperative interorganizational relationships. *Academy of Management Review,19,* 90–118.

Sandler, T. (1992). *Collective action: Theory and applications.* Ann Arbor, MI: The University of Michigan Press.

Sethi, S .P. (1995). Introduction to AMR's special topic forum on shifting paradigms: Societal expectations and corporate performance. *Academy of Management Review, 20,* 18–21.

Sharma, S., Pablo, A. L., & Vredenburg, H. (1999). Corporate environmental responsiveness strategies: The importance of issue interpretation and organizational context. *The Journal of Applied Behavioral Science, 35*(1), 87–108.

Shrivastava, P. (1995). Environmental technologies and competitive advantage. *Strategic Management Journal, 16,* 183–200.

Stopford, J. M., & Baden-Fuller, C. W. F. (1994). Creating corporate entrepreneurship. *Strategic Management Journal, 15,* 521–536.

Venkataraman, S. (1997). The distinctive domain of entrepreneurship research. *Advances in Entrepreneurship, Firm Emergence and Growth, 3,* 119–138.

Venkataraman, S. (2004). Regional transformation through technological entrepreneurship. *Journal of Business Venturing, 19*, 153–167.

Von Hippel, E., & Von Krogh, G. (2003). Open source software and the "private-collective" innovation model: Issues for organization science. *Organization Science,* 14(2), 209-223.

Wadhwa, A., & Kotha, S. B. (2006). Knowledge creation through external venturing: Evidence from the telecommunications equipment manufacturing industry. *Academy of Management Journal, 49*(4), 819–835.

Zahra, S. A., & Garvis, D. M. (2004). International corporate entrepreneurship and firm performance: The moderating effect of international environmental hostility. *Journal of Business Venturing, 15*, 469–492.

Zahra, S. A. (1993). New product innovation in established companies: Associations with industry and strategy variables. *Entrepreneurship Theory and Practice, 18*(2), 47–69.

CHAPTER FIFTEEN

Voices from the Field: The Green Economy Partnership Process

JACK HOGIN AND GEORGIA RUBENSTEIN

Over the past several years, building a green economy has increasingly become a priority for numerous governments, private companies, nonprofit organizations, and other entities throughout the world. This diverse interest in green economic development has arisen from the belief that a greener economy will yield economic, ecological, and social benefits for all stakeholders. In Minnesota, interest in developing the state's green economy also has increased in recent years. Diverse stakeholders have joined together in various partnerships to enhance different elements of the green economy in the state. Growing the green economy in Minnesota is seen as an opportunity to promote economic development, create jobs, benefit the environment, and improve the standard of living.

Funding for the Green Economy Partnership Process (GEPP) was given to the Minnesota Environmental Initiative (MEI) by the Minnesota Department of Commerce and the Blue Green Alliance, both as part of the second phase of the Mayors' Initiative on Green Manufacturing and to continue the work begun by the Minnesota Green Jobs Task Force and other recent green economic development initiatives throughout Minnesota. MEI formed a 28-member stakeholder Work Group composed of diverse representatives from business, finance, labor, environmental organizations, higher education, workforce development, and economic development agencies to develop the recommendations for GEPP. Louis Smith, attorney with Smith Partners PLLP, was retained by MEI to chair GEPP and facilitate each of the Work Group meetings. GEPP Work Group member organizations included:

- 3M
- Alliance for Metropolitan Stability
- BioBusiness Alliance of Minnesota

- Blue Green Alliance
- Center for Energy and Environment
- City of Minneapolis
- City of Saint Paul
- Great River Energy
- HIRE Minnesota
- International Brotherhood of Electrical Workers
- JetE Corporation
- Meier Tool
- Metropolitan Economic Development Association
- Minneapolis Regional Chamber of Commerce
- Minnesota AFL–CIO
- Minnesota Chamber of Commerce
- Minnesota Department of Commerce
- Minnesota Department of Employment and Economic Development
- Minnesota State Colleges and Universities
- Mortenson Construction
- Perkins + Will
- Piper Jaffray
- Remmele Engineering
- SAGE Electrochromics
- Saint Paul Port Authority
- Saint Paul Building and Construction Trades Council
- Solar Skies
- University of Minnesota

MEI and the Work Group agreed to a set of ground rules for GEPP discussions and strategy development. MEI was responsible for the design, management, and facilitation of GEPP. In facilitating the process, MEI focused discussions, assured a fair opportunity for stakeholders to participate in the meetings, and resolved any conflicts that arose. MEI identified members for the Work Group and made final determinations if and when new individuals could be added, if it was determined that essential stakeholder interests were not represented by the existing participants. All Work Group meetings were open to the public.

Work Group participants were expected to attend all meetings, keep their member organizations and constituencies informed, and bring their views to the discussions. Each Work Group member was asked to designate an alternate representative for his or her organization or constituency. One designated representative or alternate, but not both, had a seat at the table and participated in decisions at each meeting. All participants agreed to act in good faith in all aspects of GEPP. Members were expected to present their own opinions based on their experience, perspective, and training, and they agreed to participate actively, constructively, and cooperatively in the process. Participants agreed to be forthcoming about potential conflicts that arose during the process. Disagreements had to be identified and shared with the group as early as possible. All participants were expected to act as equals during GEPP and to respect the experience and perspective of the other participants. As much as possible, decisions were based on consensus of the group. Participants agreed to be supportive of GEPP, but were allowed the ability to disagree with specific decisions or outcomes of GEPP. When making statements about GEPP or its outcomes in public, Work Group participants agreed to make clear that they spoke on their own behalf, and did not necessarily represent the opinions of other participants or MEI.

GEPP used existing definitions from Minnesota statute and recent initiatives for green economy terms. Minnesota state statute 116J.437 defines green economy to include products, processes, methods, technologies, or services intended to do one or more of the following:

1. Increase the use of energy from renewable sources, including through achieving Minnesota's renewable energy standard;
2. Achieve the statewide energy-savings goal, including energy savings achieved by the conservation investment program;
3. Achieve greenhouse gas emission reduction goals; this includes strategies that reduce carbon emissions, such as utilizing existing buildings and other infrastructure, and utilizing mass transit or otherwise reducing commuting for employees;
4. Monitor, protect, restore, and preserve the quality of surface waters;
5. Expand the use of biofuels.

The Minnesota Green Jobs Task Force defined green jobs as the employment and entrepreneurial opportunities that are part of the green economy, as defined in Minnesota statute 116.437, including the four industry sectors of green products, renewable energy, green services, and environmental conservation.

The Mayors' Initiative on Green Manufacturing's definition of green manufacturing encouraged manufacturers to:

- Consider environmental factors and/or reduce environmental impacts;
- Maximize use of local suppliers;
- Reduce transportation impacts, including workers' transportation and goods' shipment;
- Reduce energy use in its product and process;
- Use renewable energy sources;
- Manage water in industrial sites and in manufacturing processes appropriately;
- Maximize a healthy environment for workers; and
- Use and maintain green sites.

The Process of Developing Recommendations

In total, the Work Group met six times from October 21, 2009 to February 2, 2010 to develop their recommendations and meet the charge for GEPP. The Work Group completed four recommendation work products in GEPP:

1. Gap analysis for green business creation and expansion in Minneapolis–Saint Paul and Minnesota
2. Design for a statewide green economy partnership initiative
3. Recommendations for Minneapolis–Saint Paul to consider while developing a regional green economy partnership
4. Draft request for proposal for statewide green economy partnership

A draft gap analysis was prepared by MEI staff, based on a literature review of recent reports on growing Minnesota's green economy. The draft gap analysis was presented to the Work Group during their first meeting, and the group then began a process of refining, prioritizing, and finalizing the analysis. To prioritize the gaps, two online surveys of Work Group members were conducted between meetings, and the survey responses were then shared and discussed with members at subsequent meetings.

To design the recommended partnership model, the Work Group reviewed and discussed several different green and general economic development structures, partnerships, and strategies used in Minnesota and throughout the United States, and heard presentations on current local green and general economic development initiatives in Minnesota, including:

- Itasca Project's Job Growth Task Force
- Regional Competitiveness Project
- Minnesota Science and Technology Strategy Project
- Saint Paul Economic Development Task Force
- MetroMSP.org
- Minnesota Renewable Energy Marketplace
- BioBusiness Alliance of Minnesota
- Minnesota Climate Investment Fund

By the end of the Work Group's fifth meeting, the group had developed a set of draft documents to meet their charge, including a draft gap analysis, a recommended design for a statewide green economy partnership initiative, and a set of recommendations for Minneapolis–Saint Paul to consider while developing a regional green economy partnership. In addition, during their fifth meeting, the Work Group began to discuss how their recommendations could be implemented, including continuing on into a possible second phase of GEPP to further develop ideas to establish the partnership designed by the Work Group.

Between the Work Group's fifth and sixth meetings, MEI hosted an open online public comment period to take comments for the Work Group on their draft recommendations.

During their sixth and final meeting on February 2, 2010, the Work Group reviewed the public comments, finalized their recommendations, and continued their discussion regarding a second phase of GEPP, including the development of the Work Group's final work product: a draft request for proposal to be used to establish the statewide green economy partnership. The meeting ended with the Work Group confirming interest in an MEI-led Phase Two for GEPP, and a discussion on the potential scope of work, cost estimates, and funding strategies for the second phase.

Recommendations

The Work Group's gap analysis contains 28 gaps and barriers for green business creation and expansion in Minneapolis–Saint Paul and Minnesota. Of the many gaps identified, the Work Group selected the following as seven priority gaps for green economic development in Minneapolis–Saint Paul and Minnesota:

- Lack of an overarching, unifying, and compelling vision for Minnesota's green economy that builds on our natural resources, progressive policies, existing businesses, academic institutions, and workforce infrastructure.
- Lack of early stage capital from private investors, especially directed to funding early stage emerging technologies or businesses and environmental technology incubation.
- Lack of significant financial subsidies (grants, loans, other) or incentives from public sources, especially in comparison to other nearby states.
- Undeveloped consumer demand in certain sectors, such as energy efficiency, solar, and biomass energy, and a lack of a systematic approach to address the limited demand.
- Instability in the renewable energy market due to fluctuating pricing, incentive policies, public opinion and demand, and commercialization support.
- Regulatory environment is evolving and is slow to adapt and thus is challenging and costly for businesses to track and influence.
- Minneapolis–Saint Paul and the greater metro region lack a strong central unifying economic development entity.

To overcome these gaps, the Work Group developed a design for a statewide green economy partnership initiative for Minnesota and a set of recommendations for Minneapolis–Saint Paul to consider while creating a regional green economy partnership initiative.

The design for a statewide green economy partnership initiative is the Work Group's primary recommended model to further develop and support the Twin Cities' and greater Minnesota's green economy. Basic elements of the statewide green economy partnership model included

- Private–public partnership
- Strong private sector leadership with strategic public, nonprofit, foundation, academic, and workforce development participation
- Statewide focus to take advantage of unique resources throughout Minnesota and to lobby on statewide policy, with external or internal regional efforts to focus on local initiatives

The partnership's operational structure design calls for a leadership team to provide overall leadership and planning and for short-term task forces and long-term advisory councils, to be used as necessary for specific tasks or issue areas. Financing for emerging and existing green businesses was identified as an immediate priority issue for the partnership to address, by task force(s), by a long-term advisory council, or by some other method. Preference was for the partnership to be either an expansion of existing organization or coalition of existing efforts, rather than creating a new organization.

The formation and funding strategy involved:

- Seed money given to the partnership to fund short-term transition and start-up costs
- Partnership's leadership team generating long-term operational funding for the partnership
- Partnership assembling significant capital to be made available to existing and emerging green businesses to facilitate green business growth in Minnesota

As referenced in the design for the statewide partnership, the Work Group recommended that regional efforts be created in addition to the statewide partnership to support local initiatives and to ensure maximum growth of the green economy in all of Minnesota's regions. Given that there are several green and general economic development initiatives currently underway in the Twin Cities, the Work Group decided to not fully design a regional Minneapolis–Saint Paul partnership, but rather to recommend that these current efforts in the Twin Cities coordinate with each other, include a specific focus on the green economy, and utilize structural elements similar to those recommended in the Work Group's statewide green economy partnership design.

Green Economic Development Structures, Examples, and Activities

According to Mitchell and Welch (2009), sustainability plans, programs, and resources are now commonplace in many cities and regions. Mitchell and Welch suggest that although sustainability strategies are good development practices, on their own they do not necessarily lead to long-term job gains and many of the green jobs created by these sustainability strategies could have only a short-term growth impact unless green economic development programs are tied to local manufacturing that will lead to long-term development in the region.

Mitchell and Welch present the following possible organizational structures for green economic development:

- Public agencies—often most effective at addressing the goal of retention
- Quasi-public and public-private ventures—draw on the strengths and assets of both sectors
- Private for-profit and nonprofit organizations—more aligned with business needs
- Task forces and councils—part of any of the structures described above with a specific, discrete mission

Whatever the structure, Mitchell and Welch suggest that all strategies for green economic development be based on the same essential tenets that apply to all economic development plans: retention, creation/expansion, and attraction.

The following briefly describes Mitchell and Welch's findings about what other cities are doing to promote green economic development. (Additional information compiled by MEI and GEPP Work Group members on other existing green and broader economic development partnerships may also be found in the Appendix to this chapter.)

- San Francisco's programs provide tools to go green, in order to connect citizens and businesses with the region's sustainability efforts; expand the use of technology to communicate progress on sustainability indicators; and work to develop and encourage growth of a regional planning policy organization, integrating business, government, and academia with the region's strong neighborhood organizations.
- Cleveland's programs build and strengthen the region's green chemistry industry to create a green chemistry and biomaterials district and foster the creation of a group of business mentors from amongst the talented and bold corporate and civic leaders in the region.
- Philadelphia's programs create public sustainability forums using resources such as the LEAD Project, Chambers of Commerce, and

local sustainable business leaders; and encourage businesses to be members of a nationally recognized sustainable business certification program, such as B-Lab.
- Santa Monica's programs create a one-stop web resource for going green in the metro area and foster collaborations with area organizations like the Green Institute to provide green building and renewable energy resources for residents and professionals.

Mitchell and Welch cite a recent report by Chapple (2008) to divide green economic development activities into two categories: expansion and capacity-building. *Expansion* refers to increasing the green product market, both on a local scale and in reference to the region's global role. It also includes technical assistance for businesses, research, and development capacity, and interindustry education. Expansion activities can include the following:

- Increasing local markets through consumer education;
- "Buying local" initiatives and Environmentally Preferable Purchasing requirements;
- Establishing international trade offices to attract businesses and exports and to keep in touch with current global trends and innovations;
- Creating technical and business assistance programs that offer consulting services for greening business processes or industry-specific assistance, as well as financing assistance, access to incentives, and networking opportunities;
- Funding emerging business development through efforts such as business plan competitions and competitive grant programs that encourage start-ups and new projects.

Capacity-building refers to strengthening industry clusters, increasing competitive advantage through increased innovative capacity, and making quality of life improvements. Capacity building activities include:

- Networking, which builds regional relationships for global competitiveness;
- Developing supply chains and flexible manufacturing networks;
- Creating strong local ties between suppliers and manufacturers to promote regional cooperation, collaboration, and innovation;
- Developing partnerships for industry-specific research and development, policy, and collaboration;
- Spatial clustering of ecoindustrial, green business parks and zones that create greater interfirm interaction and increase both rivalry and networking;
- Building innovative capacity and incubation by strengthening the intersection of business and research, and connecting universities and businesses;

- Developing the workforce through green collar job training;
- Improving quality of life through such efforts as city greening, local food programs, community energy projects, green building, and affordable housing; and
- Building area reputation and a green "brand" through marketing and consumer education.

Appendix: Additional Economic Development Partnership Examples

MEI and GEPP Work Group members identified the following existing economic development partnerships in other cities, regions, and states, some with a specific green business focus and other focused on broader business development. The partnerships take various forms, involving existing public and private entities to different degrees, or creating completely new organizations. Basic information is provided on each organization's mission, history, regional area of focus, length of existence, budget size and source, staff and board structure, and successes; more detailed information can be found on the organizations' websites.

Area Partnership for Economic Expansion (APEX), Duluth, MN

Mission: APEX's mission is to marshal private-sector leadership and resources to be the most visible and effective marketing arm for developing new business within Northeast Minnesota and Northwest Wisconsin, resulting in increased regional economic growth that will drive new employment opportunities. The primary goal is to strengthen the regional economy through the creation of sustainable wealth, which will result in quality jobs for the people who live there. It accomplishes this by promoting the retention and growth of existing companies that are vital to the region's economy, and by attracting new businesses that can benefit from the area's distinct competitive advantages.

Description and history: APEX is a private, nonprofit, business and economic development organization, founded in 2003 by progressive members of the business community, who recognized that private-sector leadership was key to growing the regional economy. The idea emerged from a 2001 region-wide economic summit of business and community leaders, which identified industry sectors that showed high growth potential in Northeast Minnesota and Northwest Wisconsin, including health care, clean energy, aerospace, and transportation. Backed by the region's largest companies and most experienced corporate executives, APEX was formed to target these industries and spearhead economic development efforts for the region. APEX offers information, advice, and resources on various business development related issues, including relocation, expansion, financial incentives for business growth, and access to workforce.

Regional area of focus: Northeast Minnesota and Northwest Wisconsin

Length of existence: Since 2003

Budget size and source: APEX is a member-funded organization, supported by over 65 members including private businesses, foundations, economic development organizations, and industry leaders who seek to improve the region's business climate. Their operating budget is $600,000 and comes strictly from membership dues.

Staff and board structure: APEX has 4 staff members: president/CEO, director of business development, director of research and communication, and an office administrator.

The APEX Board of Directors is composed of executives and directors from the region's largest and most influential organizations. Board members have demonstrated strong leadership in their industry, involvement in their community, and an interest in using their knowledge and resources to grow the region's economy. APEX membership fees range from $5,000 to $15,000, which buys a board seat, with a maximum of 2 seats per organization.

Successes: APEX continues to show results, having directly affected the creation of more than 750 jobs between 2003 to 2009 with hundreds more pending. Their "Deal Sheet" has 32 businesses in play at any one time, not including new, unqualified phone inquiries that come in weekly. They also work with current businesses and conduct 20 retention calls per year as the Minnesota Chamber's Grow Minnesota arm for the region. APEX is a funding partner of the Northland Connection, a comprehensive, regionally supported Internet portal packed with economic development data for Northeast Minnesota and Northwest Wisconsin. The website (www.NorthlandConnection.com) follows data guidelines set by the International Economic Development Council and contains up-to-date, detailed statistical information about the region's real estate, workforce, industries, infrastructure, incentives, and quality of life.

Charlotte Regional Partnership (CRP)

Mission: The mission of the CRP is to promote the planned growth and prosperity of the 16-county Charlotte region (including 12 counties in North Carolina and 4 in South Carolina) by creating a public, private, and nonprofit partnership to market and promote Charlotte USA as a highly competitive, vibrant region with an increasingly attractive quality of life.

Description and history: Because Charlotte is near the border between North and South Carolina, it had a strong sense of regionalism as far back as the 1970s. In the 1980s, six "founding fathers" (CEOs of the largest banks, the power company, and key industries), came together to create the Greater Charlotte Economic Development Council as a program of the

Charlotte Chamber of Commerce that grew into the Carolina Partnership for Counties on both sides of the border. In 1992, the Charlotte Regional Partnership was formed as its own stand-alone public, private, nonprofit partnership that by 1994 had grown to represent 16 counties on both sides of the border. In 1994, the State of North Carolina recognized that regional partnerships were productive and agreed to help fund seven such partnerships around the state. CRP focuses on three interrelated themes as its regional brand, which it promotes with its own unique trademark.

- Business: Charlotte has a diverse industry base, with strong manufacturing, technology, tourism, and banking sectors.
- Accessibility: Serving over 22 million people a year, the Charlotte/Douglas International Airport ranks first in per capita per day flights among the top 30 metro regions in the country. It has daily nonstop service to London, Frankfurt, and Munich.
- Quality: Charlotte's climate, medical facilities, cultural and regional attractions, professional sports, safe neighborhoods, and beautiful natural setting in the Appalachian Mountains give it a high quality of life.

To assist in site selection services, the 16 counties develop and maintain the database of over 800 available industrial and commercial buildings and over 650 available sites. In addition, CRP has assembled all of the state and local tax data and incentive information on one web page

Regional area of focus: 16 counties in the two-state area

Length of existence: Since 1992

Budget size and source: The annual budget is about $3 million. CRP has investors, not members, with 135 private corporate investors and 20 public investors (cities, counties, and the state). The minimum investment is not defined, but a small investment is about $1,500 a year, and there is no upper limit. Since 1994 about 25 percent of the organization's annual budget has come from the state, 25 percent from public investors (who invest more according to population), and the rest from private sector investors.

Staff and board structure: CRP has a staff of 16. CRP staff is assisted by economic development staff in the 16 member counties. These county staff maintain the real estate databases for each county, which includes listings sent in by commercial real estate brokers.

CRP has a board of 82 members, composed of public, private, and nonprofit investors. Although public partners (some of whom are elected officials) can sit on the board, they are not eligible to be officers.

Successes: The region is home to the headquarters of nine Fortune 500 companies, and there are currently more than 750 foreign-owned companies in the Charlotte region, representing 46 countries. The region has a strong connection to the film industry, and CRP houses the Charlotte Regional Film Commission, which provides filmmakers with a range of services that includes location scouting, permitting, incentives, and interaction with local and state government authorities. The Partnership's Regional Film Commission was singled out for recognition in 2007 by CoreNet, the worldwide association of corporate real estate professionals, which named the commission a finalist for its Strategies and Innovations Award. In 2007, the Charlotte Regional Partnership was recognized by Site Selection with an honorable mention as one of the nation's top economic development organizations.

Kansas City Area Development Council (KCADC)

Mission: The mission of KCADC is to represent the economic interests of the entire 2-state, 18-county region of Greater Kansas City by branding, creating positive perceptions of, and promoting the region as a location to choose for growing companies. The KCADC helps the region to compete for investment against other major metropolitan areas, supports all regional communities equally, assists companies from outside the region find the best regional location, and facilitates final negotiations between the company and its selected community. If it generates the lead, the KCADC shops it around to their community partners. If the community sources the lead by itself, KCADC provides that community with supporting materials.

Description and history: In 1971, about 20 public and private leaders in Kansas City launched a major, ongoing public relations campaign called PrimeTime. In 1976, the Chamber of Commerce created an independent but affiliated economic development organization called the Kansas City Area Development Council to take PrimeTime to the next level and compete directly with other regions for firms in search of a place to grow. KCADC is a private, nonprofit organization aimed at business retention, expansion, and attraction in the Kansas City region. In their efforts to brand and promote the region, they provide comprehensive information on their website aimed both at site selection professionals and at individuals interested in relocating to the area. These pages include a real estate database and information on the local tax structure, financial incentives offered by local and state governments, , career opportunities, schools, and neighborhoods.

Regional area of focus: Two-state, 18-county Greater Kansas City region

Length of existence: Since 1976

Budget size and source: The annual budget for KCADC is about $4 million. The organization is supported financially by more than 250 businesses and more than 50 cities and counties in the region. These supporters are called "investors" instead of "members," and make a minimum investment (not dues) of $7,500 a year for a corporation or a community. Some large corporations invest $250,000 a year.

Staff and board structure: KCADC has a staff of about 20. The organization has about 40 board members, who serve 4-year terms. The board is drawn from private sector CEOs and presidents, who are not allowed to send lower-level representatives in their place, so attendance is very good. Though there is some representation from local universities and hospitals, it is a private sector board, with no representation from, or leadership by, local elected officials. Four "community partner" members (also from the private sector) are elected to 1-year terms on a rotating basis to represent specific communities in the area.

Successes: To date, KCADC has brought more than 500 companies to the Kansas City area. In 1994, KCADC launched the SmartCities campaign, marketing the region as a telecom and high-tech hub. It set the standard across the nation for how to brand a city. In 2001, KC SmartPort was created jointly by the KCADC, the Greater Kansas City Chamber of Commerce, and the Mid-America Regional Council as a nonprofit, investor-based, public–private organization to grow the Kansas City area's transportation industry by attracting businesses with significant transportation and logistics elements and to make it cheaper, faster, more efficient, and more secure for companies to move goods to and from the area. It does not supplant the Kansas City Port Authority. In 2005, KCADC launched its current OneKC and ThinkKC campaigns. The campaigns took city branding to a whole new level, as more than 250 companies and communities used the brand in their own marketing efforts. Other initiatives include:

- Marketing the region's concentration of advanced manufacturing industries as the Alliance for Innovation In Manufacturing–Kansas City (AIM-KC)
- Marketing the advantages of the Kansas City metro area for a variety of transportation and logistics activities as Kansas City SmartPort, Inc.

Metro Denver Economic Development Corporation (Metro Denver EDC)

Mission: Metro Denver EDC works to create a competitive environment that attracts companies and jobs to the region. Each of Metro Denver EDC's economic development partner organizations is committed to the economic vitality of the entire region and is ready and able to communicate the benefits of Metro Denver first and individual communities second. They provide all information that companies or site selectors need to make an informed decision.

Description and history: Metro Denver EDC is a public–private regional economic development nonprofit organization made up of 70 cities, counties, and economic development organizations throughout the region. It was previously a program of the Denver Metro Chamber of Commerce, but was founded as a separate affiliate of the Chamber in 2003, as a "renewed commitment to private sector leadership in economic development." The organization's initial goal was to raise money for an aggressive five-year economic development program; the goals of this five-year campaign were surpassed and another five-year plan was developed in 2009. The organization provides a wide range of programs and services to assist with site selection, expansion, and market decisions, including a website with regional information and resources, demographics, business info, access to international trade, etc; data services for regional market research and analysis; connections to elected officials and business representatives in various sectors; and advocacy and assistance in community involvement for companies newly entering the marketplace.

Regional area of focus: The seven-county Metro Denver and two-county Northern Colorado region.

Length of existence: Since 2003

Budget size and source: The organization's work is largely supported by the region's business community. Most funding is from private-sector investors and the Denver Metro Chamber of Commerce, supplemented by member cities and counties. Strategic initiatives are developed among partners, and a board of investors has final decision-making authority. Metro Denver EDC raised $13.3 million in 2003 for the five-year BreakThrough! Denver campaign, and successfully completed a plan in 2009 to raise an additional $5 million over the next five years.

Staff and board structure: Highest staff (president/CEO, executive vice president, executive assistant) are shared with Chamber of Commerce. Seven other staff specialize in various areas of economic development (investor relations, marketing, etc). Executive committee members represent companies that invest $25,000+ annually, and oversee Metro Denver EDC's budget, work plan, and programs. Board of Governors members invest $10,000+ annually, help develop work plans, and serve on committees and special task forces. Supporters invest $5,000 to $10,000 annually, serve on committees, and support program progress.

Successes: Metro Denver EDC raised $13.3 million in 2004, from private business investors and the Denver Chamber of Commerce, as part of its BreakThrough! Denver campaign. The campaign surpassed its goal of creating 100,000 new jobs in 2008; by 2007, 101,000 new jobs had been created. From 2008–2009 they:

- Successfully completed a five-year capital campaign to broaden business leadership base and raise an additional $5 million over five years for programs to grow the economy.
- Continued to work with Vestas Wind Systems on its $700 million expansion in the state, including two new plants in Brighton.
- Hosted the Fifth Annual Report to Investors and Awards Luncheon in June, highlighting investors' support and work to make Metro Denver a competitive place for business.

- Continued to grow an industry-ready workforce in Metro Denver through the Metro Denver WIRED Initiative, which released the first-ever Workforce Study in 2008.
- Published an annual Industry Cluster Study, detailing seven top industries for job growth in Metro Denver and Northern Colorado.

Prosperity Partnership, Central Puget Sound Region

Mission: The Prosperity Partnership is a coalition of over 300 government, business, labor, and community organizations from King, Kitsap, Pierce, and Snohomish counties, dedicated to developing and implementing a common economic strategy. The shared goal is twofold: long-term economic prosperity and 100,000 new jobs for the central Puget Sound region.

Description and history: The coalition's strategy development began with a region-wide economic analysis. Then, in November 2004, the Partnership convened a first-ever economic summit. Over 1,100 individuals attended. A theme emerged: participants said the Puget Sound region tends to talk about problems and not act on them, but the time for talk is over. The people at the summit made a strong commitment to action, and the Partnership followed through by asking the 2005 Legislature for policy changes and infrastructure funding, even as the regional strategy was being crafted. The Partnership chose five clusters: aerospace, clean technology, information technology, life sciences, and logistics and international trade. These were chosen on the basis of an economic analysis that ranked clusters by their ability to sustain and grow jobs. Additional clusters will be selected for similar analysis in future years. Each cluster working group identified actions, investments, and public policy changes that would strengthen their ability to grow and compete. In addition to focusing on cluster-specific needs, the working groups identified concerns shared by all industry sectors about the foundations of the region's economy. These six foundations are human resources, technology, access to capital, business climate, physical infrastructure, and quality of life and social capital. The groups regularly updated coalition members on their efforts and provided opportunities for comment. The Partnership has recently added tourism and military to the list of clusters.

Regional area of focus: The central Puget Sound Region covering King, Kitsap, Pierce, and Snohomish counties.

Length of existence: Since 2004

Staff and board structure: The Prosperity Partnership is led by government, business, labor, and community leaders from King, Kitsap, Pierce, and Snohomish counties, and supported by over 300 diverse organizations throughout the region. The Partnership is led by a board of nine co-chairs, who are senior-level representatives from regional businesses, economic development organizations, and educational institutions. The Partnership's staff is provided by the regional Economic Development District, the federally designated economic development district for the central Puget Sound Region covering King, Kitsap, Pierce, and Snohomish counties. Another partner, the Puget Sound Regional Council, develops policies and coordinates decisions about regional growth and transportation planning in King, Pierce, Snohomish, and Kitsap counties. The Council is composed of over 80 county, city, port, transit, tribal, and state agencies. Over 300 organizations have signed on as a partner of the organization to help develop and implement the Prosperity Partnership's regional economic strategy. They have also agreed to help promote the development of the strategy to their members and/or constituents.

Successes: The 2005 creation and adoption of the Regional Economic Strategy for the Central Puget Sound Region, a set of detailed action initiatives focused on strengthening the region's leading industry clusters and rebuilding the foundations of the economy. Working groups have successfully raised funds, helped to create industry organizations and influenced policies relevant to their cluster-specific projects. For example, the Washington Clean Tech Alliance, one of the first statewide cleantech industry associations, was created out of the Prosperity Partnership's Clean Tech Working Group.

East Bay Economic Development Alliance and Green Corridor Partnership

Mission: To make the East Bay a center of environmental innovation, emerging green businesses and industries, green jobs, and renewable energy.

Description/history: The East Bay Economic Development Alliance (East Bay EDA) is a public/private partnership serving the San Francisco East Bay (Alameda and Contra Costa Counties) whose mission is to establish the East Bay as a world-recognized location to grow businesses, attract capital, and create quality jobs. The organization was founded in 1990 by Alameda County, its 14 cities, and special districts, as the Economic Development Advisory Board. The organization's name was changed in 1996 to the Economic Development Alliance for Business, as Contra Costa County Cities and County asked to join the organization. In June 2006, the name was changed to the East Bay Economic Development Alliance to better reflect the bi-county mission of the organization. UC Berkeley Chancellor Robert Birgeneau initiated the East Bay Green Corridor Partnership, together with the East Bay EDA and other partners as a way to attract, retain, and grown green businesses in the East Bay area, with an emphasis on businesses launched by entrepreneurial scientists at UC Berkeley and Lawrence Berkeley National Laboratory.

Regional area of focus: East Bay, California
Length of existence: Since 2007

Budget size and source: Each partner contributes $10,000 annually for staff and marketing.

Staff and board structure: Executive Director (new position as of fall '09). Partners include:

- East Bay Economic Development Alliance, an alliance of public and private organizations in Alameda and Contra Costa counties (hosts the organizations offices)
- Eight East Bay cities (Berkeley, Oakland, Richmond, Emeryville, El Cerrito, Alameda, Albany, San Leandro)
- Educational Institutions (UC Berkeley; Peralta and Contra Costa Community Colleges)
- Lawrence Berkeley National Laboratory

Successes: The Partnership received $75M in Recovery Act funds. BP is investing $350 million over 10 years with the Partnership to create a biofuels lab at Berkeley, the Joint Biofuels Energy Institute, which will train 170 researchers (undergraduate through post-doctorate) in biofuels development.

City of Portland—Bureau of Planning and Sustainability (BPS)

Mission: To promote integrated land-use planning and development based on sustainability principles and practices. BPS also develops and implements policies and programs that provide environmental, economic, and social benefits to residents, businesses, and government, which strengthen Portland's position as an international model of sustainable practices and commerce.

Description/history: Portland merged its Bureau of Planning with the Office of Sustainable Development, creating one unified office, the Bureau of Planning and Sustainability. BPS seeks to integrate sustainability practices into all aspects of planning and economic development.

Length of existence: Since 2008

Budget size and source: Funded by City of Portland

Staff and board structure: Staff are City of Portland employees. Advisory committees include the Portland Planning Commission, consisting of nine members appointed by the mayor and confirmed by the city council; the Sustainable Development Commission and Portland/Multnomah Food Policy Council, both citizen advisory councils; and the Peak Oil Task Force.

Regional area of focus: City of Portland, Oregon

Successes: Sustainability-related economic development initiatives include:

- PDX Lounge, a collaborative network of local government, businesses and nonprofit organizations advancing sustainable industry in Oregon.

The Lounge has traveled the country, holding networking events at various green conferences. In support of these activities, the Portland Development Commission is broadening the network to foster sustainability initiatives in the Portland metropolitan region.
- Portland + Oregon Sustainability Institute, a major new initiative founded to implement an aggressive sustainability roadmap in the Portland region. It was created in 2008 by 50 organizations that came together to confront the complex and connected issues presented by the changing environment, climate, economy, and social landscape. The City of Portland provided start-up funds to support a year's salary for the institute's director.
- Oregon Sustainable Wood Products Forum, which took place on Tuesday, May 13, 2008. Approximately 40 people in the wood products sector were in attendance. Participants represented sustainable wood products suppliers, retailers, wholesalers, manufacturers, and assistance providers from around the state, although the Portland Metropolitan region did see a higher proportion of attendance.
- BEST Business Center is a "one-stop shop" for businesses in Portland that want to become greener and more profitable. The Center provides free tools and advice to help businesses in Portland become more profitable and sustainable. It is a partnership of city and regional government programs and energy utilities, including BPS, City of Portland Water Bureau, Metro, Pacific Power, Portland Development Commission, and Portland General Electric.

References

Chapple, K. (2008). *Defining the green economy: A primer on green economic development.* Berkeley, CA: University of California, Center for Community Innovation.

Mitchell, K., & Welch, A. (2009). *Current structures, strategies, and examples for green economic development.* Minneapolis, MN: Blue Green Alliance.

CHAPTER SIXTEEN

Moving the Green Economy Forward: Conclusions from Research and Practice

ALFRED MARCUS

The Nobel Prize winning economist Doug North (2005) has established that major economic leaps forward do not rest on technology alone. They require a sound institutional foundation, but the nature of this institutional foundation and how it works is not well-known. If movement toward a strong institutional foundation is firmly in place, then the chances of breakthroughs occurring in such important areas of the green economy as renewable energy production, storage, and utilization and/or next-generation feedstocks like algae systems or biological or chemical catalysis are likely to be greater. This chapter provides nine conclusions from academic research about what is needed for the green economy to gain momentum. It then turns to a practitioner's perspective and provides six key challenges that practitioners typically believe must be confronted and overcome.

Nine Conclusions from Research

What is needed to create a strong institutional foundation for a green economy? Here are some conclusions from academic research.
 Conclusion 1. The number and type of organizations needed to induce takeoff must be large, their activities many, and the relationships among them dense and complicated. Types of supporting organizations that are involved include public agencies, quasi-public and public–private ventures, private for profit and nonprofit organizations, task forces and councils, trade offices and/or technical and business assistance organizations, eco-industrial and green business parks, green zones or incubators, university technology centers, etc. (Mitchell & Welch, 2009). The activities in which these organizations engage are many. They include providing for consumer education, creating forums for business networking, establishing

standards for quality, certifying products and services, funding, facilitating supply chain development, engaging in workforce improvement, building an area's reputation and "green" brand, attracting dedicated venture capital, and facilitating flexible manufacturing. How many of these organizations must be in place, how diverse do they have to be, how dense must be their relationships, and what are optimal ties among them? Theoretically, the problem of their organization can be conceptualized as a collective-action one (Olson, 1965; Ostrom, 2000), for which Axelrod's (1997) solution assumes spontaneous self-organization that reduces or eliminates a need for hierarchy. However, for such a solution to be realized, these organizations have to be well-informed, able to recognize the moves of other organizations, and have access to the history of these moves, conditions often not met. Although such weakness can be mitigated over an extended period (Hargrave & Van de Ven, 2006), often through such means as government intervention, in which governments disseminate information, encourage cluster formation and/or subsidize research and development (Bianco, Lynch, Miller, & Sened, 2007), this does not always happen.

Conclusion 2. The influence of government, social movements, and natural capital is large. The organizations involved do not exist in a vacuum. A broader milieu of governments, social movements, and natural capital surrounds them. For breakthroughs to take place, for instance, it is not clear what is more important: the natural capital, that is the availability of resources like the wind or sun; social movements; or government? Russo's (2003) empirical analysis of wind energy generation in California found that an abundance of wind determined the rate at which wind projects were done. In contrast, Sine and Lee's work (2009), empirical analysis of wind projects across a number of U.S. states found that large-scale social movements had greater influence. According to Sine and Lee's (2009), social movements help to solve collective action problems. The role of government is very important (York & Lenox, 2009). To what extent are governments, along with social movements and natural capital, decisive factors in bringing about breakthroughs in sustainable renewable energy and energy conservation technologies?

Conclusion 3. Organizations have to engage in joint problem-solving. McEvily and Marcus (2005) suggest that the organizational ties that facilitate the commercialization of green technologies must rely on joint problem-solving if they are to be successful. The preconditions for joint problem-solving are trust and information-sharing. Jacobsson and Bergek (2004) maintain that joint problem-solving, trust, and information-sharing arise in a system made up not only of entrepreneurs and established companies, but also of the providers of such services as engineering, legal, and accounting, and finance (banks and venture capitalists). Different relationships prevail among entrepreneurs, established firms, and service firms (Marcus, Triemstra, & Miel, 2009). Firms transfer knowledge about what is possible and how the future is likely to look (Marcus & Anderson, forthcoming). Exchange of knowledge flourishes in such ties as supplier–customer alliances, licensing agreements, and research consortia (Arikan, 2009). Formal institutions like governments provide norms to guide, direct, and to govern knowledge-sharing (Burer & Wüstenhagen, 2008). According to Bell, Tracey, and Heide (2009), the conditions needed in a cluster to achieve complex problem-solving to confer competitive advantage on cluster depends on how the exchange of knowledge takes place, whether it is decentralized and cooperative and has a foundation of dense social ties or it is formal, hierarchical, and centralized.

Conclusion 4. There must be positive feedback loops. Jacobsson and Bergek (2004) propose that an underlying wave of market and technological opportunities by itself is insufficient for success, because blocking factors in the commercialization of sustainable businesses and technologies are so strong. Thus, long time-spans may exist before alternatives to conventional production to take off (Marcus & Anderson, forthcoming). Factors that prevent progress and keep momentum from building include market uncertainty, lack of legitimacy, opposition from incumbents, and inconsistent government support. To overcome these factors, protected market niches must come into being. These are spaces where critical exploratory work can be done, successful experiments are possible, and experience is accumulated. Within the protected market spaces, different specialties and subspecialties within areas such as renewable energy production, storage, and utilization and/or next generation feedstocks like algae systems or biological or chemical catalysis can flourish. Niches of excellence can develop. Coalitions then can spring up that support the niches and create positive impressions so that citizens come on board and government assistance is available. A chain reaction comes into being in which the start-ups in these niches of excellence enjoy success, which stimulates more start-ups in the niches, funding grows, and customer acceptance sets in. The pool of labor expands, and additional labor becomes more qualified and easier to attract.

Conclusion 5. Both new entrepreneurs and incumbent firms play important roles. Hockerts and Wüstenhagen (2009) describe the dynamic in which high-echelon innovation at levels of "breakthrough," "creative destruction," or the "next industrial revolution" take place among the start-ups, while incumbent firms initially respond to green challenges by optimizing current business models and tightening up existing systems to extract more efficiency. Once start-ups create future market spaces and carve out new customer demand, incumbents switch direction and are apt to follow. Start-ups are not held back by a concern about cannibalizing present business models and are more focused on single issues, which gives them a head start. However, the incumbents often are quick to catch up as they often are fast followers that rapidly introduce copycat products. Sine, Haveman, and Tolbert (2005) describe the challenge of field emergence as hostile criticism and skepticism from financial backers, suppliers, customers, the general public, and employees. To increase new entry and facilitate start-up formation, regulative, cognitive, and normative challenges must be overcome. Otherwise perceived risk overwhelms sense of perceived reward. Sine, Haveman, and Tolbert (2005) find two different processes in which legitimization takes place. General institutions legitimate an entire sector and stimulate start-ups across the entire range of new field, but their effect is especially felt among startup firms with novel technologies, but trade-specific institutions legitimate the existing sector made up of incumbents and established technologies. More-advanced firms relate differently to the opportunity to forge voluntary agreements with governments than less-advanced firms do.

Conclusion 6. The burden on entrepreneurs to develop markets is high. Santos and Eisenhardt (2009) detail the difficult tasks in which successful entrepreneurs have to engage in order to construct markets in emerging fields like the green economy. In a number of in-depth and thoroughly researched case studies they show that successful entrepreneurs have to engage in processes of claiming, demarcating, and controlling market spaces. Mostly they do so by means of alliances and exercising soft power and subtle persuasion to influence the behavior of others in ways that are favorable to them. As Santos and Eisenhardt (2009) understand nascent fields, they are incompletely defined and are missing in primary definitions about matters as elementary as markets and lacking in dominant action logics. Hence, there is great ambiguity, which means outcomes are hard to predict and development costs are largely unknown. To be successful, entrepreneurs therefore must take very vigorous steps, much of it at the symbolic level, to fill this type of void.

Conclusion 7. By themselves, governments do not have the power to induce a long-term takeoff. Marcus and Geffen (1998) argue that opposing logics in governments and markets, plus firms' capacities, have to be harmonized if sustainable business development is to occur. York and Lenox (2009) empirically show that in the case of green building, government incentives by themselves did not provide a sufficient stimulus for takeoff. The effectiveness of government incentives required a rich set of ventures, quasi-public and public–private, for–profit and not-for-profit, that provided technical and business assistance. In the case of green building, there was an important role played by such bodies as the American Institute of Architects, the U.S. Green Building Council, and the LEED Accredited Practitioner Program.

Conclusion 8. Venture capital (VC) can be very influential. Though governments alone may not be able to induce takeoff of green businesses, they can be influential with regard to VC funding (Burer & Wüstenhagen, 2008). One of the most important factors in preventing VCs from investing in new fields like the green economy is their perceived risk. Consistent government support over time in the form of regulatory devices, tax incentives, investment credits, public equity, renewable energy goals and standards may lower the perception of this risk. Burer and Wüstenhagen (2008) argue that the German Electric Feed-In Law of 1991 had an especially strong impact in guaranteeing a preferred rate for selling electricity generated in alternative ways, and that this helped to kick-in a virtuous cycle of volume increase and cost reduction. Recognizing the important role of government, Burer and Wüstenhagen (2008) maintain that some VC firms try to influence legislation and in other ways engage in active risk-management vis à vis government, while other VC firms diversify their country and technology portfolios and in other ways engage in passive risk management strategies vis a vis government.

Conclusion 9. Managerial perceptions are important. The burden is large on managers in emerging business sectors like energy efficiency and renewable energy to calculate the impact of public policies. Government policies may influence their firms' performance in many ways, but they are less likely to know for sure what the ultimate impact will be, in comparison to firms in a more mature industry like oil and gas that have had many more years of experience in dealing with government. Public policies can alter the distribution of power among substitutes, rivals, suppliers, and customers, and can create barriers that slow industry entry, but in an emerging sector like energy efficiency and renewable energy, it is not so clear how this will play itself out. Examples of policies that government uses are many and are likely to be somewhat confusing to conceptually constrained and resource-constrained managers in an emerging sector. What are they to make of the mixture of price supports, certification requirements, and investment subsidies that their companies might be offered, which might be one day on the public policy agenda, and the next day pulled from it? Such policies influence the stability of the industries in which managers in emerging business sectors compete, but the policies themselves tend to be understood as unstable in the minds of the managers and therefore as potentially unsettling rather than stability-creating. These policies, and the impact of demand growth, price competition, and other factors that have important consequences for investment decisions and competition, may be hard for managers in an emerging sector to control, and if they can control them, it may require considerable effort to do so, effort that they may believe should be reserved for their marketplace activities.

Six Challenges from Practitioners

Practitioners point to various challenges that must be managed in order for a strong institutional foundation for a green economy to emerge. These practitioners' perspectives are a summary of comments made at the Cross-Sector Leadership for the Green Economy Conference that was sponsored by the University of Minnesota's Center for Integrative Leadership. This conference was held on April 29-30, 2010 (see http://www.leadership.umn.edu/news/annual_conferences.html)

Challenge #1: Cooperation and Conflict. The costs of coordination are high in both time and money. When efforts are made to increase cooperation, there are power dynamics, value differences, and conflicts of interest. To bring people together is important, but it can be difficult. Trust is a major factor. Other collaboration barriers may include: a lack of transparency and openness, different interpretations about the facts and what people say, and moving forward when discourse is insufficiently clear. Collaboration does not occur naturally. It often involves conflict. Thus, there must be a safe place to communicate as well as clear goals. To enhance cooperation:

- Rules of engagement are needed so that there is open discussion and each party is respected.
- Forums are needed to connect people who have similar interests.
- Companies need to understand not only what they are doing but also what their competitors are doing.
- There must be opportunities to engage in common efforts.
- Big business as well as small business must be part of the discussion.
- Advocacy organizations as well as organizations that represent businesses must be represented.
- Regular meetings must be held, with progress updates.
- A leader or leadership group must take initiative, bear risk, enforce rules, and assist the parties in cooperating.

Leaders with vision who can identify and clarify the problems need to step forward. Often, important perspectives are not heard. Some stakeholders are underrepresented. There needs to be dialogue, better understanding of the risks, better interpersonal ties, and networking, which take time and effort. None of these mechanisms necessarily overcomes the fear of being involved in new and risky market opportunities.

Challenge #2: Focus vs. Diversity. Does the pursuit of many diverse opportunities undermine focus? Though diversity exists, it would be wrong to move toward premature closure. There are legitimate disparate paths. No type of activity should be excluded. Many solutions are required, even if they are small-scale. Transition to a new economy cannot necessarily be orchestrated. It is likely to be emergent and unplanned. The endpoint is not known. The journey is ongoing and therefore it is critical to be flexible and to be able to adapt.

Challenge #3: Government vs. Business vs. Academia. Cross-sector collaboration is needed because of the complexity of the issues and the amount of information to be absorbed, communicated, and transferred. There is a need to bridge gaps between government, industry, and academia.

- Government can help bring academic and university ideas to industry. Unfortunately, it does not have sufficient money or motivation to push clean technology on its own. A large part of the burden lies with private industry to change.

- Universities must accelerate their patenting and licensing, but they also must understand customers and markets better.
- Technical schools must be involved in training green collar workers.
- Utilities and people in energy businesses must learn to share their expertise.

Research partnerships among government, business, and universities are important but these partnerships often are less successful than they should be. Universities must develop and test models of cleantech incubation and economic development. They must pursue comparative research that verifies what works.

Government support is needed, but government should not and cannot pick the winners. Academia's role is to provide alternatives, while business' role is to choose alternatives. Government has many functions it can serve:

- It can mandate energy efficiency in buildings.
- It can establish programs that promote sustainable consumer products.
- If it gets the incentives right, for example through applying full cost accounting energy pricing, it will raise energy costs, and big jumps will be made in energy efficiency, but such policies are likely to alienate consumers and businesses, which will be burdened by higher prices.
- It can reduce complexity in policy and regulation so that it is easier to get projects done.

Venture capital is critical. It is affected by public policies. The "valley of death" for many technologies is the gap between initial funding and seed funding. Start-ups need funding for further growth. Without additional capital, it is hard for start-ups to take the next step. Funding and policy support for innovative start-ups and small businesses needs to be maintained and expanded. When there is success, it is a result of a virtuous cycle of venture capital, technology, and policy development.

Government can play a role in helping entrepreneurs who seek venture capital to overcome barriers. It can help find funding for industry to do research and adopt new practices. It can provide assistance to small businesses in achieving "green" certification so they are able to supply larger organizations. It is not an issue of more government involvement, but clarity about what the government does and assurances that whatever it does, it will do well. Sometimes the public officials' attitudes are more important than policies. They should lay out a vision for growing a green economy.

Challenge #4: Egotism vs. Altruism. Though altruism is a starting point, profit is still the bottom line in most personal and business decisions. There is a need for green businesses to understand their customers if they wish to be profitable. They have to understand why people are motivated to consume green products and services. Changes in behavior and attitude must go hand-in-hand with changes in technology. New technology that comes without changes in attitudes and behavior will not work.

- Consumers have to have reasons to change their consumption. It cannot be that it is just "better" for the environment. Incentives for doing the "right" thing must be in place. Incentive programs, like rebates for green products, help spark interest, but the existence of too many programs of this type is confusing and causes the public to lose interest.
- There is a need for research on green economy selling points that attract diverse customers. These selling points could extend from national security to health (air quality) to aesthetics.
- In the end, businesses must demonstrate that "green" is profitable. Many indicators of "greenness" are at an abstract level; there needs to be a way to make these indicators more relevant for small private businesses. For small private businesses, the biggest barrier is lack of understanding of the costs vs. benefits. There are large amounts of risk and uncertainty for small private businesses.

Some see the "green" economy as an opportunity, but many regard it as replacing an old economy that worked relatively well. For businesses, it is hard to make the change to green technology, because they need resources. They need not just money and capital, but, more important, they need knowledge resources. They need experts in green technology to whom they can talk.

Many businesses would be very willing to make the transition to green practices, but they just don't know how or where to get help. Even though there is knowledge about the best ways to approach many "green" issues, long-held behaviors, habits, and beliefs continue to be big barriers.

Green economy successes are not known enough outside of the people who have been involved. Some see these businesses as elitist and restricted to a small and marginal slice of the population, an environmental issue, and policy whose broader implications are not well appreciated. Many resent the subsidies and government intervention.

Challenge #5: Local vs. National vs. International. The U.S. may lack the advantages that many countries have for making rapid advances. Switzerland's energy consumption is lower to begin with, and the infrastructure already exists (public transportation, for example).

- In China, the political system is very different; NGOs do not really have a say. The central government of China drives things, and state-owned operations are influential. China is often ranked first in clean technology. China is building ecoindustrial parks.
- Canada has allocated a great deal of money at the provincial and national level for technology transfer and commercialization.

- The European Union sets targets to which everyone is supposed to conform, but this requires a lot of front-end negotiations.
- Europe is more willing to set long-term plans and stick with them than the U.S. is.
- Germany has many incentives for small businesses.
- Israel is an example of how national pride can be mobilized for clean energy.

The decentralized political process in the United States makes a coherent, stable energy policy difficult to achieve. It does, however, allow for the public's voice to be heard more than in other countries, specifically China and Switzerland. There are many overlapping national, state, and local programs, and there is a lack of clarity about the different roles they play in stimulating green economic activity.

- National investment in green tech R&D is essential to give a boost to the market
- Nonetheless, national agencies like the Environmental Protection Agency (EPA) and the Department of Energy (DOE) cannot handle everything. What works in each region or locality is likely to be different than what works in another.

Challenge #6: Long-term vs. Short-term Thinking. Poor economic conditions lead businesses to focus on short-term economic results. Short-term thinking, characteristic of Americans, is hindering the "green economy." Europeans and Asians tend to think about longer-term paybacks; Asians can think as far as 100 years ahead.

- The United States should set 10– to 15-year goals and meet these goals. However, it is difficult to get people in the U.S. to make long-term decisions. Five years ahead may not be manageable.
- To rise to the challenge, Americans must be given a reason to act now. With this motivating them, they might be capable of great achievements, as they have shown in the past in sending a man to the moon, harnessing nuclear energy, building a transcontinental railroad, etc.
- However, recent shocks to the economy keep encouraging short-term thinking, and successful green policies require long-term thinking, supported by a diverse set of stakeholders.
- Research should be carried out on how to manage the long-range/short-range conundrum. Lessons should be learned from the examples of how other technological fields were commercialized.
- The way out of this dilemma is to stick to constant, steady, incremental change. Green is a process that grows incrementally though this obviously has drawbacks in that it can get in the way of systemic change.

References

Arikan, A. T. (2009). Interfirm knowledge exchanges and the knowledge creation capability of clusters. *Academy of Management Review, 34*(4), 658–676.

Axelrod, R. (1997). *The Complexity of cooperation.* Princeton, NJ: Princeton University Press.

Bell, S. J., Tracey, P., & Heide, J. B. (2009). The organization of regional clusters. *Academy of Management Review, 34*(4), 623–642.

Bianco, W., Lynch, M., Miller, G., & I. Sened, I. (2007). The constrained instability of majority rule. *Political Analysis,* September.

Burer, M. J., & Wüstenhagen, R. (2008). Cleantech venture investors and energy policy risk: An exploratory analysis of regulatory risk management strategies. In R. Wüstenhagen, J. Hammischmidt, S. Sharma, and M. Starik (Eds.), *Sustainable innovation and entrepreneurship* (pp. 290–309). Northampton, MA: Edgar Elgar.

Drori, I., Ellis, S., & Shapira, Z. (forthcoming) *The evolution of Israeli hi-tech: A genealogical approach.* Stanford University Press.

Hargrave, T., & Van de Ven, A. (2006). A collective action model of institutional innovation. *Academy of Management Review, 31*(4), 864–888.

Hockerts, K., & Wüstenhagen, R. (2009). Greening Goliaths vs. emerging Davids—Incumbents and new entrants in sustainable entrepreneurship. Paper presented at the Academy of Management Annual Meeting.

Jacobsson, S., & Bergek, A. (2004). Transforming the energy sector: The evolution of technological systems in renewable energy technology. *Industrial and Corporate Change, 13*(5), 815–849.

Marcus, A., & Geffen, D. (1998). The dialectics of competency acquisition. *Strategic Management Journal, 19,* 1145–1168.

Marcus, A., & Anderson, M. (forthcoming) Commitment to an emerging field: An enactment theory. *Business and Society.*

Marcus, A., Triemstra, K., & Miel, D. (2009). *Directory, market analysis, and employment opportunities in the Twin Cities green marketplace.* Twin Cities Blue Green Alliance, June 25.

McEvily, B., & Marcus, A. (2005). Embedded ties and the acquisition of competitive capabilities. *Strategic Management Journal, 26,* 1033–1055.

Mitchell, K., & Welch, A. (July, 2009). *Current structures, strategies, and examples for green economic development.* Blue Green Alliance Twin Cities.

North, D. (2005). *Understanding the process of economic change.* Princeton, NJ: Princeton University Press.

Olson, M. (1965). *The logic of collective action.* Cambridge, MA: Harvard University Press.

Ostrom, E. (2000). Collective action and the evolution of social norms. *Journal of Economic Perspectives, 14*(3), 137–158.

O'Rourke, A. (2009). *The emergence of cleantech.* Yale dissertation (committee: D. Esty, M. Chertow, B. Gentry).

Russo, M. (2003). The emergence of sustainable industries: Building on natural capital. *Strategic Management Journal, 24,* 317–331.

Santos, F. & Eisenhardt, K. (2009). Constructing markets and shaping boundaries, *Academy of Management Journal, 52,* 643–671.

Sine, W., Haveman, H. & Tolbert, P. (2005). Risky business? Entrepreneurship in the new independent-power sector. *Administrative Science Quarterly, 50,* 200–232.

Sine, W., and Lee, B. (2009). Tilting at windmills? The environmental movement and the emergence of the U.S. wind energy sector. *Administrative Science Quarterly, 54,* 123–155.

York, J., & Lenox, M. (2009). It's not easy building green: The interaction of private and public institutions in the adoption of voluntary standards. Unpublished working paper., University of Virginia. Darden School.

CONTRIBUTORS

About the Editors

Alfred Marcus teaches and conducts research in strategic management, macroeconomics, business ethics, and business and the natural environment. He has been chair of the Strategic Management and Organization Department at the Curtis L. Carlson School of Management of the University of Minnesota (Minneapolis); director of the Carlson School's Strategic Management Research Center; and a visiting professor at MIT's Sloan School of Management, the Technion, and the Norwegian School of Management. Marcus earned a PhD from Harvard University and undergraduate and master's degrees from the University of Chicago. From 1975–76, he was a research consultant at the National Academy of Sciences, Washington, D.C. He is the author or editor of fourteen books, including *Reinventing Environmental Regulation* (with Donald Geffen and Ken Sexton). His articles have appeared in such journals as *Academy of Management Journal, Strategic Management Journal, Academy of Management Review,* and *The Journal of Forecasting.*

Paul Shrivastava is currently the David O'Brien Distinguished Professor and director of the David O'Brien Centre for Sustainable Enterprise at the John Molson School of Business, Concordia University, Montreal. He also serves as senior advisor at Bucknell University and at the Rajiv Gandhi Indian Institute of Management (IIM), Shillong, India, and he leads the International Chair for Arts and Sustainable Enterprise at the ICN Business School, Nancy, France. Dr. Shrivastava was part of the team of professionals who helped to found Hindustan Computer Ltd., one of India's largest computer companies. He founded the nonprofit Industrial Crisis Institute, Inc., in New York, and published the *Industrial Crisis Quarterly.* He founded *Organization and Environment* (published by Sage Publications). He was founding president and CEO of eSocrates, Inc., a knowledge management software company, and the founding chair of the ONE (Organizations and the Natural Environment) Division of the Academy of Management. Dr. Shrivastava received his PhD from the University of Pittsburgh. He was a tenured associate professor of management at the Stern School of

Business, NYU, and held the Howard I. Scott Chair in Management at Bucknell University.

Sanjay Sharma recently became the dean of the School of Business Administration at University of Vermont, after completing a term as the dean of the John Molson School of Business, Concordia University, Montreal, the largest English-language business school in Canada. Before joining Concordia he was the Canada Research Chair of Organizational Sustainability, a professor of strategy and sustainability and the director of a cross-university Centre for Responsible Organizations at the School of Business and Economics, Wilfrid Laurier University, Waterloo, Canada. His expertise is in the area of helping organizations to develop internal motivation and build capacity and capabilities to reconcile their economic, social, and environmental performance and generate competitive advantage via sustainable business models in developed and developing nation contexts. He has published six books and over 60 articles on research in corporate sustainability. He has been involved in the foundation and governance of the Organizations and the Natural Environment Division at the Academy of Management. He also founded GRONEN (Group on Organizations and the Natural Environment), the international academic think-tank on corporate sustainability that brings together top North American and European scholars to define the academic field of research in sustainability. Sanjay hosted and chaired the 2007 "Greening of Industry Network International Conference" at Waterloo. Sanjay consults for several large multinational corporations, governments, and international organizations. Before pursuing an academic career, Sanjay was a senior manager and CEO of multinational corporations for 16 years.

Stefano Pogutz is a tenured researcher and associate professor of management in the Department of Management and Technology at Università Bocconi, Milan, Italy. He is the director of Bocconi's postgraduate master's program in Green Management, Energy and Corporate Social Responsibility and he is chair of the faculty group Business and the Environment of the CEMS (The Global Alliance in Management Education) Masters in International Management. His expertise and research are in the field of sustainability and innovation, green technologies, environmental management, and corporate social responsibility.

Contributors

Javier Aguilera-Caracuel is a researcher in the management department of the University of Granada (Spain) and member of the research group Innovation, Sustainability and Business Development (ISDE). He received his PhD from the University of Granada in 2010. His research interests include environmental management in multinational enterprises and corporate social responsibility.

Amrou Awaysheh is an assistant professor of operations management at the IE Business School, Instituto de Empresa, in Madrid, Spain. Professor Awaysheh's research focuses on how firms attempt to manage socially responsible practices in their operations and supply chains. His research examines the drivers of these practices and how they impact performance. His research has appeared in journals such as *Journal of Operations Management* and *International Journal of Operations and Production Management*. He has taught the core operations management and supply chain management courses in the MBA program at the IE Business School. He also developed an elective for the MBA program titled "Sustainable Development and Corporate Social Responsibility."

Michelle Bernard earned a BA in political science with a minor in religious studies from Washington University in St. Louis in 2010. Michelle is also a Hirsh Family Fellow for CNISS (the Center for New Institutional Social Sciences) at Washington University in St. Louis, and she aided in the creation of comprehensive global maps of institutional ecosystems that support renewable energy. She earned an Undergraduate Research Award for the analysis of the data, and received a grant to travel to Minnesota to present this research at the Green Economy Conference. Following graduation, she dedicated one year to teaching in a charter school in a high-needs area of Memphis, Tennessee.

Oana Branzei is the David G. Burgoyne Faculty Fellow and associate professor of strategy at the Richard Ivey School of Business, the University of Western Ontario. Her research explores the prosocial functions of business and explains how socio-emotional resources and relational capabilities pattern the creation, capture, conversion, and distribution of value. Her research interests include positive social change, social innovation, and hybrid organizing.

Timo Busch teaches courses on sustainable finance and corporate strategy at ETH Zurich, the Swiss Federal Institute of Technology in Zurich, Switzerland, and at the Duisenberg School of Finance, the Netherlands. His research interests include corporate strategies for a low-carbon economy, organizational adaptation to climate change, and the business case for corporate environmental sustainability. Previously, Timo worked at the Wuppertal Institute for Climate, Environment and Energy, in Germany.

Eva Carmona-Moreno is an assistant professor of strategic management at the University of Almería, Campus de Excelencia Internacional Agroalimentario (ceiA3), Almería, Spain. Her current research interests focus on environmental management and strategic human resource management.

José Céspedes-Lorente is a professor of management at the University of Almeria, Campus de Excelencia Internacional Agroalimentario (ceiA3), Almería, Spain. He received his PhD from the UNED (Universidad

Nacional de Educación a la Distancia, Madrid, Spain). His current research interests are strategic management, environmental management, organizational learning, and entrepreneurship.

Susan K. Cohen received her PhD from the University of Minnesota in 1998 and has taught at the University of Pittsburgh's Katz Graduate School of Business since then. Her research examines how firms enhance their performance (innovation, output and quality, survival, and profitability) through research and development activities. She is particularly interested in how firms manage tensions between acquiring, protecting, and leveraging their technological knowledge by structuring internal and external capabilities. Susan's research has been published in the *Strategic Management Journal, Academy of Management Review, Journal of Economic Behavior and Organization*, and *Advances in Strategic Management*.

Eulogio Cordón-Pozo is an associate professor of management in the Management Department of the University of Granada (Spain) and member of the research group group Innovation, Sustainability and Business Development (ISDE). He received his PhD from the University of Granada in 2004. His research interests concern innovation and strategic management .

Michael Craig graduated summa cum laude from Washington University in St. Louis in 2010 with a bachelor of arts in environmental studies and a minor in environmental engineering. In addition to his research on the interaction of institutions and renewable energy, he completed a senior thesis in ecology, characterizing patterns of seed predation around habitat corridors. He has written on energy innovation issues for Americans for Energy Leadership, and currently works as an energy analyst in Washington, D.C., at an environmental nonprofit.

Deborah E. de Lange is assistant professor of international business at Memorial University of Newfoundland, Canada. After many years in industry, Dr. de Lange obtained her PhD at the University of Toronto and published her PhD dissertation as a book entitled *Power and Influence: The Embeddedness of Nations*. She has written a second book, *Research Companion to Green International Management*. She is finishing a third book about corporate governance, called *Cliques and Capitalism: A Modern Networked Theory of the Firm*. Her research interests include trade, foreign direct investment, diplomacy, international business and organizations, corporate governance, sustainability, and network and embeddedness theories. She has taught globalization, sustainable strategy, and high-technology strategy and currently teaches international business at graduate and undergraduate levels.

Israel Drori is a professor at the School of Business Administration, College of Management Academic Studies, Rishon Lezion, Israel, and a visiting professor at Tel Aviv University. He has visited at a number of

U.S. and Canadian universities, including, in 2011–2012, the University of Michigan's Ross School of Business. His PhD is in anthropology from UCLA and he is the winner of Clifford Geertz Prize for Best Article in Cultural Sociology for "Repertoires of Trust: The Practice of Trust in Multinational Corporation amid Political Conflict," *American Sociological Review,* 72: 143–161. Along with S. Ellis and Z. Shapira, he is the co-author of the book *The Evolution of Israeli Hi-Tech* (forthcoming, Stanford University Press).

Shmuel Ellis received his PhD in social psychology from Tel Aviv University in 1985. During the years 1985 to 1988, he was on the faculty of the Open University of Israel and from 1990 to 1992 he was a visiting assistant professor at the Sloan School of Management, MIT. Shmuel Ellis is currently the chairperson of the Department of Undergraduate Management at Tel Aviv University. His main areas of interest are learning from experience, organizational learning, knowledge management, organizational genealogy, and social networks. Along with I. Drori and Z. Shapira, he is the co-author of the book *The Evolution of Israeli Hi-Tech* (forthcoming, Stanford University Press).

M. Ángeles Escudero-Torres is an assistant professor of management in the Management Department of the University of Granada (Spain) and member of the research group Innovation, Sustainability and Business Development (ISDE). Her research focus is on internationalization and innovation of firms .

Emilio Galdeano-Gómez is a professor of applied economics at the University of Almería, Campus de Excelencia Internacional Agroalimentario (ceiA3), Almería, Spain. His research topics of interest include environmental economics, agricultural economics, and regional development. He is an editing board member of the *International Journal of Sustainable Society.*

Jörg H. Grimm is a research associate at the Chair of Logistics Management at the University of St. Gallen, Switzerland. His research focuses on the management of firms' multi-tier value chains, and in particular on how to ensure that their suppliers and subsuppliers comply with firms' sustainability standards. Before joining the University of St. Gallen, he worked as a consultant for supply chain management and strategic sourcing with Capgemini Consulting in Munich, Germany, and Bucharest, Romania, with an industrial focus on oil and gas exploration and production.

Jens Hamprecht is a senior researcher at the Chair for Sustainability and Technology at ETH Zurich (the Swiss Federal Institute of Technology, Zurich). His research focuses on the market strategies and the institutional strategies of corporations in the context of greenhouse gas reduction. His research has been published in the *Journal of Management Studies,* the *Journal of Supply Chain Management, Business & Society,* and in several edited books.

Before joining ETH Zurich, he gained his PhD at the University of St. Gallen, Switzerland.

Volker H. Hoffmann is an associate professor for sustainability and technology in the Department of Management, Technology, and Economics of ETH Zurich, the Swiss Federal Institute of Technology. His work concentrates on corporate climate strategies and the relation between climate policy and innovation. He has published in journals such as the *Journal of Management Studies, Long Range Planning, Climate Policy, Energy Policy, Ecological Economics,* and *Energy Economics.* Before joining the faculty of ETH Zurich, he was a project manager at McKinsey & Company.

Joerg S. Hofstetter is assistant professor of management at the University of St. Gallen, Switzerland, vice director of the university's Chair of Logistics Management, and he co-leads the International Center for Sustainability in Value Chains (ISVC). He holds a BSc and MSc from the University of Stuttgart and a PhD from the University of St. Gallen. His research interests concern the management of multi-tier value chains, and in particular the ability of brand owners to ensure value chain partners' compliance with their sustainability standards. Prior to his academic career he worked for ITT Automotive in North America and Lufthansa Cargo in the Far East.

Jack Hogin was associate director of environmental projects at the Minnesota Environmental Initiative (MEI), a nonprofit organization that builds innovative partnerships among the business, government, and nonprofit communities to develop collaborative solutions to Minnesota's environmental problems. Jack's work at MEI focused primarily on energy and green/sustainable business. He led several MEI stakeholder processes, including the Green Economy Partnership Process. Jack has a bachelor of arts degree in economics from Carleton College, Minnesota.

Nuria Esther Hurtado-Torres is an associate professor of international management in the Management Department of the University of Granada (Spain) and member of the research group Innovation, Sustainability and Business Development (ISDE). She received her PhD from the University of Granada in 2000. Her research interests include internationalization and innovation of firms.

Robert D. Klassen is a professor of operations management and Magna International Inc. Chair in Business Administration at the Richard Ivey School of Business, University of Western Ontario, Canada. Klassen's research explores the multifaceted linkages between the natural environment, social issues, and firm performance, termed the triple bottom line, with a particular emphasis on the pivotal role of manufacturing operations and supply chains. He has over two dozen refereed publications, including widely cited articles in such journals as *Management Science, Journal of Operations Management, Academy of Management Journal, Production and*

Operations Management, and *International Journal of Operations and Production Management.* At Ivey, he has taught the core operations management course in the MBA and EMBA programs, as well as electives in sustainable development, services management, operations strategy, and technology management. He has coauthored two textbooks, *Foundations of Operations Management* (Canadian edition), and *Cases in Operations Management: Building Customer Value through World-Class Operations.* He has also written over two dozen teaching cases and simulation exercises in the areas of operations strategy, process analysis, and sustainable development.

Moritz Loock is a senior researcher at the IWÖ-HSG (Institute for Economy and the Environment at the University of St. Gallen, Switzerland) and programme manager for the Advanced Studies Programme in Renewable Energy Management (REM-HSG) at the University of St. Gallen. His main fields of research are strategic management in emerging markets and discontinuous change and consequences for management and management education. Dr. Loock focuses on topics such as business model design, management skills, and decision-making. He obtained the title of *doctor oeconomiae* from the University of St. Gallen for his thesis titled "Business Models for Renewable Energy: Determinants of Financial Performance and Investment Decisions." It includes papers on strategic business model design in different renewable energy fields.

Florian Lüdeke-Freund is a research assistant at the Centre for Sustainability Management (CSM), Leuphana University of Lüneburg, Germany, in the field of corporate sustainability management, where he focuses on sustainable business models, strategic sustainability management, and entrepreneurship. His PhD project deals with these topics in the context of renewable energies, especially regarding the German photovoltaic industry. Currently, he is working on a follow-up study to the research presented in this book.

Sonja Lüthi is a postdoctoral researcher at University of St. Gallen's Good Energy Chair for Renewable Energy Management (Switzerland). She holds a PhD degree in economics from the University of St. Gallen and a master's degree in geography with a minor in environmental economics from the University of Fribourg, Switzerland. In her cumulative PhD thesis, she investigated factors in effective policy design from the point of view of solar and wind energy project developers. From June 2009 to September 2010, she was a visiting scholar, first at University of California Davis and later University of California Berkeley. Since 2011 she has been a project leader for the energy department of the Canton of St. Gallen, Switzerland.

Joel Malen is a PhD student at the Carlson School of Management, University of Minnesota. His research looks at business and public policy, focusing on the relationship between environmental policy and firm strategies related to clean technology development. He holds a master's degree in international relations from Johns Hopkins University.

Alfred Marcus. *See biography with editors, above.*

Martina Müggler holds a master of arts in international affairs and governance from the University of St. Gallen, Switzerland, and a CEMS master's degree in international management. After the completion of her studies, she joined SBB Cargo AG, a Swiss rail freight provider, as a management trainee. Her research analyzes the practices that companies implement to ensure compliance with their sustainability standards in the supply chain.

Nils J. Peters is consultant with McKinsey & Company in Zurich, Switzerland. He mainly focuses on portfolio strategy, growth strategy, and corporate finance-related projects for industrials in the metals and mining and pharmaceutical sectors. Before joining McKinsey, he achieved his doctoral degree at the University of St. Gallen, Switzerland, empirically analyzing sustainability strategies for companies in the consumer goods, chemical, and cosmetics industry.

Kim Poldner is oikos PhD Fellow at the University of St. Gallen in Switzerland. Her dissertation engages the major discourses of ecological, social, cultural, and economic sustainability in the ethical fashion industry. She works with multisensorial ethnographies, including video material of interviews and fashion shows, artifacts, and web-based sources. Kim is also co-founder of http://www.ecofashionworld.com

Georgia Rubenstein is senior environmental project associate at the Minnesota Environmental Initiative (MEI), a nonprofit organization that builds innovative partnerships among the business, government, and nonprofit communities to develop collaborative solutions to Minnesota's environmental problems. Georgia works on outreach and communications at MEI and leads events that bring together diverse stakeholders to discuss and collaborate around specific environmental issues. She currently serves on steering committees for both her neighborhood association and the Land Stewardship Project, an organization that focuses on sustainable farming and food systems. Georgia has a bachelor of arts degree in international development studies from McGill University, Montreal, Canada.

Itai Sened is a professor of political science and the director of the Center for New Institutional Social Sciences (CNISS) at Washington University in St. Louis. He is the author of *The Political Institution of Private Property* (1997); *Political Bargaining: Theory, Practice and Process* (2001, with Gideon Doron); *Multiparty Democracy* (2006, with Norman Schofield); and coeditor of *Explaining Social Institutions* (1995, with Jack Knight); and of *Economic Institutions, Rights, Growth, and Sustainability: The Legacy of Douglass North* (forthcoming, with Sebastian Galiani). He published numerous articles in the top-refereed journals in political science, including *The American Political Science Review, The American Journal of Political Science, The Journal of Politics, The British Journal of Political Science,* the *European Journal for Political Research,* and the *Journal of Theoretical Politics.*

David C. Sprengel obtained his PhD from the Department of Management, Technology, and Economics of ETH Zurich, the Swiss Federal Institute of Technology. His research analyzes corporate responses to direct and indirect climate change effects. He thereby focuses on empirically identifying response strategies to regulatory and stakeholder pressures to reduce direct CO_2 emissions across different industries in the context of climate change. Currently, David works as a project manager with McKinsey & Company.

Chris Steyaert is professor of organizational psychology at the University of St. Gallen in Switzerland. His current interests concern creativity, multiplicity, and reflexivity in organizing change, intervention, and entrepreneurship. His latest books are *The Politics and Aesthetics of Entrepreneurship* (2009, co-edited with Daniel Hjorth) and *Relational Practices, Participative Organizing* (2010, co-edited with Bart Van Looy).

Bryan Stinchfield is an assistant professor of organization studies at Franklin & Marshall College in Lancaster, Pennsylvania. His research interests include how organizations and entrepreneurs use innovation to increase their own, and their society's, level of sustainability. Bryan teaches courses in strategy, organizational behavior, and entrepreneurship. Bryan earned his PhD from Southern Illinois University in Carbondale, Illinois.

Kathleen M. Sutcliffe (PhD, University of Texas at Austin) is the Gilbert and Ruth Professor of Management and Organizations at the Stephen M. Ross School of Business at the University of Michigan. For the past decade, her research has been aimed at understanding how organizations and their members cope with uncertainty and unexpected events, and how complex organizations can be designed to be more reliable and resilient. She is currently investigating these issues in wild land firefighting and health care. She has published widely in management and organization journals as well as in health care. A recent book is titled *Managing the Unexpected: Resilient Performance in an Age of Uncertainty* (2007, co-authored with Karl Weick).

Hildy Teegen is dean of the Darla Moore School of Business, University of South Carolina, in Columbia, South Carolina. Prior to coming to the University of South Carolina in September 2007, she was director of the George Washington University's Center for International Business Education and Research (CIBER) in Washington, D.C. Dr. Teegen earned bachelor's degrees in Latin American studies and international business and finance from the University of Texas at Austin in 1987. In 1993, she received her PhD in international business, also from the University of

Texas at Austin. Dr. Teegen has written extensively about global business, most recently about interactions among firms, governments, and non-governmental organizations. Her research has been widely published in such journals as the *Journal of International Business Studies* and *The Journal of Business Ethics*.

Gurneeta Vasudeva is an assistant professor at the Carlson School of Management, University of Minnesota. Her research interests lie in the areas of interorganizational alliances, technological innovation, and comparative national institutional contexts. Her research has appeared in top management journals. Prior to her current position, she taught at the Indian School of Business, in Hyderabad, India and worked in the area of international development as an energy policy analyst in Washington, D.C. She has a PhD from the George Washington University, and MSc and BSc degrees in applied mathematics from Delhi University.

Matthew S. Wood is an assistant professor of entrepreneurship at the University of North Carolina, Wilmington. His research interests include the role of innovation, entrepreneurial cognition, and decision-making, and academic entrepreneurship. Matthew earned his PhD from Southern Illinois University in Carbondale, Illinois.

Rolf Wüstenhagen is the Good Energies Professor for Management of Renewable Energies and a director of the Institute for Economy and the Environment at the University of St. Gallen, Switzerland. He has held visiting faculty positions at the University of British Columbia, and Copenhagen Business School and National University Singapore. His research focuses on decision-making under uncertainty by energy investors, consumers, and entrepreneurs, and how such choices are influenced by energy policy. He embarked on his academic career after retiring from one of the leading European energy venture-capital funds.

INDEX

ACA, *See* adaptive conjoint analysis
ACBC, *See* adaptive choice-based conjoint experiment
adaptive choice-based conjoint experiment (ACBC) (PV finance), 108–9, 114–17
   and graph, 117
   and Hierarchical Bayes (HB) Estimation, 116
   and method, 114–15
   and Prospect Theory, 119–20
   and results, 116–19
   and sample, 115–16
adaptive conjoint analysis (ACA) approach (PV market), xviii, 37–51
   attributes and attribute levels used, 43–50
   average part-worth utility estimates, 46
   and consumer theory, 40–1
   data analysis and results, 44–50
   and decision-making, 40
   and discrete choice theory, 41
   and feed-in tariff (FIT), 38–9, 42, 45–7, 49–50
   and investment, 49
   and location attributes, 40
   and optimal features of projected projects, 39–40
   and potential attributes for survey, 42
   and questionnaire, 41–4
   and random utility theory, 41
   and sample and questionnaire, 41–4
   and SMRT Simulation, 47, 49–50
   and theory, 40–1
   and willingness to accept policy, 47–8
Africa, 160, 205, 249–50
Aiken, L. S., 86
Alliance for Metropolitan Stability, 277
Alliance to Save Energy (ASE), 60
America's Energy Future, 5
American Institute of Architects, 298
American Recovery and Reinvestment Act (2009), 14, 53–4
American Wind Energy Association (AWEA), 6
Anderson, J. C., 251
APEX, *See* Area Partnership for Economic Expansion
Apple, 132–3, 142
Area Partnership for Economic Expansion (APEX), 285–6
Asia, 24, 27–8, 205, 249, 250, 303
ASE, *See* Alliance to Save Energy
Australia, 129, 249
Austria, 115–16, 250
AWEA, *See* American Wind Energy Association
Axel Springer, 179, 183, 186–7
Axelrod, R., 296

Bansal, P., 83, 247
Barkema, H. G., 243
Barron, R. M., 91, 93–4, 102
Barsky, D., 245
Bell, S. J., 297
Bergek, A., 296–7
Bhargava, A., 233
BioBusiness Alliance of Minnesota, 277
biofuels, 54, 125–31, 134–7, 279, 293
biomass, 8, 11, 71, 281
Blue Green Alliance, 277–8
Blundell, R., 235
Bond, S., 235
Böttcher, J., 110–11
BP, 264–6, 269, 293

BPS, *See* Bureau of Planning and Sustainability
Brammer, S., 97
Brazil, 20, 201, 269
Brouthers, K. D., 245
Brundtland Report (1987), xv, 82, 99
Bureau of Planning and Sustainability (BPS) (Portland), 293–4
Burer, M. J., 298–9
Bush, George W., 20

Calantone, R. J., 246
Canada, 1, 129, 203, 205–6, 209, 213, 249–50, 265, 302
CapX2020, 9
CO2 Capture Project (CCP), xxi, xxv, 263–74
  advisory board of, 265
  benefits of, 270–2
  and board membership, 269
  and cooperation, 263–74
  and corporate venture creation, 269–70
  costs of carbon capture and storage (CCS), 263, 267
  and decision-making, 265
  and executive board, 265–6
  and exploratory approach, 266
  and hierarchy of benefit types, 272
  and interviews, 266–74
  and market opportunities, 267–8
  organization of, 268
  and public perception, 267–8
  reasons for, 267
  specific benefits list, 271
  and technology, 265–6
carbon emissions, 5, 26, 89, 101, 133, 264, 268, 273, 279
CCP, *See* CO2 Capture Project
Center for Integrative Leadership, 299
CFP, *See* corporate financial performance
Chapple, K., 284
Charlotte Regional Partnership (CRP), 286–8
Chevron, 265
China, 129, 205, 213, 302–3
Christmann, P., 247
Chung, Y., 227
Cixous, H., 160
Cleantech Group, 128–9, 131

cleantech venture capital (VC), xxii, xxiv, 79, 125–38, 142, 146–51, 270, 296, 298, 310
  and economic conditions, 132–3
  and energy efficiency (EE), 125–8
  environment of, 128–30
  and exit mechanisms, 133–4
  exploratory analysis of, 134–7
  and factors that affect path dependence, 131
  and fifteen most active countries, 129
  and growth, 134–5
  and ideology and culture, 130–1
  initial conditions of, 130–3
  and initial public offerings (IPOs), 125
  and intellectual and technical capital, 132
  and large-scale social movements, 130
  and mergers and acquisitions (M&As), 125
  and past deals, 133
  and path creation, 125
  and path dependence, 125–38
  and physical and social capital, 131–2
  and prior VC experience, 132–3
  and public policies, 132–3
  and renewable energy (RE), 125, 130
  summary of, 137–8
Cleveland, 283
climate change, xi, 3, 23–5, 83, 154, 162, 241
coal energy, 4–5, 13, 54, 126, 270
COC, *See* codes of conduct
codes of conduct (COC), 197–8, 202, 204, 208–12, 215
Coffee and Farmer Equity (CAFE), 195
Combs, J. G., 90
Competitive Renewable Energy Zones (CREZs), 10
Compustat, 88
conjoint experiment methods, 39–40
  See also adaptive conjoint analysis
ConocoPhillips, 265
consumer theory, 40–1
Control Center for Renewable Energies, 13
corporate financial performance (CFP), xviii–xix, 81–101
corporate social performance (CSP), 86, 91

corporate social responsibility (CSR), 86
CREZs, *See* Competitive Renewable Energy Zones
Cronbach's alpha, 25–6, 204–5, 255–6
Crook, T. R., 90
cross-sector cooperation, xxi–xxii, 261, 295–303
  and the future, *See* green economy, future of
  and the green economy, *See* Green Economy Partnership Process
  and leadership, xii–xiii, xiv
  and private firms, *See* CO2 Capture Project
Cross-Sector Leadership for the Green Economy Conference, 299
CRP, *See* Charlotte Regional Partnership
CSP, *See* corporate social performance
CSR, *See* corporate social responsibility

Data Envelopment Analysis (DEA) technique, 227
DEA, *See* Data Envelopment Analysis
Debt Service Cover Ratio (DSCR), xix, 111, 113, 116–21
decision-making, 1–2, 22, 25, 37–8, 40, 42, 44, 49–51, 54–9, 66, 68–74, 101, 109–10, 114–15, 119–21, 125, 127, 137, 145, 152, 164, 169, 175, 211, 245, 252, 265, 269–70, 278–9, 289–92, 299–303
Deepwater Horizon oil spill (2010), 3
Denmark, 1, 129, 250
Dillman, D. A., 203
DOE, *See* U.S. Department of Energy
Dow Jones 2500 global index, 24
DSCR, *See* Debt Service Cover Ratio
Dun & Bradstreet (D&B), 249
Dyer, J., 90

East Bay Economic Development Alliance, 292–3
Economics of Ecosystem and Biodiversity (TEEB) Report (2010), xi
Ederington, W. J., 223–4
EFC, *See* efficiency change
efficiency change (EFC), 227–8, 230–4
egotism, 301–2
EIA, *See* U.S. Energy Information Administration

Eisenhardt, K., 298
Eisenmann, T., 90
electricity, 4–15, 19, 39, 107, 110, 117, 121n4, 142, 147, 299
EMS, *See* environmental management systems
EnCana, 165
Energy Reliability Council of Texas (ERCOT), 9–10, 13
ENI, 265
entrepreneurs, xxii, 64, 69, 73, 79, 87, 146, 157–72, 178, 183–4, 188, 247, 270, 272, 279, 292, 296–8, 301
  and fashion, *See* ethical fashion industry
environmental and social performance (ESP), xix, 81–101
environmental management systems (EMS), 226–7
ERCOT, *See* Energy Reliability Council of Texas
ESP, *See* environmental and social performance
ethical fashion industry, xx, xxv, 157–72
  and art, 169–71
  and awareness, 163–9
  and biomimicry, 160, 162, 168
  and duty of care, 158–9, 162, 171
  and ecosystem design stages, 158, 162
  and effect, 169
  and feminism, 160–2
  and findings, 162–71
  implications of, 171–2
  and methods, 160–2
  micro- and macro-practices of, 164–8
  practices of ecosystem design, 170
  and shecopreneurs, xx, xxv, 79
  and social imaginaries, 159–60
  and voice, 160
EU ETS, *See* European Union Emission Trading Scheme
EUREP-GAP code, 226–7
European Union Emission Trading Scheme (EU ETS), 23
exporting, xxi, xxv, 221–38
  conclusions on, 236–7
  and data envelopment analysis technique, 227
  and definitions of variables, 231
  and descriptive statistics, 232
  discussion and conclusions on, 236–7

exporting—*Continued*
  and efficiency change, 227–8, 230–4
  and endogeneity, 222, 230
  and environmental management systems, 226–7
  and environmental productivity, 227–35
  and export performance model, 227–8
  and "factor endowment hypothesis," 223–4
  and generalized method of moment, 230, 234–6, 238n4
  and heteroskedasticity, 230
  and hypotheses, 225–6
  and intermediate imported inputs, 230
  and knowledge, 225–6
  and methods, 226–30
  and "pollution haven hypothesis," 237n1
  and productivity, 222–6, 233
  and results, 230–6
  and technological change, 227–9
  total factor productivity index, 222, 227, 229, 231
  and wage rate, 228–31, 234
Exxon, 20

Fabozzi, F. J., 110
Fairtrade certification, 198
Färe, R., 227
fashion industry, *See* ethical fashion industry
Federal Energy Regulatory Commission (FERC), 8
feed-in tariff (FIT), xviii, 38–9, 42, 45–7, 49–50
FERC, *See* Federal Energy Regulatory Commission
Ferrell, O. C., 250
Filatotchev, I., 178
finance
  and fashion, *See* ethical fashion industry
  and high-tech, *See* high-tech cluster revolution
  and innovation, *See* innovation and finance
  and photovoltaic projects, *See* photovoltaic projects and finance
  and venture capital, *See* cleantech venture capital

Finland, 129, 250
firm–government interactions, 56–7
firms and political activity (figure), 55
FIT, *See* feed-in tariff
food industry, 248–9
Fowler, S. J., 87
France, 44–5, 129, 250
Frankel, J., 224
Freeman, R. E., 25
future of green economy, *See* green economy, future of

game theory, 22, 30, 199
gas, 4, 263–4, 267–9, 272–3, 279, 299
Geffen, D., 298
George, E., 22
geothermal energy, 11, 54
GEPP, *See* Green Economy Partnership Process
Gerbing, D. W., 251
German Electric Feed-In Law (1991), 132, 298–9
German Renewable Energy Sources Act (EEG), 108, 110, 121n4,5
Germany, 10–15, 20, 37, 44–5, 48–9, 108, 114–15, 122n8, 129, 250, 303
  and cleantech VC, 129
  and greenhouse gas emissions, 20
  and installed wind power capacity, 10–15
  and PV panels, 37, 44–5, 48–9
  and PV project financing, 108, 114–15, 122n8
Glick, W., 66
global financial crisis (2007–present), 81, 107–8, 129
Global Reporting Initiative (GRI), 197
global supply chains, xiv, xx–xxi, 175
  and exporting, *See* exporting
  and internationalizing, *See* internationalization of SMEs
  and sustainable practices, *See* proactive supply chain sustainability standards (PSCSS)
  and sustainability standards, *See* sustainability standards
global warming, 20, 264, 267
  See also climate change
GLS Gemeinschaftsbank eG, 109
Gomez-Mejia, L. P., 90
Goodstein, J. D., 21, 25

INDEX

Graves, S. B., 86
Greece, 44–5, 49, 250
Green Corridor Partnership, 292–3
green economy, xi–xxvi
   and business schools, xvi
   and creating knowledge, xvi
   and cross-sector cooperation, *See* cross-sector cooperation
   and cross-sector leadership, xii–xiii, xiv
   future of, *See* green economy, future of
   and global governance, xv
   and global impacts, xv
   and global supply chain, *See* global supply chain
   and innovation, *See* innovation
   and institutions, *See* institutions
   issues in moving toward, xiii–xvi
   and methodology, xvi–xxvi
   partnership process, *See* green economy partnership process
   and path dependence, xiv–xv
green economy, future of, xxii, xxv, 295–303
   and challenges, 299–303
   and cross-sector cooperation, 296–301
   and egotism, 301–2
   and entrepreneurs, 297–8
   and focus, 300
   and governments, 296, 298–301
   and long-term thinking, 303
   and managerial perceptions, 299
   and organizations, 295–8
   and positive feedback loops, xxii, 297
   and public interest, 296
   research conclusions, xxii, 295–9
   and scope, 302–3
   and technology, 295
   and venture capital, 298–9
Green Economy Partnership Process (GEPP), xxii, xxv, 277–94
   Area Partnership for Economic Expansion (APEX), 285–6
   Bureau of Planning and Sustainability (BPS), 293–4
   and capacity-building, 284–5
   Charlotte Regional Partnership (CRP), 286–8
   developing recommendations, 280–1
   development partnership examples, 285–94
   and development structures, 283–5
   East Bay Economic Development Alliance, 292–3
   economic development partnership examples, 285
   and expansion activities, 284
   Green Corridor Partnership, 292–3
   green economy definition, 279
   green manufacturing definition, 279
   Kansas City Area Development Council (KCADC), 288–9
   Metro Denver EDC, 289–91
   Minnesota Environmental Initiative (MEI), xxii, 277–81, 283, 285
   Prosperity Partnership, 291–2
   recommendations, 281–2
   work group member organizations, 277–8
Green Economy Partnership Process Work Group, 277–85
greenhouse gas (GHG) emissions, and regulation, xvii, xxiv, 19–33, 101
   and climate change, 23
   greenhouse-gas-intensive product lines, 19
   hypothesis testing, 23–6
   and institutional pressure, 25–30, 32
   level of perceived regulatory uncertainty, 26–32
   nine sources of pressure to reduce, 32
   and organizational response strategy, 22–4, 33
   questionnaire, *See* GHG regulations questionnaire
   responses to institutional pressures, 20–3
   and results, 26–9
Greenhouse Gas Program, 264
greenhouse gas (GHG) regulations questionnaire, 24–33
   companies studied, 27
   data sample, by industry, 24
   items measuring perceptions, 32
   organizational response strategy, 33
   results of the regression analysis, 28
Grell, A., 111–12
GRI, *See* Global Reporting Initiative
grid, 1, 5, 8–10, 12–13, 15, 39, 43, 110, 113, 133

Grosskopf, S., 227
growth, xiv, 1, 7–9, 14–15, 56–64, 71, 81, 83, 90, 100, 107–8, 126, 128, 130–2, 134–6, 142–3, 145–8, 150–3, 169, 227–8, 230, 234–6, 264, 267–8, 282–3, 285–6, 291–2, 299, 301

Hamprecht, J., 178
Hart, D., 68
Hart, S. L., 83, 87
Haveman, H., 298
Heide, J. B., 297
heuristics, 55–7, 68–70
Hierarchical Bayes (HB) Estimation, 45–6, 116
high-tech cluster revolution, xix–xx, xxiv, 141–54
  and contingency theory, 143
  and "coopetition," 149–52
  and density dependence principle, 147
  and double liabilities, 149–50
  and funding, 146–50
  and generalist organizations, 145
  and governments, 142, 146–7, 151–4
  and location, 145–7
  old and new firm industry interactions, 149–53
  and organizational ecology, 143–5
  and random variation, 144
  and "resource partitioning," 144–8
  Route 128, 141–2
  and the sales market, 150
  Silicon Valley, 141–3
  and social networks, 145
  and specialist organizations, 144–5
  and specialized employees, 145–50
  start of a cluster, 146
  and structural inertia, 144
  two scenarios for, 145–52
Hitt, M. A., 244
Hockerts, K., 142, 297
Hoffmann, V. H., 178
Hofstetter, J. S., 178
Holzinger, I., 68–9
Hope, C., 87
Hoskisson, R. E., 178, 244
Hull, C. E., 86, 99
Hult, G. T. M., 242, 246, 250
Hurley, R. F., 242, 246, 250
hydroelectricity, 11, 54

India, 129, 171, 205
information technology (IT)
  clusters, 142
  venture capitalists, 132–3
innovation, xiii, xviii–xx, 79
  and fashion, *See* ethical fashion industry
  and finance, *See* innovation and finance
  and high-tech, *See* high-tech cluster revolution
  and internationalization, *See* internationalization
  and photovoltaic projects, *See* photovoltaic projects, and finance
  and venture capital, *See* cleantech venture capital
innovation and finance, xviii–xix, xxiv, 79, 81–101
  analysis and results, 91–101
  and balanced constructs, 82–5
  the complementary hypothesis, 87–8
  and Compustat, 88
  contrasting short-term versus long-term results, 96
  and "corporate environmental and social performance," 83
  and corporate financial performance (CFP), xviii–xix, 81–101
  corporate social performance (CSP) 86, 91
  and corporate social responsibility (CSR), 86
  and the corporation, 85–8
  descriptive statistics and correlations, 92
  and environmental and social performance (ESP), xviii–xix, 81–101
  ESP–CFP relationships, 84–97
  and GICS code, 91
  and KLD Research and Analytics, Inc., 88–9
  limitations and future research, 100–1
  and managerial implications, 100
  mediated model of ESP, innovation, and long-term CFP, 95
  moderating effect of innovation on ESP and CFP, 87
  regression results for short-term CFP, 93
  and research and development (R&D), 86, 90

INDEX

and return on assets (ROA), 90
sample and data collection, 88–91
and short-term financial
  performance, 92
and the social performance construct,
  82–5
and the substitution
  hypothesis, 86–7
Time-dependent inverse-U-shaped
  relationship, 98
and Tobin's q, 90
institutional pressure, 25–32
institutions, xiii, xvii–xviii, 1–2, 295–8
  and the future, 295–8
  and greenhouse gas, *See* greenhouse gas
    emissions, and regulation
  and managers, *See* managerial
    perceptions
  and regulation, *See* regulation
  and wind energy, *See* wind energy
Intel, 132–3
International Energy Agency
  (IEA), 4–5, 264
International Roundtable on Sustainable
  Palm Oil (RSPO), 182
International Study Group on Exports
  and Productivity (2008), 223
internationalization of small and medium
  enterprises (SMEs), xxi, xxv, 241–57
  and construct reliabilities, 252
  and descriptive statistics and
    correlations, 251
  discussion of, 252–5
  and distribution, 244
  and diversification, 244–5
  and food industry, 248–9
  and hypotheses, 245–8
  and improvement, 243–4
  and "innovation capability," 242
  and innovativeness, 242–50, 256–7
  and international diversification,
    244, 250
  and knowledge, 242–8
  and learning orientation, 245–6,
    250, 256
  and the Likert scale, 249–50
  and measurement of constructs, 255–7
  and methods, 248–50
  and proactive environmental strategy,
    241–3, 246–9, 252, 254, 255–6

and research and development, 247–8
and resources, 245
and results, 250–2
and sources of international
  knowledge, 248
and Spain SMEs, 243, 248–9
and standardized factor loadings, 251
sand structural models, 252–3
Integrated Production, 226–7
Internet, xv, 90, 115, 199, 203, 286
iPods, 142
Ireland, 129
Ireland, R. D., 244
ISO 14001, 197, 226–7
ISO 26000, 197
Israel, 129, 133, 303
Italy, 44–5, 250

Jacobsson, S., 296–7
Japan, 142, 152–3, 164, 213, 250

Kahneman, D., 119
Kansas City Area Development Council
  (KCADC), 288–9
Karemera, D., 223–4
Katz, J. P., 90
KCADC, *See* Kansas City Area
  Development Council
Kenny, D. A., 91, 93–4, 102
KLD Research and Analytics,
  Inc., 88–9
Knight, 242
Koch, J., 127–8, 138
Kyoto Protocol, xvii, 20, 23, 31–2

Lancaster, K. J., 41
Lang, T., 111–12
Larsen, T., 85
Larson, A., 87
Latin America, 183, 205, 249–50
Lattas, J., 160
Lawrence, T. B., 178
learning orientation, 245–6, 250, 256
Lee, B., 296
LEED Accredited Practitioner
  Program, 298
Lenardič, D., 110–11
Levi's, 177
Levinson, A., 224
Li, L., 244

living standards, xi
long-term thinking, 303
Lounsbury, M., 21, 31
Lyles, M. A., 250

Madsen, P. M., 222
Malmquist Total Factor Productivity (TFP), 227
managerial perceptions of public policy, xviii, xxiv, 53–74, 299
  cognitive template for, 61–3
  and demand fluctuations, 58
  and description of the sample, 70–1
  and descriptive statistics, 65
  and firm–government interactions, 56–7
  and government assistance, 53
  and heuristics, 55–7, 68–70
  implications of, 67–70
  importance of, xxii
  and instability, 57–61, 64, 71
  and interviews with managers, 60–3
  and lawyers, 59
  and market–government interactions, 53
  and mental model of managers, 57–9
  and motivations for political activity, 55
  and questionnaire, 60, 63–7, 70–4
  and results of the regression analyses, 67
  survey items of, 71–4
  and the tilt away from government, 63–4
  and the US 2009 Economic Stimulus Act, 54
Managi, S., 223–4
Marcus, A., 16, 53, 54, 76, 78, 104, 139, 275, 296, 298
Marshall, A., 145
Martin, J., 70
mathematical psychologists, 39
Mattel, 177
Mayors' Initiative on Green Manufacturing, 277
McEvily, B., 296
McWilliams, D., 86
MEI, *See* Minnesota Environmental Initiative
methodology, xvi–xxvi
Metro Denver EDC, 289–91
Microsoft, 132–3

Midwest Independent Transmission System Operator (MISO), 8–9
Migros, 179, 182–7
Miller, D., 22, 25, 30
Millington, A., 97
Minier, J., 223–4
Ministry of Economy, Trade, and Industry (METI), 142, 153
Minnesota Environmental Initiative (MEI), 277–81, 283, 285
Minnesota Green Jobs Task Force, 277
Minnesota Public Utility Commission (PUC), 9
Minnesota wind power capacity (2007–2009), 6–10
MISO, *See* Midwest Independent Transmission System Operator
Mitchell, K., 283–4
Musgrave, 179, 182–5, 187

Nachum, L., 243
National Research Council (NRC), 5
natural gas, 4–5, 13, 126
neoclassical economic theory, 3–4
Netherlands, 129, 250
Nevitt, P. K., 110
New Zealand, 249
Nike, 177, 198
nongovernmental organizations (NGOs), xv–xvi, 21, 23, 32, 63, 175, 177, 181, 185–6, 189, 195–6, 199–200, 211, 261, 266, 268–9, 273, 302, 313
North, D. C., 3–4
North, Doug, 295
Norway, 129, 250, 265
NRC, *See* National Research Council
nuclear energy, 5, 89, 126, 303

Obama, Barack, 14
OECD, *See* Organisation for Economic Co-operation and Development
Office of Energy Security, 9
oil, 3, 21, 24, 134–6, 171, 182, 263–5, 267–73, 299
Oliver, C., 21, 29–30, 68–9
Olson, M., 56, 132, 269, 273, 296
Organisation for Economic Co-operation and Development (OECD), 4–5, 14–15
organizational response strategy, 22–4, 33

INDEX

path dependence, xiv–xv, xxiv, 37, 125–38, 185
Pearce, J., 68–9
Peng, M. W., 178
PES, *See* proactive environmental strategy
Peters, N., 178
Petrobras, 265, 269
petroleum, *See* oil
photovoltaic (PV) business models, 112–14
photovoltaic (PV) market, xviii, xxiv, 37–51, 107–21, 122n8
  adaptive conjoint analysis (ACA) of *See* adaptive conjoint analysis (ACA)
  barriers to, 37
  and finance, *See* photovoltaic projects and finance
  and Germany, 37, 44–5, 48–9
  high cost of, 37
  and incentives, 37–8; *See also* feed-in tariff (FIT)
  and legal factors, 39
  and policy, 39
  project developers, 38–9
  and solar irradiation, 39
  and Spain, 37, 44–5, 49–50
  and technology, 37
photovoltaic (PV) projects and finance, xix, 107–21
  and adaptive choice-based conjoint (ACBC), *See* adaptive choice-based conjoint (ACBC)
  and brands, 110, 113, 115–21
  and business models, 112–14
  and conjoint experiment results, 116
  and debt capital, 107–8, 110–11, 121
  and Debt Service Cover Ratio (DSCR), xix, 111, 113, 116–21
  discussion of, 119–21
  and economic requirements, 111
  and Germany, 108
  and maintenance concept, 111
  and method, 114–15
  and off-balance-sheet financing, 109–10
  and part-worth utilities, 114
  and power plant capacity, 110–11
  and project development, 109–13, 115
  and project financing, 109–13, 114–19
  research approach, 108–9
  and sample, 115–16
  and Spain, 108
  and sponsor types, 111–12
  and technical components, 110
  and US investment, 108
Pickle, S., 38
Porter, M. E., 87, 97, 223–4
Porter hypothesis, 223–4
Portugal, 44–5, 250
positive feedback loops, xxii, 297
Prashant, K., 90
Pricewaterhouse Coopers, 265
private investment in public goods, *See* CO2 Capture Project
proactive environmental strategy (PES), 241–3, 246–9, 252, 254
proactive supply chain sustainability standards (PSCSS), xx, xxv, 177–90
  and Axel Springer, 179, 183, 186–7
  and capabilities, 178–81
  and case selection and analysis, 179–82
  conclusions on, 188–90
  and continuous improvement, 187, 190
  and cross-functional integration, 186–7, 190
  and external stakeholder collaboration, 185–6, 189
  and health, safety, and environmental risks, 187–9
  and institutional entrepreneur, 178, 183–5, 188
  institutionalization of, 182–8
  and interfirm dialogue, 183–4, 189
  and the media, 177
  and Migros, 179, 182–7
  and Musgrave, 179, 182–5, 187
  and NGOs, 175, 177, 181, 185–6, 189
  and resource-based view, 178–9, 183, 185, 187–8
  and risk management, 184–5
  and Smurfit Kappa Group, 179, 183–4, 186
  and theme analysis of capabilities, 180–1
Production Tax Credit (PTC) (U.S.), 14
Prospect Theory, 119–20
Prosperity Partnership (Central Puget Sound Region), 291–2
PTC, *See* Production Tax Credit
PUC, *See* Public Utility Commission(s)

Public Utility Commission (PUC), 9–10
Purchase Likelihood (SMRT Simulation), 47, 49–50
PV technology, *See* photovoltaic (PV) projects and finance

Qian, G., 244

Rands, G. P., 82
recycling, xi, 89, 129, 147, 160–1, 241, 249, 255–6
Red Eléctrica de España (Spain), 13
regional transmission operator (RTO), 8, 10, 13
regulation and greenhouse gas (GHG) emissions, xvii, xxiv, 19–33
  and climate change, 23
  and game theory, 22
  and hypothesis testing, 23–6
  and institutional pressures, 20–3
  and the Kyoto Protocol, 20
  and level of perceived regulatory uncertainty, 26–32
  and organizational response strategy, 22–4, 33
  and questionnaire, *See* GHG questionnaire
  and results, 26–9
  and strategy, 21–3
Renewable Energy Plan (Spain), 12
Renewable Energy Policy Project (U.S.), 14
Renewable Energy Sources Act (Germany), 12
Renewables Portfolio Standard (RPS), 8–10, 12–13
Repsol, 265
"resource partitioning," 144–8
Reuter, A., 110
Roath, A. S., 250
Rose, A., 224
Roth, K., 247
Rothenberg, S., 86, 99
RPS, *See* Renewables Portfolio Standard
RSPO, *See* International Roundtable on Sustainable Palm Oil
RTO, *See* regional transmission operator
Russia, 183, 205, 250
Russo, M., 296

SA8000, 198
San Francisco, 283, 292
Santa Monica, 284
Santos, F., 298
Sargan, J., 233–4
Sawtooth Software, 43, 47, 114, 117
Scheryögg, G., 127–8, 138
Schumpeter, J. A., 85
Schwenk, C. R., 250
Scott, W. R., 25
SDA Bocconi School of Management (Milan), xiii
Sharma, S., 178
shecopreneurs, 157–72
  See also ethical fashion industry
Shell, 265
Shook, C. L., 90
Shrivastava, P., 247
Siegel, D., 86
Sine, W., 296–8
Singh, J., 90
Sinkovics, R. R., 250
SKG, *See* Smurfit Kappa Group
small and medium enterprises (SMEs), xxi, xxv, 241–57
  See also internationalization of SMEs
SMEs, *See* small and medium enterprises
Smith, Louis, 277
Smith Partners PLLP, 277
Smurfit Kappa Group (SKG), 179, 183–4, 186
social imaginary, 159–60, 171–2
solar energy, xviii–xix, xxiv, 1, 8, 11, 15, 37–51, 53–4, 107–21, 122n8, 125–30, 134–6, 142, 147, 151, 281
  and cleantech VC, 125–30, 134–6
  uneven path in developing, 130
  US investment in, 54, 108
  See also adaptive conjoint analysis approach; photovoltaic market; photovoltaic projects and finance
Spain, xxi, 1–3, 5–6, 10–13, 15, 37, 44–5, 49–50, 108, 129, 221, 226, 237, 243, 248–9, 255, 265
  agri-food industry, 226
  and installed wind power capacity, 10–11
  and SMEs, 243, 248–9
Starbucks, 195
Starik, M., 82

Statoil, 265
Sterzinger, George, 14
Stigler, G. J., 56–7
Suarez, S., 70
Sun Microsystems, 132–3
Suncor, 265
SunEnergy Europe GmbH, 109
supply chains, *See* global supply chains
"sustainable development," xv
sustainable practices and global supply chains, xx, xxv, 195–216
   and codes of conduct, 197–8, 202, 204, 208–12, 215
   and Coffee and Farmer Equity, 195
   and conceptual model of supply chain, 202
   and confirmatory factor analysis, 204–5
   and Cronbach's alpha, 204–5
   and customers, 196–7
   and dependency, 199–200, 216
   and descriptive statistics, 207
   and discussion, 209–11
   and distance, 200–2
   and Fairtrade certification, 198
   and hypotheses, 198–201
   and ISO 14001, 197
   and ISO 26000, 197
   and limitations, 212–14
   and managerial implications, 211–13
   and NGOs, 195–6, 199–200, 211
   and primary supplier practices, 215
   and research methods, 202–6
   and results, 206–8
   and SA8000, 198
   and social issues, 196–206
   and socially responsible suppliers, 215
   and technology, 195
   and transparency, 198–9
   and triple bottom line (TBL), 196
Sweden, 129, 250
Switzerland, 44–5, 115–16, 129, 302–3
Sydow, J., 127–8, 138

tables listing, ix–x
Tapscott, D., 209
Taylor, M. S., 224
TBL, *See* triple bottom line
Texas Public Utility Commission (PUC), 9–10
Texas wind power capacity (2007–2009), 6–10
TFP, *See* total factor productivity index
3M, 277
ThyssenKrupp, 20
Ticoll, D., 209
TNS survey company, 249
Tolbert, P., 297–8
Tomas, M., 250
Tosi, H. L., 90
total factor productivity (TFP) index, 222, 227, 229, 231
Toulouse, J. M., 25, 30
Tracey, P., 297
transaction cost economics, 22
triple bottom line (TBL), 196
Tversky, A. 119
Twin Cities, 7–9, 282

UmweltBank AG, 109
UN Environment Program, xi–xii, 107
UN Global Compact, 182
UNEP, *See* United Nations Environment Program
United Kingdom, 129, 182, 250, 265
University of Minnesota, 278, 299
U.S. Department of Energy (DOE), 4, 6–7, 9, 14, 264–5, 303
U.S. Economic Stimulus Act (2009), *See* American Recovery and Reinvestment Act
U.S. Energy Information Administration (EIA), 5
U.S. Green Building Council, 298
U.S. machine tool industry, 58
U.S. National Academy of Sciences, 5

Van Beers, C., 224
Van den Bergh, J. C. J. M., 224
Van der Linde, C., 87, 223–4
venture capital, xxii, xxiv, 79, 125–38, 142, 146–51, 270, 296, 298, 310
   See also cleantech venture capital
Vermeulen, F., 243
Volkswagen, 20

Waddock, S. A., 86
Wagner, M., 223
WCED, *See* World Commission on Environment and Development

Wecker, Claus, 110
Welch, A., 283–4
Werner, S., 90
West, S. G., 86
wind energy, xvii, xxiv, 1, 3–15, 53–4,
    71, 108, 120, 125–7, 129–31, 134–7,
    147, 151, 296
  and analytic narrative, 4, 14
  cost-competiveness of, 4–5
  growth of, 134
  and institutional theory, 3–4
  and institutions, 3–15
  and Minnesota–Texas analysis, 6–10
  and resource availability, 5–14
  US investment in, 108
  U.S.–Spain–Germany analysis, 10–14
Windwärts Energie GmbH, 109
Wiser, R., 38
World Bank, 248–50
World Business Council for Sustainable
    Development, 100
World Commission on Environment
    and Development (WCED),
    xv, 82, 99
World Soccer Cup, 120
World Resource Fund, xv
World Wildlife Fund (WWF), 182
Wright, M., 178
Wüstenhagen, R., 142, 298–9
WWF, *See* World Wildlife Fund

Xcel Energy, 8–9

Yeoh, P. L., 244
Yingli, 120

Zaheer, S., 243
Zimmerman, M. A., 245